Algorithms, Methods, and Applications in Mobile Computing and Communications

Agustinus Borgy Waluyo
Monash University, Australia

A volume in the Advances in Computer and
Electrical Engineering (ACEE) Book Series

Published in the United States of America by
  IGI Global
  Engineering Science Reference (an imprint of IGI Global)
  701 E. Chocolate Avenue
  Hershey PA, USA 17033
  Tel: 717-533-8845
  Fax: 717-533-8661
  E-mail: cust@igi-global.com
  Web site: http://www.igi-global.com

Library of Congress Cataloging-in-Publication Data

Names: Waluyo, Agustinus Borgy, 1973- editor.
Title: Algorithms, methods, and applications in mobile computing and
  communications / Agustinus Borgy Waluyo, editor.
Description: Hershey, PA : Engineering Science Reference, [2018] | Includes
  bibliographical references.
Identifiers: LCCN 2017051711| ISBN 9781522556930 (h/c) | ISBN 9781522556947
  (eISBN)
Subjects: LCSH: Mobile computing. | Mobile communication systems.
Classification: LCC QA76.59 .A44 2018 | DDC 004--dc23 LC record available at https://lccn.loc.gov/2017051711

This book is published in the IGI Global book series Advances in Computer and Electrical Engineering (ACEE) (ISSN: 2327-039X; eISSN: 2327-0403)

British Cataloguing in Publication Data
A Cataloguing in Publication record for this book is available from the British Library.

All work contributed to this book is new, previously-unpublished material. The views expressed in this book are those of the authors, but not necessarily of the publisher.

For electronic access to this publication, please contact: eresources@igi-global.com.

Advances in Computer and Electrical Engineering (ACEE) Book Series

Srikanta Patnaik
SOA University, India

ISSN:2327-039X
EISSN:2327-0403

MISSION

The fields of computer engineering and electrical engineering encompass a broad range of interdisciplinary topics allowing for expansive research developments across multiple fields. Research in these areas continues to develop and become increasingly important as computer and electrical systems have become an integral part of everyday life.

The **Advances in Computer and Electrical Engineering (ACEE) Book Series** aims to publish research on diverse topics pertaining to computer engineering and electrical engineering. **ACEE** encourages scholarly discourse on the latest applications, tools, and methodologies being implemented in the field for the design and development of computer and electrical systems.

COVERAGE

- Programming
- Algorithms
- Qualitative Methods
- Applied Electromagnetics
- Analog Electronics
- Electrical Power Conversion
- VLSI Fabrication
- Circuit Analysis
- Optical Electronics
- VLSI Design

IGI Global is currently accepting manuscripts for publication within this series. To submit a proposal for a volume in this series, please contact our Acquisition Editors at Acquisitions@igi-global.com or visit: http://www.igi-global.com/publish/.

Titles in this Series

For a list of additional titles in this series, please visit: www.igi-global.com/book-series

Electronic Nose Technologies and Advances in Machine Olfaction
Yousif Albastaki (University of Bahrain, Bahrain) and Fatema Albalooshi (University of Bahrain, Bahrain)
Engineering Science Reference • copyright 2018 • 318pp • H/C (ISBN: 9781522538622) • US $205.00 (our price)

Quantum-Inspired Intelligent Systems for Multimedia Data Analysis
Siddhartha Bhattacharyya (RCC Institute of Information Technology, India)
Engineering Science Reference • copyright 2018 • 329pp • H/C (ISBN: 9781522552192) • US $185.00 (our price)

Advancements in Computer Vision and Image Processing
Jose Garcia-Rodriguez (University of Alicante, Spain)
Engineering Science Reference • copyright 2018 • 322pp • H/C (ISBN: 9781522556282) • US $185.00 (our price)

Handbook of Research on Power and Energy System Optimization
Pawan Kumar (Thapar University, India) Surjit Singh (National Institute of Technology Kurukshetra, India) Ikbal
Ali (Jamia Millia Islamia, India) and Taha Selim Ustun (Carnegie Mellon University, USA)
Engineering Science Reference • copyright 2018 • 500pp • H/C (ISBN: 9781522539353) • US $325.00 (our price)

Big Data Analytics for Satellite Image Processing and Remote Sensing
P. Swarnalatha (VIT University, India) and Prabu Sevugan (VIT University, India)
Engineering Science Reference • copyright 2018 • 253pp • H/C (ISBN: 9781522536437) • US $215.00 (our price)

Modeling and Simulations for Metamaterials Emerging Research and Opportunities
Ammar Armghan (Aljouf University, Saudi Arabia) Xinguang Hu (HuangShan University, China) and Muhammad
Younus Javed (HITEC University, Pakistan)
Engineering Science Reference • copyright 2018 • 171pp • H/C (ISBN: 9781522541806) • US $155.00 (our price)

Electromagnetic Compatibility for Space Systems Design
Christos D. Nikolopoulos (National Technical University of Athens, Greece)
Engineering Science Reference • copyright 2018 • 346pp • H/C (ISBN: 9781522554158) • US $225.00 (our price)

Soft-Computing-Based Nonlinear Control Systems Design
Uday Pratap Singh (Madhav Institute of Technology and Science, India) Akhilesh Tiwari (Madhav Institute of
Technology and Science, India) and Rajeev Kumar Singh (Madhav Institute of Technology and Science, India)
Engineering Science Reference • copyright 2018 • 388pp • H/C (ISBN: 9781522535317) • US $245.00 (our price)

701 East Chocolate Avenue, Hershey, PA 17033, USA
Tel: 717-533-8845 x100 • Fax: 717-533-8661
E-Mail: cust@igi-global.com • www.igi-global.com

Table of Contents

Detailed Table of Contents

Section 1
Mobile Data Networking and Communication

Chapter 1

Varun G. Menon, Sathyabama University, India
Joe Prathap P. M., RMD Engineering College, India

Mobile ad hoc networks (MANETs) are a collection of wireless devices like mobile phones and laptops that can spontaneously form self-sustained temporary networks without the assistance of any pre-existing infrastructure or centralized control. These unique features have enabled MANETs to be used for communication in challenging environments like earthquake-affected areas, underground mines, etc. Mobility and speed of devices in MANETs have become highly unpredictable and is increasing day by day. Major challenge in these highly dynamic networks is to efficiently deliver data packets from source to destination. Over these years a number of protocols have been proposed for this purpose. This chapter examines the working of popular protocols proposed for efficient data delivery in MANETs: starting from the traditional topology-based protocols to the latest opportunistic protocols. The performances of these protocols are analyzed using simulations in ns-2. Finally, challenges and future research directions in this area are presented.

Chapter 2

Sander Soo, University of Tartu, Estonia
Chii Chang, University of Tartu, Estonia
Seng W. Loke, Deakin University, Australia
Satish Narayana Srirama, University of Tartu, Estonia

The emerging Internet of Things (IoT) systems enhance various mobile ubiquitous applications such as augmented reality, environmental analytics, etc. However, the common cloud-centric IoT systems face limitations on the agility needed for real-time applications. This motivates the Fog computing architecture, where IoT systems distribute their processes to the computational resources at the edge networks near data sources and end-users. Although fog computing is a promising solution, it also raises a challenge in mobility support for mobile ubiquitous applications. Lack of proper mobility support will increase the latency due to various factors such as package drop, re-assigning tasks to fog servers, etc. To address the challenge, this chapter proposes a dynamic and proactive fog computing approach, which improves the task distribution process in fog-assisted mobile ubiquitous applications and optimizes the task allocation based on runtime context information. The authors have implemented and validated a proof-of-concept prototype and the chapter discusses the findings.

Chapter 3

Heterogeneous networks are comprised of dense deployments of pico (small cell) base stations (BSs) overlaid with traditional macro BSs, thus allowing them to communicate with each other. The internet itself is an example of a heterogeneous network. Presently, the emergence of 4G and 5G heterogeneous network has attracted most of the user-centric applications like video chatting, online mobile interactive classroom, and voice services. To facilitate such bandwidth-hungry multimedia applications and to ensure QoS (quality of service), always best-connected (ABC) network is to be selected among available heterogeneous network. The selection of the ABC network is based on certain design parameters such as cost factor, bandwidth utilization, packet delivery ratio, security, throughput, delay, packet loss ratio, and call blocking probability. In this chapter, all the above-mentioned design parameters are considered to evaluate the performance of always best-connected network under heterogeneous environment for mobile users.

Chapter 4

Wireless mesh network (WMN) is a widely accepted network topology due to its implementation convenience, low cost nature, and immense adaptability in real-time scenarios. The components of the network are gateways, mesh routers, access points, and end users. The components in mesh topology have a dedicated line of communication with a half-duplex radio. The wireless mesh network is basically implemented in IEEE 802.11 standard, and it is typically ad-hoc in nature. The advantageous nature of WMN leads to its extensive use in today's world. WMN's overall performance has been increased by incorporating the concept of multi-channel multi-radio. This gives rise to the problem of channel assignment for maximum utilization of the available bandwidth. In this chapter, the factors affecting the channel assignment process have been presented. Categorizations of the channel assignment techniques are also illustrated. Channel assignment techniques have also been compared.

Chapter 5

Next-generation network promises to integrate cross-domain carriers; thus, infrastructure can be provided as a service. 5G-PPP's vision is directed toward solving existing 4G LTE mobility challenges that congest core networks, disrupt multimedia and data transfer in high mobility situations such as trains or cars. This research adopts 5G methodology by using software-defined networking (SDN) to propose a novel mobile IP framework that facilitates seamless handover, ensures session continuity in standard and wide area coverage, and extends residential/enterprise indoor services across carriers under service level agreement while ensuring effective offload mechanism to avoid core network congestion. Performance excels existing protocols in setup and handover delays such as eliminating out-band signaling in bearer setup/release and isolating users' packets in virtual paths. Handover across cities in wide area motion becomes feasible with lower latency than LTE handover inside city. Extending indoor services across carriers becomes equivalent to LTE bearer setup inside a single carrier's PDN.

Section 2
Mobile Data Access and Management

Chapter 6

Pawan Kumar Verma, Ambedkar National Institute of Technology, India
Rajesh Verma, Raj Kumar Goel Institute of Technology and Management, India
Arun Prakash, Motilal Nehru National Institute of Technology, India
Rajeev Tripathi, Motilal Nehru National Institute of Technology, India

This chapter proposes a new hybrid MAC protocol for direct communication among M2M devices with gateway coordination. The proposed protocol combines the benefits of both contention-based and reservation-based MAC schemes. The authors assume that the contention and reservation portion of M2M devices is a frame structure, which is comprised of two sections: contention interval (CI) and transmission interval (TI). The CI duration follows p-persistent CSMA mechanism, which allows M2M devices to contend for the transmission slots with equal priorities. After contention, only those devices which have won time-slots are allowed to transmit data packets during TI. In the proposed MAC scheme, the TI duration follows TDMA mechanism. Each M2M transmitter device and its corresponding one-hop distant receiver communicate using IEEE 802.11 DCF protocol within each TDMA slot to overcome various limitations of TDMA mechanism. The authors evaluate the performance of the proposed hybrid MAC protocol in terms of aggregate throughput, average transmission delay, channel utility, and energy consumption.

Chapter 7

Tamer Emara, Shenzhen University, China

The IEEE 802.16 system offers power-saving class type II as a power-saving algorithm for real-time services such as voice over internet protocol (VoIP) service. However, it doesn't take into account the silent periods of VoIP conversation. This chapter proposes a power conservation algorithm based on artificial neural network (ANN-VPSM) that can be applied to VoIP service over WiMAX systems. Artificial intelligent model using feed forward neural network with a single hidden layer has been developed to predict the mutual silent period that used to determine the sleep period for power saving class mode in IEEE 802.16. From the implication of the findings, ANN-VPSM reduces the power consumption during VoIP calls with respect to the quality of services (QoS). Experimental results depict the significant advantages of ANN-VPSM in terms of power saving and quality-of-service (QoS). It shows the power consumed in the mobile station can be reduced up to 3.7% with respect to VoIP quality.

Chapter 8

A very large number of broadcast items affect the access time of mobile clients to retrieve data item of interest. This is due to high waiting time for mobile clients to find the desired data item over wireless channel. In this chapter, the authors propose a method to optimize query access time and hence minimize power consumption. The proposed method is divided into two stages: (1) The authors present analytical models and utilize the analytical models for both query access time over broadcast channel and on-demand channel; (2) they present a global index, an indexing scheme designed to assist data dissemination over multi broadcast channel. Several factors are taken into account, which include request arrival rate, service rate, number of request, size of data item, size of request, number of data item to retrieve, and bandwidth. Simulation models are developed to find out the performance of the analytical model. Finally, the authors compare the performance of the proposed method against the conventional approach.

Section 3
Mobile Data Visualization

Mobile money transfer services (MMTS) are widely spread in the countries lacking conventional financial institutions. Like traditional financial systems they can be used to implement financial frauds. The chapter presents a novel visualization-driven approach to detection of the fraudulent activity in the MMTS. It consists in usage of a set of interactive visualization models supported by outlier detection techniques allowing to construct comprehensive view on the MMTS subscriber behavior according to his/her transaction activity. The key element of the approach is the RadViz visualization that helps to identify groups with similar behavior and outliers. The scatter plot visualization of the time intervals with transaction activity supported by the heat map visualization of the historical activity of the MMTS subscriber is used to conduct analysis of how the MMTS users' transaction activity changes over time and detect sudden changes in it. The results of the efficiency evaluation of the developed visualization-driven approach are discussed.

Chapter 10

3D maps have become an essential tool for navigation aid. The aim of a navigation aid is to provide an optimal route from the current position to the destination. Unfortunately, most mobile devices' GPS signal accuracy and the display of pathways on 3D maps in the small screen of mobile devices affects the pathway architectural from generating accurate initial positions to destinations. This chapter proposed a technique for visualizing pathway on 3D maps for an interactive user navigation aid in mobile devices.

This technique provides visualization of 3D maps in virtual 3D workspace environments which assists a user to navigate to a target location. The Bi-A* path-finding algorithm was used for establishing dynamic target location in Voronoi diagram/Delaunay triangulation. This approach could navigate more than two users in a 3D walk-space and at the same time showing their whereabouts on 3D projections mapped. The map shows the users' location in the scene to navigate from source to the target and the target also moves to the source to meet on the same physical location and image plane.

Preface

Driven by the proliferation of mobile devices and wireless communication technology, coupled with the increasing pace of mobile adoption in the social community around the globe, it is becoming a new norm where every individual is equipped with one or more mobile devices. Accordingly, the number of mobile users has now exceeded the traditional desktop users. Realizing the demand and size of mobile market worldwide, mobile device providers have been competing in releasing their latest version of their products to the market. Aligned with this phenomenon, wireless standard committee and telecommunication industry have worked together to continuously evolve the network in delivering high-speed internet connectivity which will help unlock new applications and deliver sophisticated mobile experience to the users.

However, despite the opportunities, there exist constraints, challenges and complexities in realising the full potential of mobile computing such as limited power and storage restriction, frequency of disconnection, security, asymmetric communications costs and bandwidth, and small screen size - all of which are required to be effectively addressed to support the enormous potential of mobile computing applications.

This book aims to dedicate to the latest research and development in algorithms, methods, and applications in the area of mobile computing and communication with the purpose to advance and disseminate the most recent research under this theme.

The chapters in this book are divided into three domains, namely (1) Mobile Data Networking and Communication, (2) Mobile Data Access and Management, and (3) Mobile Data Visualization.

Section 1 (Mobile Data Networking and Communication) comprises five chapters in relation to the most recent work in advanced mobile network and data communication techniques. Chapter 1 presents state of the art review of routing protocols in ad-hoc networks ranging from the traditional topology based protocols, geographic protocols and opportunistic routing protocols, followed by some performance evaluations. This survey can serve as valuable starting point to learn about wireless routing protocol for mobile ad-hoc networks (MANET). Chapter 2 introduces a mobility-aware framework for proactive Fog service provisioning while taking into account the dynamic characteristics of mobile distributed computing. The study provides new insights about how distributed systems can achieve the high-performance process migration in the edge networks. Chapter 3 discusses various predicting parameters required for estimation of always best-connected networks in a heterogeneous environment. In this chapter, the behaviour of networks is studied over some average time instants such as busy/peak hour, business hour and festival hours. In Chapter 4, channel assignment techniques in wireless mesh networks is presented. Several factors affecting the channel assignment process such as, interference, connectivity, fault tolerance, load balancing, fairness is discussed. These factors informed the design of the channel assignment technique. Chapter 5 proposes a novel network based Mobile IP (MIP) framework using SDN, called

SMARC, which guarantees seamless accessibility to indoor services with an effective offload mechanism inside and across carriers under service level agreement (SLA). The framework provides a unique unified architecture that can be adopted over any IP infrastructure such as WiFi, LAN, and WiMAX.

Section 2 (Mobile Data Access and Management) is dedicated to various approaches for access control, data processing and optimization in mobile computing environment. Chapter 6 studies a hybrid-MAC protocol for the densely deployed M2M networks, considering a number of factors, including; aggregate throughput, average transmission delay, channel utility, and energy consumption. Chapter 7 concerns with power saving mechanism based on artificial neural network for VoIP services over WiMAX systems. The proposed mechanism considers two states of VoIP conversation, namely; mutual silent state and talk-spurt state. Chapter 8 addresses optimal broadcast channel models, and Global indexing to determine the optimum number of database items to be broadcast in a channel with a view to minimising energy utilisation of mobile users.

Section 3 (Mobile Data Visualization) contains two chapters, discussing visualization advancement in mobile space. Chapter 9 demonstrates interactive visual representation of the MMTS subscribers allowing detection of the groups of users with similar behavior and outliers. This chapter contains detailed description of the proposed approach, including its enhancement targeted to detect short-term types of the behavior frauds (such as theft of mobile phones), suggested analysis workflow as well as discussion of the introduced modification and its influence on the efficiency of the proposed approach. Lastly, Chapter 10 describes pathway analysis for 3D mobile interactive navigation aid using Bent function, Voronoi diagram and its dual Delaunay triangulation. The proposed approach aims to offer the most favourable locations and path with 3D maps in a mobile device.

I hope this book is able to provide collective insights to readers especially on the recent advancements in various aspects of mobile computing ranging from mobile data communication and networking, to mobile data access, management and visualization.

Agustinus Borgy Waluyo
Monash University, Australia

Section 1
Mobile Data Networking and Communication

Chapter 1
Moving From Topology–Dependent to Opportunistic Routing Protocols in Dynamic Wireless Ad Hoc Networks:
Challenges and Future Directions

Varun G. Menon
Sathyabama University, India

Joe Prathap P. M.
RMD Engineering College, India

ABSTRACT

Mobile ad hoc networks (MANETs) are a collection of wireless devices like mobile phones and laptops that can spontaneously form self-sustained temporary networks without the assistance of any pre-existing infrastructure or centralized control. These unique features have enabled MANETs to be used for communication in challenging environments like earthquake-affected areas, underground mines, etc. Mobility and speed of devices in MANETs have become highly unpredictable and is increasing day by day. Major challenge in these highly dynamic networks is to efficiently deliver data packets from source to destination. Over these years a number of protocols have been proposed for this purpose. This chapter examines the working of popular protocols proposed for efficient data delivery in MANETs: starting from the traditional topology-based protocols to the latest opportunistic protocols. The performances of these protocols are analyzed using simulations in ns-2. Finally, challenges and future research directions in this area are presented.

DOI: 10.4018/978-1-5225-5693-0.ch001

INTRODUCTION

Recent advances in wireless technology have led to the exponential growth and usage of wireless mobile devices worldwide. Today billions of wireless devices are connected with the help of infrastructure like access points and base stations. These infrastructure supported wireless networks provide an increasing number of wireless local area network (LAN) hot spots, allowing travelers and users with portable laptops and mobile phones to surf the Internet from hotels, airports, railway stations, coffee shops and other public locations. However, these infrastructure supported wireless network comes with a number of limitations. They consume plenty of time and money for installation and maintenance; have constraints in flexibility, suffer from low utilization of local wireless resources and are particularly vulnerable to natural disasters and unpredicted failures. To overcome these limitations, self-sustained, infrastructure-less and decentralized wireless networks have been proposed, known as mobile ad hoc networks (Giordano and Lu, 2001; Chlamtac et al., 2003; Menon & Prathap, 2016).

Mobile ad hoc networks (MANETs) are a collection of wireless devices like mobile phones, laptops, PC's and iPads that can form instantaneous temporary networks without the support of any pre-existing network infrastructure or centralized control. It works as an autonomous system of mobile hosts connected by wireless communication links. The network is configured in a way that all the devices can dynamically join or quit the network at any time without disrupting communication between other devices. Every device in the network plays the dual role of a router and a host, cooperates and coordinates with each other to make routing decisions in the network. Data is transmitted in the network in a store and forward manner from the source node to the destination node via the intermediate nodes. Ease of deployment, speed of deployment and the ability to self-organize and self-adapt without the help of any underlying infrastructure has contributed to the growing popularity of MANETs in research as well as in industry. Today MANETs are used for communication and resource sharing in numerous challenging environments like earthquake and volcano affected areas (Mase, 2011; Menon et al., 2016), underground mines, battlefields etc. Figure 1 shows an example MANET used in disaster recovery operations

Figure 1. MANETs in disaster recovery operations

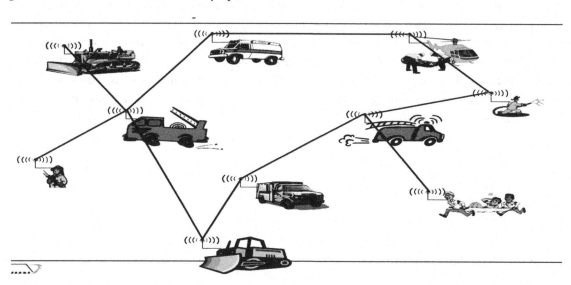

One of the major challenges in these highly dynamic networks is to efficiently deliver data packets from the source to the destination device. Ensuring reliable and continuous communication between the devices is yet another major challenge in these networks. Over these years a number of routing protocols have been proposed for data delivery and communication in MANETs. Figure 2 gives the taxonomy of all the protocols proposed for MANETs. Recent advancements in wireless technology have enabled mobile devices in MANETs to move freely with higher speeds in random directions. The mobility and speed of these wireless devices have become highly unpredictable and is increasing day by day. Also the number of connected devices in the network is increasing rapidly leading to highly dense and scalable ad hoc networks. As the mobility and number of devices increases in the network the performance of most of the existing routing protocols comes down drastically leading to low transmission efficiency and reduced Quality of Service. Very few researches have been done to identify the reasons behind this performance degradation.

This paper presents the design and working of popular routing protocols from three major categories proposed for MANETs; traditional topology based protocols, geographic routing protocols and opportunistic routing protocols. Further the research analyses the working and behavior of these protocols in highly dynamic ad hoc networks and discusses the reasons for performance degradation. The performances of the three categories of protocols are evaluated and compared in highly dynamic MANETs using simulations in Network Simulator-2.34 (Saha et al., 2013) with Random Way Point Mobility Model (Navidi and Camp, 2004). Finally the paper discusses the challenges and future research directions existing in this research area.

TRADITIONAL TOPOLOGY DEPENDENT PROTOCOLS

The earliest proposed routing protocols for MANETs depended on pre-determined routes to the destination for data delivery. These protocols were thus known as traditional topology dependent protocols. They are classified into three types based on their working; reactive protocols, proactive protocols and hybrid protocols.

Figure 2. Taxonomy of protocols proposed for MANETs

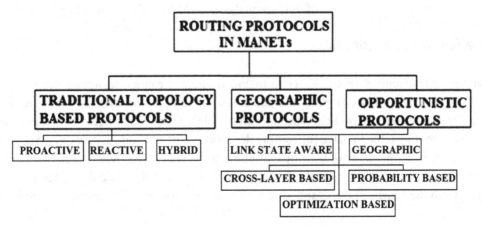

Reactive Routing Protocols

Reactive on demand routing protocols created routes only when they were requested by the source device. Source device initiated the route discovery process once it wanted to send a data packet to a destination device. This helped to reduce the overhead of maintaining the route table and route information throughout the network lifetime. In this type of protocols, once a route was established, it was maintained until the destination become unreachable or the route was no longer required.

Dynamic Source Routing (DSR) (Johnson and Maltz, 1996) was one of the first standard reactive protocols proposed for MANETs. DSR used a source routing strategy in which, when a source device needs to send a data packet to the destination device, it computes a possible route to the destination. This is done only on demand, which is only when packets are needed to be sent to a particular destination device. This route is maintained by the source device as long as it is working and data packets can be delivered using the particular path. Once it becomes unreachable, the source device searches for new routes. Due to unpredictable and continuous movement of devices, source initiated routes become obsolete frequently and this leads to the poor performance of DSR in dynamic MANETs.

Ad hoc On Demand Distance Vector (AODV) was a reactive protocol proposed by Perkins and Royer (1999). This protocol became popular and was well accepted by the research community. This protocol tried to combine the advantages offered by the proactive Destination-Sequenced Distance-Vector (DSDV) protocol (Perkins and Bhagwat, 1994) and the reactive Dynamic Source Routing (DSR) routing protocol. AODV protocol minimized the number of route broadcasts done in DSDV method by sending route information to other devices only on demand. The route request and route reply messages were used by AODV in route creation and maintenance. This protocol offered much better performance compared to all previous protocols and was used in a variety of applications. This protocol could only be used in symmetric links and had a number of performance issues in dynamic MANETs.

Temporally Ordered Routing Algorithm (TORA) (Park and Corson, 1997) was an improved reactive routing protocol proposed for MANETs. This protocol offered multiple routes for every source-destination device pairs and focussed on loop free operation in dynamic MANETs. This protocol worked in three phases; route creation, route maintenance and route erasure. One unique feature of this protocol was the broadcast of clear packet message to remove the stale routes in the network. Although the protocol gave better performance compared to previous protocols, there were a number of performance issues with high mobility of devices in dynamic MANETs. Also the implementation of the protocol was highly complex and incurred high overhead. Thus most of these reactive on demand protocols suffered from performance degradations with increasing mobility of devices in dynamic MANETs and was highly unsuitable.

Proactive Routing Protocols

Table driven proactive routing protocols try to maintain updated route information from every device to the other devices. The methodologies used are similar to the conventional distance vector and link state routing strategies. They periodically send routing table updates in the network and are completely dependent on the routing table to make the routing decisions in the network. Every device acknowledges the topology changes in the network by propagating route updates throughout the network to maintain a consistent network view. The major disadvantage of these protocols is the dependence on predetermined routes from the source to the destination device in the network. This contributes to its very low performance in networks with mobile devices. This routing strategy is predominantly suitable for wired networks and static wireless networks.

Destination-Sequenced Distance-Vector (DSDV) (Perkins and Bhagwat, 1994) is one of the first standard table driven proactive routing protocols proposed for ad hoc networks. In this distance vector protocol, a routing table is maintained by every device in the network. This protocol uses Bellman Ford Algorithm (Bellman, 1958; Yen, 1970) to calculate the shortest path from the source device to the destination based on the shortest number of hops. The major advantage of this routing mechanism is that it tries to avoid loops in the network using sequence numbers generated by destination devices. Whenever a modification happens in the neighborhood, these sequence numbers are increased. One of the major challenges in this protocol was in periodically updating and maintaining the correct routing table information with all the devices in the network.

An improved routing protocol, Optimized Link State Routing (OLSR) was proposed by Jacquet et al. (1998). The main contribution of this protocol was the introduction of the Multipoint Relay concept or MPR. Using this concept, instead of having only one next forwarder device in the routing table, OLSR allowed every device in the network to discover multiple relay devices for forwarding data packets. OLSR was one of the first protocols to exploit multipath forwarding in the network. Although it improved performance on previous protocols, OLSR suffered from many drawbacks. OLSR used large amount of bandwidth in the network to propagate the updates on MPR and to compute the best paths to each device. Due to heavy power and network resource usage, OLSR could not give desired performance in dynamic MANETs

Hybrid Routing Protocols

Another class of traditional topology based protocols, hybrid protocols tried to combine the advantages from proactive and reactive classes of protocols. The Zone-Based Hierarchical Link State Routing protocol (ZRP) (Haas and Pearlman, 1997) is the most referenced protocol in this category. Zone-Based Hierarchical Link State Routing Protocol (ZRP) defined a zone for each device in the network. The routing technique then used both proactive and reactive routing strategies to route the data packet inside and outside the zone. This protocol had considerable overhead and remained unsuitable for dynamic MANETs. This protocol too suffered from serious performance degradations with increasing speed of mobile devices in the network. It could not adapt to the dynamic wireless environment variations and was unable to provide the expected Quality of Service (QoS) in the dynamic MANETs.

GEOGRAPHIC ROUTING PROTOCOLS

To overcome the limitations of traditional topology dependent protocols, geographic routing protocols (Karp and Kung, 2000; Mauve et al., 2001; Menon et al., 2013; Cadger et al., 2013) were proposed for dynamic MANETs. Geographic routing protocols used location information to route the packets in a hop by hop fashion from the source to the destination devices in the network. This section discusses the working of major geographic routing protocols proposed for dynamic MANETs.

DREAM (Basagni et. al, 1998) was one of the earliest protocols to give the concept of using location information in routing the data packets in the network. This protocol used the information on the location of the devices positioned in the network for making routing decisions. New concept known as the distance effect was introduced by this protocol. The distance effect concept proposed that the relative velocity between the devices is lower if they are located at a greater distance. DREAM protocol selects the shortest hop path for data transmission. The major drawback of this protocol was the difficulty of obtaining the positional information of the devices in the network and in propagating this information and its updates to the neighbour devices. Also with this protocol, when the destination device moved away from the current position, frequent data loss occurred in dynamic MANETs.

Greedy Perimeter Stateless Routing (GPSR) (Karp and Kung, 2000) is the most referenced geographic routing protocol. This protocol selects the device that had maximum progress to the destination (nearest to the destination) as the best forwarder to forward the data packet. When this strategy was not possible in some region in the network, GPSR used a technique of routing around the perimeter of the region known as face routing. But the major problem with this protocol in dynamic MANETs is when the best forwarder device moved away from the current location and it became impossible to forward the data packet leading to low performance.

Secure Position Aided Ad hoc Routing (SPAAR) protocol (Carter and Yasinsac, 2002) tried to enhance the security of data transmission in the network. A private key-public key encryption method was introduced to improve the security of incoming and outgoing messages in the network. As the prime focus of this protocol was on security, this protocol couldn't improve the performance of data transfer in dynamic MANETs. Also this protocol had higher overhead compared to the previous routing protocols and was unsuitable for networks with extremely mobile devices.

Location Aided Knowledge Extraction Routing for Mobile Ad Hoc Networks (LAKER) protocol (Li and Mohapatra, 2003) tried to combine the advantages of source initiated protocol DSR and geographic protocol LAR. LAKER protocol was designed to learn the density of devices deployed in the network for better routing decisions. LAKER could be efficiently used in networks with non-uniform distribution of devices. The major drawback of LAKER was the low delivery rate obtained in data transmission in extremely dynamic ad hoc networks.

Blind Geographic Routing (BGR) protocol (Witt and Turau, 2005) was aimed to reduce the energy consumption of the devices in the network. This protocol used the concept of forwarding area in the network. It was a beacon-less geographic routing protocol and thus saved much network resource and energy. The protocol was designed in a way that the devices in the network battle with each other to become the forwarding devices. This protocol also suffered from low data delivery rate in extremely dynamic ad hoc networks and had high overhead due to the calculation of forwarding area.

Geographic Landmark Routing (GLR) protocol (Na and Kim, 2006) was an improvised geographic routing protocol that was aimed to solve the communication void problem in dynamic MANETs. This protocol solved many problems that existed in previous geographic routing protocol but incurred high overhead in operation in dynamic MANETs. Greedy Other Adaptive Face Routing Plus (GOAFR+) (Hwang et al., 2009) was yet another geographic routing protocol using the greedy forwarding and face routing strategies in MANETs.

Energy Aware Geographic Routing (EAGPR) protocol (Elrahim et al., 2010) offered better Quality of Service compared to previous geographic routing protocols in dynamic MANETs. The forwarding strategy of this protocol was based on the position information and the energy level of the devices. The algorithm was mainly aimed for working in wireless sensor networks. The performance of this algorithm came down with increasing mobility of devices in ad hoc networks and was unsuitable for extremely mobile environments.

OPPORTUNISTIC ROUTING PROTOCOLS

Opportunistic Routing protocols (OR) utilized the reception of the same broadcasted packet at multiple devices in the network and selected one best forwarder device dynamically from the set of multiple receivers. The most important advantage of this class of protocols is that they do not commit to a fixed route before data transmission. The next forwarder device and the route are only determined dynamically based on current network conditions and thus leads to its better performance compared to all previous classes of routing protocol proposed for dynamic MANETs (Yang et al., 2009; Menon and Prathap, 2016; Menon et al., 2016)

When a sender device wants to send a data packet to a particular destination device, it broadcasts the data packet to a list of candidate devices that are in its transmission range. Now these candidate relay devices are prioritized based on some metric like Expected Transmission Count (ETX) (Biswas and Morris 2005) or Expected Transmission Time (ETT) (Lee et al., 2013) calculated dynamically from the network. The candidate devices that receive the data packet run a coordination scheme to determine the best forwarder for the current data packet. Thus the forwarder device is selected dynamically from the network based on current network characteristics. The data packet is then forwarded by the best forwarder device and this opportunistic routing strategy continues till the data packet reaches the destination.

Opportunistic Routing protocols are classified into five categories based on their working; link state topology based protocols, geographic distance based protocols, cross layer based protocols, probability based protocols and optimization based opportunistic routing protocols. The first category, link state topology based opportunistic routing protocols uses link delivery probabilities for candidate list generation and in making decisions on best forwarder device in the network. Geographic distance based uses the information on the distance between the devices in the network to make routing decisions. Cross layer based opportunistic routing protocols make use of information from MAC and Physical layers in making routing decisions in the network. Probability based opportunistic routing protocols use delivery probabilities in the network for making routing decisions and optimization based opportunistic routing protocols tries to optimize the candidate set using machine learning approach, graph theory etc. Figure 3 presents the taxonomy of entire opportunistic routing protocols proposed for dynamic MANETs.

Link State Topology Based Opportunistic Routing Protocols

Link state topology based opportunistic routing protocols make use of a link state style updating mechanism for the calculated metric in the network. Also they use link delivery probabilities as the decision making metric in the network. Further they try to notify each device with the delivery probability of every link in the network using the link state type topology and updating mechanism.

The first OR protocol proposed in this category was Extremely Opportunistic Routing (ExOR) protocol (Biswas and Morris, 2005). This protocol introduced the batching systems in which a group of 10 to 100 packets were broadcasted by the source device. This broadcasted group of data packets also consisted of information on the potential forwarder devices. The priority of the devices was decided using the Expected Transmission Count (ETX) metric which calculated the expected number of transmissions for successful delivery of a packet over a link in the network. The major disadvantage of ExOR is that it uses a link state topology updating scheme. ExOR requires periodic network wide measurement of ETX value which is very difficult in dynamic MANETs with extremely mobile devices. Moreover,

Figure 3. Taxonomy of opportunistic routing protocols proposed for dynamic MANETs

Link State Topology Based ORPs	Geographic Distance Based ORPs	Probability Based ORPs	Cross Layer Based ORPs	Optimization Based ORPs
ExOR (Biswas and Morris, 2005)	**ROMER** (Yuan et al., 2005)	**FPOR** (Conan and Friedman, 2008)	**PRO** (Lu et al., 2009)	**LCOR** (Dubois-Ferriere et al., 2007)
OAPF (Zhong et al., 2006)	**OPRAH** (Westphal, 2006)	**Delegation Forwarding** (Erramilli et al., 2008)	**ILOR** (Bletsas et al., 2010)	**OMNC** (Zhang and Li, 2009)
MORE (Chachulski et al., 2007)	**DTRP** (Nassr et al., 2007)	**OR-Flooding** (Guo et al., 2009)	**SPOR** (Lee and Haas, 2011)	**Consort** (Fang et al., 2011)
Code OR (Lin et al., 2008)	**GOR** (Keng et al., 2007)	**OPF** (Lu and Wu, 2009)	**EEOR** (Mao et al., 2011)	**AdaptOR** (Bhorkar et al., 2012)
XCOR (Koutsonikolas et al., 2008)	**DICE** (Zhang and Li, 2008)	**EBR** (Nelson et al., 2009)	**CORMAN** (Wang et al., 2012)	**PLASMA** (Laufer et al., 2012)
Economy (Hsu et al., 2009)	**POR** (Yang et al., 2009)	**MaxOpp** (Bruno and Conti, 2010)	**QOR** (Lampin et al., 2012)	**TOUR** (Xiao et al., 2013)
SOAR (Rozner et al., 2009)	**MGOR** (Zeng et al., 2009)		**Parallel-OR** (Shin and Lee, 2013)	**ORL** (Tehrani et al., 2013)
Slide OR (Lin et al., 2010)	**TLG-OR** (Zhao at al., 2013)		**MTOP** (Lee et al., 2013)	**MAP** (Fang et al., 2013)
O3 (Han et al., 2011)	**XLinGo** (Rosario, et al., 2014)		**CAOR** (Zhao et al., 2014)	**LOR** (Li et al., 2013)
			ORW (Ghadimi et al., 2014)	

communication and coordination between the candidate devices generated duplicate transmissions when they were connected with links of low quality.

Opportunistic Any Path Forwarding (OAPF) protocol (Zhong et al., 2006) improved on ExOR protocol with a new metric known as Expected Any Path Transmissions (EAX) which calculated the expected number of transmissions for successful delivery of data packet between a pair of devices in the network. This metric was used by OAPF for candidate list generation and prioritization of the forwarder device. This protocol also required network wide periodic measurement of EAX and continuous updating which was quite impossible in extremely mobile environments. Thus this protocol too suffered from performance degradations in dynamic MANETs.

MAC-Independent Opportunistic Routing and Encoding (MORE) protocol (Chachulski et al., 2007) tried to increase the throughput of the network by integrating network coding into OR. This protocol too used the batch mechanism in its operation and obtained better performance than ExOR protocol in dynamic MANETs. MORE was one of the first protocols to use network coding as the coordination method between the candidate devices in the network. This protocol used ETX as the metric to generate the candidate set and to prioritize the potential forwarder devices.

Code OR protocol (Lin et al., 2008) is another link state OR protocol that combined OR with segmented network coding. This protocol too used ETX for candidate list generation and prioritization. Although this protocol offered better throughout, it suffered from many problems in dynamic MANETs. It was quite difficult to determine the optimal segment size of the data packet with this protocol and this contributed to the increased overhead in data transmission in the network.

Economy protocol (Hsu et al., 2009) was proposed to reduce the duplicate transmission in extremely mobile networks caused by previous OR protocols. Economy protocol introduced a new concept which removed numerous unused and unreachable devices from the candidate list and reduced duplicate transmissions. Token passing method was used by this protocol for coordination between the forwarder devices in transmitting a data packet in the network. Economy gave better throughput in the network compared to previous OR routing strategies, but incurred high overhead in data transmission and it remained unsuitable for extremely dynamic MANETs.

Simple Opportunistic Adaptive Routing (SOAR) protocol (Rozner et al., 2009) was an improved version of ExOR protocol and it used the same batching mechanism to transmit the data packets in the network. Design and working of this protocol was uncomplicated and new techniques could be easily integrated to the protocol. Although it offered better performance compared to ExOR and other OR protocols, SOAR also suffered from the problem of periodic updating of the ETX metric in dynamic MANETs.

Cumulative Coded Acknowledgments (CCAK) (Koutsonikolas et al., 2010) and Slide OR (Lin et al., 2010) used similar network coding strategies with OR in dynamic MANETs. Slide OR used a segmented coding mechanism and combined the packets belonging to different overlapping segments to increase the throughput of data transmission in the network. Both the protocols tried to improve the reliability of data transmission in the network and achieved higher throughput compared to previous OR protocols in dynamic MANETs. But both these protocols suffered from performance degradations with increase in mobility of devices in dynamic MANETs.

Optimized Overlay-based Opportunistic routing (O3) (Han et al., 2011) was one of the advanced OR protocols using network coding proposed in this category. The main objective of this protocol was to introduce a standard in the number of optimal coded packets that needs to be transmitted at a time in the network. This protocol solved some of the issues that existed in OR protocols with network coding. Although this protocol offered better throughout compared to the previous OR protocols, this protocol too suffered from performance degradations with increased mobility of devices in dynamic MANETs.

Geographic Opportunistic Routing Protocols

Geographic opportunistic routing strategies used the location information of the devices to generate the candidate list and to prioritize the set of forwarder devices. This type of OR protocols were much more flexible and dynamic compared to other categories of OR protocols and offered better performance.

Resilient Opportunistic Mesh Routing (ROMER) (Yuan et al., 2005) was one of the first proposed geographic OR protocols. This protocol used location information of the devices in the network with probabilities in data transmission to prioritize the forwarder devices in the network. Using ROMER protocol, forwarder devices located in the shortest paths were assigned a probability of one in data transmission. This protocol helped to reduce the occurrences of packet dropping attacks in the network with extremely mobile devices. One of the major drawbacks with this protocol was the increasing number of duplicate transmissions with rising mobility of devices in the network. This protocol was therefore seldom used in dynamic MANETs.

OPRAH protocol (Westphal, 2006) used a number of positive techniques from the earlier proposed AODV routing protocol. The main feature of this protocol was in maintaining more than one route to the destination device in the network. Route with minimum number of hops was selected as the best route to the destination device. The major advantage of this protocol was that it was less complex and had low overhead. The major drawback of OPRAH protocol was that it suffered from duplicate data packet transmissions in the network. Moreover, often this protocol was unable to discover the optimal path to the destination, resulting in higher timing overhead and low performance in dynamic MANETs.

Directed Transmission Routing Protocol (DTRP) (Nassr et al., 2007) is another geographic OR protocol that used the transmission probabilities similar to ROMER protocol in dynamic MANETs. Similar to ROMER, all the forwarder devices in the shortest path of transmission was assigned a probability one. All the remaining devices that took part in data transmission were assigned a different probability value calculated based on the current network scenario. The protocol used beacon messages for transfer of location information between the devices in the network. The major drawbacks of this protocol were high energy consumption and overhead in data transmission.

Geographic Opportunistic Routing (GOR) (Keng et al., 2007), was one of the earliest protocols to use the timer based coordination scheme among the various forwarder devices in the candidate set. Timer based coordination technique was much simpler and efficient compared to many previous methods used in Dynamic MANETs. This protocol used the Expected One Hop Throughput (EOT) metric that used the delay caused by the coordination process among the devices to make routing decisions in the network. GOR used the neighbor overhearing method to avoid packet retransmissions by lower priority forwarders in dynamic MANETs with extremely mobile devices. Although the timer based coordination methods was better compared to the previous methods, GOR suffered from duplicate data transmissions in dynamic MANETs.

Position Based Opportunistic Routing (POR) (Yang et al., 2009) and Multi-rate Geographic Opportunistic Routing (MGOR) (Zeng et al., 2009) protocols used the information on the position of the devices in the network to generate the candidate set and also to prioritize the devices in the candidate set. MGOR protocol was an improved version of the GOR protocol and used the OEOT metric for candidate set generation and prioritization. Both the protocols achieved better performance compared to all previous protocols in dynamic MANETs. Both protocols achieved higher throughput and lower delay compared to the previous protocols in dynamic MANETs. The major issue with POR protocol was buffer occupancy in the devices. MGOR suffered from duplicate data transmissions in dynamic MANETs.

TLG-OR (Zhao at al., 2013) combined geographic location information with details of link quality between the devices and the remaining energy information of devices to improve the QoS for video traffic in dynamic MANETs. Link quality and energy of the devices were the two most important parameters used by this protocol in deciding the forwarder devices in the network. This protocol however had higher overhead in data transmission and did not have any provision to handle communication voids in dynamic MANETs with extremely dynamic devices. Also the protocol had serious performance degradations in wireless networks with interference.

XLinGo (Rosario, et al., 2014) protocol was also aimed to improve the quality of video transmission in dynamic MANETs. This protocol also aimed at reducing the energy usage by the devices in routing of data packets in the network. This protocol offered better performance in video transmission compared to TLG-OR protocol in dynamic MANETs. This protocol too had higher overhead in data transmission and did not have any provision to handle communication voids in dynamic MANETS. Also this protocol was not suitable for use in dynamic MANETs with extremely dynamic devices.

Probability Based OR Protocols

Probability based OR protocols used various probabilities of data transmission and delivery in the network as the main metric in candidate set calculation. A number of OR protocols depending on probability of data delivery, links and transmissions have been proposed over these years.

Fixed Point Opportunistic Routing (FPOR) protocol (Conan and Friedman, 2008) tried to reduce the delay experienced by the data packets in the network. This device utilized the probability of devices coming in contact with each other to generate the candidate relay set. Contact probabilities of every device were estimated in the network and this was given the prime importance in FPOR protocol. This protocol suffered from low performance in dynamic MANETs with extremely mobile devices. Also the protocol was unable to efficiently manage the communication holes in the network.

Delegation Forwarding protocol (Erramilli et al., 2008) also used the contact probabilities of devices in the network to make the routing decisions. It worked on the theory that frequently encountered devices would be better forwarders in the network. This protocol was less complex and reduced some of the overhead caused by earlier OR protocols. This was one of the better protocols used in communication in disaster recovery operations. The protocol however had issues with bandwidth usage and storage in the network. Duplicate messages generated in the network reduced the performance of this protocol in dynamic MANETs with highly dynamic devices.

OR-Flooding (Guo et al., 2009) protocol was designed to work in low duty cycle networks. Delay information at the devices was used by this protocol to make forwarding decisions in the network. The major advantage of this protocol was in low energy consumption compared to previous OR protocols. Also this flooding technique had less delay compared to all previous flooding techniques in dynamic MANETs with dynamic devices. However, this protocol could only be used in duty-cycled stationary networks.

Encounter Based Routing (EBR) protocol (Nelson et al., 2009) set an upper limit on the amount of duplicate copies that can be generated from a data packet. The main objective of this protocol was to solve the major issue of redundant data packets in the networks. Working of EBR is based on the encounter probabilities of devices in the network and this protocol assumes that the devices that encounter frequently are better forwarders for any data packet in the network. EBR had better performance compared to Delegation Forwarding in dynamic MANETs. Although EBR was able to limit the number of duplicate packets generated in the network, it was unable to improve upon the QoS of data transmission in networks with extremely mobile devices. One of the latest OR protocol in this category, MaxOpp was proposed by Bruno and Conti, (2010). Although this protocol was a much improved version compared to all previous protocols, this protocol too could not offer better performance in terms of Quality of Service in dynamic MANETs.

Cross Layer Based Opportunistic Routing Protocols

Cross layer based OR protocols utilized information from Network, MAC and Physical layer to improve the efficiency of OR protocols in dynamic MANETs with extremely mobile devices. This information was used to generate the candidate sets and to prioritize the forwarder devices. Some of these cross layer based protocols offered better performance compared to the previous three categories of OR protocols in dynamic MANETs.

PRO protocol (Lu et al., 2009) used information from the network layer along with data from MAC and Physical layers to improve the QoS of OR in dynamic MANETs. The protocol used Link Quality Indicator (LQI) and Received Signal Strength Indicator (RSSI) as the major indicators for candidate set selection and prioritization. The protocol measured the quality of various links in the network and eliminated the low quality links. This protocol utilized overhearing property of the neighboring devices as the coordination mechanism among the devices in the network. The transmission link that had the maximum RSSI value was then selected to forward the data packet to the destination. PRO protocol did not offer techniques to handle communication voids in the network and had high overhead in dynamic MANETs.

ILOR protocol (Bletsas et al., 2010) used the information on link quality towards the destination in prioritizing the candidate set of devices and in selecting the best forwarder device in the network. The best forwarder device was selected using the link quality information towards the destination in this protocol. The major issue with this protocol was that the data packets can be relayed only up to two hops in the network. This protocol was thus highly unsuitable for large ad hoc networks with extremely mobile devices.

Simple and Practical Opportunistic Routing (SPOR) (Lee and Haas, 2011) was an interference aware cross layer based OR protocol that used acknowledgements for each data transmission to avoid duplicate packet retransmissions in extremely mobile networks. The major limitation with SPOR is that the performance of this protocol comes down if the path of data transmission consists of more than four hops. This protocol was therefore highly unsuitable for large ad hoc networks with many mobile devices. Numerous limitations prevented this protocol from being used in real time applications.

Energy-Efficient Opportunistic Routing protocol (EEOR) (Mao et al., 2011) was aimed at minimizing the energy usage of devices in the network. This protocol used information about transmission power of devices in the network and was mainly used in wireless sensor networks. This protocol suffered from low packet delivery rate in the network. This protocol could not offer better QoS in dynamic MANETs with extremely mobile devices.

Cooperative Opportunistic Routing protocol (CORMAN) protocol (Wang et al., 2012) used information about the position and speed of devices in the network to improve on previous OR protocols in dynamic MANETs. The protocol operated in similar fashion to the earlier proposed cross layer based PRO protocol. This protocol used Link Quality Indicator (LQI) and Received Signal Strength Indicator (RSSI) as the major indicators for candidate set selection and prioritization. This protocol made use of a realistic propagation model in the network. CORMAN achieved better throughput compared to previous OR protocols, but its performance came down with increasing mobility of devices in the network.

The protocols Parallel-OR (Shin and Lee, 2013) and QoS Oriented Opportunistic Routing protocol (QOR) (Lampin et al., 2012) tried to increase the throughput of data transmission in the network by using information on signal power from lower layers. Both the protocols exploited multiple paths to the destination. QOR protocol used a token based coordination method among the devices in the network. The major drawback of these protocols was that they never took account of the increasing signaling overhead in data packet transmission in the network and gave moderate performance with increasing mobility of devices in the network. They worked well with sensor networks but could not offer better QoS in dynamic MANETs.

Context Aware Opportunistic Routing (CAOR) protocol (Zhao et al., 2014) offered much better performance compared to all cross layer based protocols in dynamic MANETs with extremely mobile devices. CAOR used coding gain information to increase the delivery rate in the network. The coordination mechanism used was based on packet overhearing in the network. CAOR achieved higher data delivery rate compared to all the previous protocols in dynamic MANETs with extremely dynamic devices. But the performance of this protocol came down with increasing mobility of devices in dynamic MANETs. One of the latest cross layer based protocol ORW, (Ghadimi et al., 2014) was mainly targeted to achieve energy efficiency for wireless sensor networks.

Optimization Based Opportunistic Routing Protocols

Optimization based ORPs tried to optimize the candidate set selection and prioritization in OR using various mathematical techniques like graph theory, machine learning etc. Each of the protocol proposed in this category tried to improve on the basic building blocks of OR in dynamic MANETs.

Least Cost Opportunistic Routing (LCOR) (Dubois-Ferriere et al., 2007) is one of the best optimized protocols proposed in this category. This protocol helped to find the optimal candidate set for any source device in dynamic environments. But the major problem with this protocol was the increased number of duplicate data transmissions in dynamic MANETs with extremely mobile devices.

Consort (Fang et al., 2011) and OMNC (Zhang and Li, 2009) optimization based OR protocols also tried to find out the optimal candidate set in the network. These protocols also aimed to reduce the time overhead caused in the network. But both the protocols couldn't offer optimal paths to the destination and often suffered from serious performance issues with rising mobility of devices in the network.

Adapt-OR (Bhorkar et al., 2012) was an adaptive ORP that introduced a dynamic learning approach to improve the performance of OR in extremely dynamic ad hoc networks. The protocol introduced a new learning frame work for the network through which the details of various connections could be learned. The protocol had major issues with the management of control packets in the network and was often unable to discover the best path to the destination device in dynamic MANETs

Optimization based OR protocol, PLASMA (Laufer et al., 2012) targeted to improve the performance of routing in wireless mesh networks. PLASMA ensured that every device in the network was linked to more than one gateway in the network. PLASMA then used a variation of the Bellman Ford algorithm to compute the optimal paths from the source to the destination device in extremely dynamic networks. PLASMA improved on the throughput offered in the network but was unsuitable for highly scalable networks. Time-sensitive Opportunistic Utility-based Routing protocol (TOUR) (Xiao et al., 2013) aimed at reducing the delay of data transmission in the network. The protocol selected potential paths that offered much lesser delay in data transmission. But the protocol could not improve on the delivery rate and throughput of data transmission in the network and suffered from many problems in dynamic MANETs.

Multi-constrained Any Path (MAP) protocol (Fang et al., 2013) used optimization techniques similar to the Dijkstra's algorithm to compute the optimal path to the destination. Although this protocol had a number of advantages compared to the previous protocols, it could not guarantee required QoS in dynamic MANETs with extremely mobile devices. Localized Opportunistic Routing (LOR) protocol (Li et al., 2013) offered good performance compared to previous protocols in dynamic MANETs with extremely mobile devices. LOR protocol divides the entire network into smaller sub networks based on graph theory. It then used different routing strategies within the smaller sub networks and between these sub networks to achieve better routing performance in highly scalable and mobile ad hoc networks. LOR too could not guarantee required Quality of Service in dynamic MANETs with extremely mobile devices.

PERFORMANCE ANALYSIS

Performance analysis of various categories of routing protocols is done using simulations in Network Simulator 2.34. NS-2 has very high simulation credibility because it can generate various network scenarios required by the user easily and accurately (Saha et al., 2013). The simulator is installed on Core i7, Windows 10, 4 GB, 2.20 GHz system. The study analyses the performance of the different protocols using the most important Quality of Service parameters; Packet Delivery Ratio (PDR) and Average End to End Delay. PDR is one of most important performance metric that assesses the success of a routing protocol in dynamic MANETs. PDR helps to assess the number of successful data transmissions by the protocol and the effective data delivery rate in the network. Average End to End Delay measures the delay incurred by the packet in reaching the destination device. The average delay of all transmissions from various sources to destinations is calculated. This metric is very vital in determining the success of a routing protocol in dynamic MANETs. Comparison of these two metrics helps to analyze the QoS support provided by each of these protocols in dynamic MANETs.

MAC protocol used for the simulation is IEEE 802.11g. 100 nodes are deployed in a network area of 1000×800 m² rectangular region. The transmission range of the nodes is set at 250 m. Constant Bit Rate (CBR) traffic is being generated from the source to the destination nodes in the network at a rate of 20 packets per second (40kbps). The size of the data packet is 512 bytes. The simulation starts at 100 seconds and ends at 900 seconds. Mobility in the network is created by varying the speed of nodes from 5 m/s to 40 m/s. Dynamic movement of nodes are generated using random way point mobility model and the performance of the routing protocols are measured. Most referenced and popular protocols from each category, ExOR (opportunistic), GPSR (geographic), AODV (traditional-reactive) and DSDV (traditional-proactive) are selected for the performance comparison and analysis.

Figure 4 shows the Packet Delivery Ratio (PDR) obtained for the four different protocols with varying speed of nodes in the network. From the results it is evident that the data delivery rate of traditional routing protocols AODV and DSDV comes down with increasing mobility of nodes in the network. This is because; more often these protocols depend on pre-determined or static routes to the destination for data delivery. It is almost impossible to have fixed routes in highly dynamic ad hoc networks. Also it is very difficult to propagate the routing table and its updates reliably to other nodes in highly dynamic ad hoc network. Figure 5 shows the average end to end delay incurred by the data packets in reaching the destination using the different routing mechanisms. From the results it is evident that the traditional routing protocols suffer from higher delay in transmission and the delay increase considerably with rising mobility of nodes. Geographic routing protocol GPSR maintains a better performance because it uses the dynamic greedy forwarding and perimeter routing strategy. GPSR forwards the data packet to nodes that are nearest to the destination. Opportunistic routing protocol ExOR achieves much better performance compared to all other category of protocols in highly dynamic ad hoc networks. This is

Figure 4. Packet Delivery Ratio vs. Speed

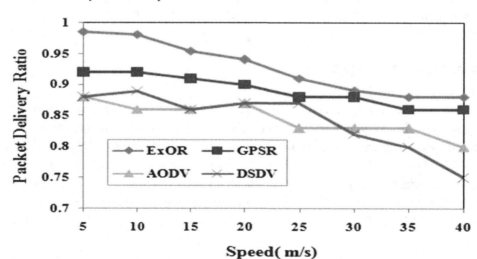

Figure 5. Average end to end delay vs. Speed

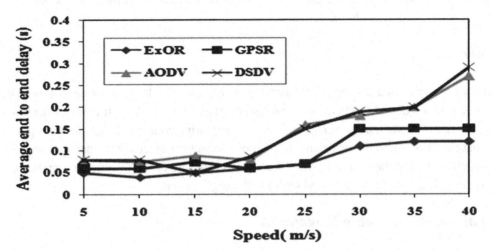

because ExOR uses a very dynamic routing strategy. It selects the transmission link and node that is having the maximum probability of data delivery in the network for forwarding the data packet. This selection is dynamic and is different for each data packet and this strategy is continues till the data packet reaches the destination. But even the performance of the opportunistic routing protocol, ExOR comes down with increasing mobility of nodes. Thus with increasing mobility of nodes, the major challenge is to design an efficient routing protocol that could give high Quality of Service (QoS) even in highly mobile environment. The next section describes the major challenges and future research direction in routing of data packets in highly dynamic ad hoc networks.

CHALLENGES AND FUTURE RESEARCH DIRECTIONS

Routing of data packets in highly dynamic ad hoc networks has become a major area of research recently. Researchers are facing a number of challenges and issues in designing efficient routing protocols for these networks. Increasing number of connected devices and its ever increasing mobility have been the major challenge in this regard. This section discusses the major challenges in designing an efficient routing protocol in highly dynamic ad hoc network and the future research directions in this area.

Quality of Service

Maintaining the required Quality of Service (QoS) is a major challenge for routing protocols in highly dynamic ad hoc networks. Users and applications choice of protocol depends on the QoS offered to them. Protocols must be designed to achieve higher data delivery rate with minimum delay in the network. This area of research is still wide open with numerous possibilities. As the numbers of connected devices are increasing day by day, the networks are expanding with limitless boundaries. Also, recent technological advancements have enabled the devices in the network to move at a much faster rate. Mobility of devices in the network has increased rapidly leading to highly dynamic networks. Routing protocols has to take care of this highly dynamic nature of the network while maintaining the required quality of service in the network. New protocols offering excellent QoS with affordable overhead in extremely dynamic environments is the need of the hour.

Scalability

Rising number of connected devices in the network is a major challenge for researchers in designing efficient routing protocols. The networks are expanding at a faster pace. Routing protocols has to ensure good performance even when more number of devices are getting connected. Applications prefer routing protocols that can accommodate maximum number of connected devices in the network and still deliver data packets efficiently at the destination nodes. Efficient routing protocols need to handle networks of variable sizes while ensuring the required QoS for the applications.

Mobility and Unpredictable Movement

Accommodating mobility has become one of the primary requirements of all new routing protocols. As the mobility of wireless devices is increasing at a rapid rate, routing protocols has to ensure minimum data loss in the network with fewer retransmissions. Data loss due to mobility has to be reduced or eliminated. Duplicate messages generated with broadcasting in wireless networks have to be controlled to minimize the traffic in the network. Handling mobility of nodes in the network and also ensuring excellent performance is the most important requirement for routing protocols in dynamic ad hoc networks.

Mobility Models

Numerous mobility models have been used to generate dynamic ad hoc networks for simulation and study. Accuracy and suitability of mobility models has been a prime area of research over these years. Larger networks with highly mobile devices have led to the need for new and advanced mobility models. Many conventional mobility models are unsuitable for the study of routing protocols in the current highly dynamic ad hoc networks. Research in this direction is progressing at a rapid rate.

Privacy and Security

Ensuring the privacy and security of data transmitted in the network is one of the most important requirements for any application in dynamic ad hoc networks. The challenge for the routing protocols is in providing security to the transmitted data while maintaining excellent QoS and minimum overhead in the network. Combining security with QoS is a major requirement and also an important challenge for new routing protocols in dynamic ad hoc networks.

Opportunistic Routing For VANETS

Vehicular Ad Hoc Networks (Li & Wang, 2007; Menon & Prathap 2017) are specific type of MANETs in which the mobile nodes are moving vehicles. In VANETs every vehicle is equipped with an on-board unit and a group of sensors. The radio interfaces or on-board unit enables short range wireless ad hoc networks to be formed. VANETs often has multiple Road Side Units (RSUs) deployed as intermediary servers near the vehicles to process the data. VANETs offer both vehicle to vehicle and vehicle to roadside unit communication. Every vehicle in the network plays the role of a sender, receiver and a router to broadcast data to the vehicular network and the roadside units which then uses the data to ensure safe and free flow of traffic. Opportunistic routing is an emerging routing technology in VANETs. Opportunistic routing can be used for efficient communication and resource sharing between the moving vehicles. This research area is still underexplored and needs more focus. This would help VANETs to be used efficiently in the design and development of Intelligent Transportation Systems (ITS) that offers improved safety and better transportation. Opportunistic routing could also help in efficient traffic monitoring and emergency services using VANETs.

Opportunistic Routing for Underwater Sensor Networks

In recent years underwater wireless sensor networks (UWSNs) (Sozer et al., 2000; Menon et al., 2016) have emerged as one of the most popular researched areas in the networking field due to its numerous applications in ocean exploration, underwater surveillance and pollution detection. With specific characteristics such as higher delay in transmission, continuously moving nodes, limited storage and battery, most of the routing protocols that work with traditional sensor networks cannot be used in underwater wireless sensor networks. It has been a challenge for researchers to route the data packets efficiently from the source to the destination. Designing efficient routing protocols for underwater sensor networks is yet another emerging area of research. Opportunistic protocols are being applied for routing data packets in underwater sensor networks and have been found to be effective.

CONCLUSION

This research paper discussed the progress of routing protocols used in ad hoc networks from the traditional topology based to the latest opportunistic protocols. The paper initially discussed the working of three major categories of routing protocols; traditional topology based protocols, geographic protocols and opportunistic routing protocols. The working and behavior of the most popular protocols from each category was presented and analyzed in detail. Further the paper discussed the issues and drawbacks of these protocols in highly dynamic ad hoc networks. The performance of the various categories of routing protocols in highly dynamic ad hoc networks was analyzed using simulations in NS-2.The research then discussed the reasons for the performance degradation of these protocols with increasing mobility of nodes in the network and proposed the need for more efficient routing protocols. The paper finally discussed the issues, challenges and future research directions in this area.

REFERENCES

Basagni, S., Chlamtac, I., Syrotiuk, V. R., & Barry, W. (1998). A Distance Routing Effect Algorithm for Mobility (DREAM). In *Proceedings of the 4th Annual ACM/IEEE International Conference on Mobile Computing and Networking* (pp.76-84). New York: IEEE. 10.1145/288235.288254

Bellman, R. (1958). On a Routing Problem. *Quarterly of Applied Mathematics, 16*(1), 87–90. doi:10.1090/qam/102435

Bhorkar, A., Naghshvar, M., Javidi, T., & Rao, B. (2012). Adaptive Opportunistic Routing for Wireless Ad-Hoc Networks. *IEEE/ACM Transactions on Networking, 20*(1), 243–256. doi:10.1109/TNET.2011.2159844

Biswas, S., & Morris, R. (2005). ExOR: Opportunistic Multi-Hop Routing for Wireless Networks. In *Proceedings of the 2005 Conference on Applications, Technologies, Architectures, and Protocols for Computer Communications (SIGCOMM '05)* (pp. 133-144). New York: ACM. 10.1145/1080091.1080108

Bletsas, A., Dimitriou, A., & Sahalos, J. (2010). Interference-limited Opportunistic Relaying with Reactive Sensing. *IEEE Transactions on Wireless Communications, 9*(1), 14–20. doi:10.1109/TWC.2010.01.081128

Bruno, R., & Conti, M. (2010). MaxOPP: A Novel Opportunistic Routing for Wireless Mesh Networks. In *Proceedings of IEEE Symposium on Computers and Communications (ISCC)* (pp. 255-260). Riccione, Italy: IEEE. 10.1109/ISCC.2010.5546793

Cadger, F., Curran, K., Santos, J., & Moffett, S. (2013). A Survey of Geographical Routing in Wireless Ad-Hoc Networks. *IEEE Communications Surveys and Tutorials, 15*(2), 621–653. doi:10.1109/SURV.2012.062612.00109

Carter, S., & Yasinsac, A. (2002). Secure Position Aided Ad hoc Routing. In *Proceedings of the IASTED International Conference on Communications and Computer Networks (CCN02)* (pp.329-334). Academic Press.

Chachulski, S., Jennings, M., Katti, S., & Katabi, D. (2007). Trading Structure for Randomness in Wireless Opportunistic Routing. In *Proceedings of the 2007 Conference on Applications, Technologies, Architectures, and Protocols for Computer Communication* (pp. 169-180). Academic Press. 10.1145/1282380.1282400

Chlamtac, I., Conti, M., & Liu, J. (2003). Mobile Ad Hoc Networking: Imperatives and Challenges. *Ad Hoc Networks, 1*(1), 13–64. doi:10.1016/S1570-8705(03)00013-1

Conan, J. L. V., & Friedman, T. (2008). Fixed Point Opportunistic Routing in Delay Tolerant Networks. *IEEE Journal on Selected Areas in Communications, 26*(5), 773–782. doi:10.1109/JSAC.2008.080604

Dubois-Ferriere, H., Grossglauser, M., & Vetterli, M. (2007). Least-Cost Opportunistic Routing. In *Proceedings of 2007 Allerton Conference on Communication, Control, and Computing* (pp. 1-8). Allerton, UK: Academic Press.

Erramilli, A. C. V., Crovella, M., & Diot, C. (2008). Delegation Forwarding. In *Proceedings of ACM International Symposium on Mobile Ad Hoc Networking and Computing* (pp. 251-260). Hong Kong: ACM.

Fang, X., Yang, D., & Xue, G. (2011). Consort: Device-Constrained Opportunistic Routing in Wireless Mesh Networks. In Proceedings of IEEE INFOCOM (pp. 1907-1915). Shanghai, China: IEEE.

Fang, X., Yang, D., & Xue, G. (2013). Map: Multi-Constrained Any Path Routing in Wireless Mesh Networks. *IEEE Transactions on Mobile Computing, 12*(10), 1893–1906. doi:10.1109/TMC.2012.158

Ghadimi, E., Landsiedel, O., Soldati, P., Duquennoy, S. & Johansson, M. (2014). Opportunistic Routing in Low Duty-Cycled Wireless Sensor Networks. *ACM Transactions on Sensor Networks, 10*(4), 67:1-67:39.

Giordano, S., & Lu, W. (2001). Challenges in Mobile Ad Hoc Networking. *IEEE Communications Magazine, 39*(6), 29–129. doi:10.1109/MCOM.2001.925680

Guo, S., Gu, Y., Jiang, B., & He, T. (2009). Opportunistic Flooding in Lowduty- Cycle Wireless Sensor Networks with Unreliable Links. In *Proceedings of IEEE/ACM Annual International Conference on Mobile Computing and Networking (MobiCom)* (pp. 133-144). Beijing, China: IEEE. 10.1145/1614320.1614336

Haas, Z.J., & Pearlman, M.R. (1997). *The Zone Routing Protocol (ZRP) for Ad Hoc Networks.* Internet Draft, Available at hdraft-haaszone-routing-protocol-00.txt.

Han, M. K., Bhartia, A., Qiu, L., & Rozner, E. (2011). O3: Optimized Overlay Based Opportunistic Routing. In *Proceedings of the ACM International Symposium on Mobile Ad Hoc Networking and Computing (MobiHoc)* (pp. 2:1-2:11). ACM.

Hsu, C., Liu, H., & Seah, W. (2009). Economy: A Duplicate Free Opportunistic Routing. In *Proceedings of the 6th ACM International Conference on Mobile Technology Application and Systems* (pp. 17:1-17:6). New York: ACM.

Hwang, H., Hur, I., & Choo, H. (2009). GOAFR plus-ABC: Geographic routing based on Adaptive Boundary Circle in MANETs. In *Proceedings of the 2009 International Conference on Information Networking* (pp. 1-3). Chiang Mai, Thailand: Academic Press.

Jacquet, P., Muhlethaler, P., & Qayyum, A. (1998). *Optimized Link State Routing Protocol.* Internet Draft, Available at draft-ietf-manetolsr-00.txt.

Johnson, D. B., & Maltz, D. A. (1996). Dynamic Source Routing in Ad Hoc Wireless Networks. *Mobile Computing, Kluwer Academic Publishers, 353*, 153–181. doi:10.1007/978-0-585-29603-6_5

Karp, B., & Kung, H. T. (2000). GPSR: Greedy Perimeter Stateless Routing for Wireless Networks. In *Proceedings of the 6th Annual International Conference on Mobile Computing and Networking* (pp. 243-254). New York: Academic Press. 10.1145/345910.345953

Koutsonikolas, D., Hu, Y., & Wang, C. (2008). XCOR: Synergistic Interflow Network Coding and Opportunistic Routing. In *Proceedings of the ACM Annual International Conference on Mobile Computing and Networking* (pp.1-3). San Francisco, CA: ACM.

Lampin, Q., Barthel, D., Aug-Blum, I., & Valois, F. (2012). QOS Oriented Opportunistic Routing Protocol for Wireless Sensor Networks. In *Proceedings of IEEE/IFIP Wireless Days* (pp. 1-6). Dublin, Ireland: IEEE. 10.1109/WD.2012.6402804

Laufer, R., Velloso, P. B., Vieira, L. F. M., & Kleinrock, L. (2012). Plasma: A New Routing Paradigm for Wireless Multihop Networks. In *Proceedings of IEEE Conference on Computer Communications (INFOCOM)* (pp. 2706-2710). Orlando, FL: IEEE. 10.1109/INFCOM.2012.6195683

Lee, G., & Haas, Z. (2011). Simple, Practical, and Effective Opportunistic Routing for Short-Haul Multi-Hop Wireless Networks. *IEEE Transactions on Wireless Communications*, *10*(11), 3583–3588. doi:10.1109/TWC.2011.092711.101713

Li, F., & Wang, Y. (2007). Routing in vehicular ad hoc networks: A survey. *IEEE Vehicular Technology Magazine*, *2*(2), 12–22. doi:10.1109/MVT.2007.912927

Li, J., & Mohapatra, P. (2003). LAKER: Location Aided Knowledge Extraction Routing for Mobile Ad Hoc Networks. In Proceedings of 2003 IEEE Wireless Communications and Networking, 2003. WCNC 2003 (pp. 1180-1184). New Orleans, LA: IEEE.

Li, Y., Mohaisen, A., & Zhang, Z. (2013). Trading Optimality for Scalability in Large-Scale Opportunistic Routing. *IEEE Transactions on Vehicular Technology*, *62*(5), 2253–2263. doi:10.1109/TVT.2012.2237045

Lin, Y., Li, B., & Liang, B. (2008). CodeOR: Opportunistic Routing in Wireless Mesh Networks with Segmented Network Coding. In *Proceedings of the IEEE International Conference on Network Protocols (ICNP)* (pp. 13-22). Orlando, FL: IEEE.

Lin, Y., Li, B., & Liang, B. (2010). SlideOR: Online Opportunistic Network Coding in Wireless Mesh Networks. In *Proceedings of IEEE Conference on Computer Communications (INFOCOM)*, (pp.171-175). San Diego, CA: IEEE. 10.1109/INFCOM.2010.5462249

Lu, M., Steenkiste, P., & Chen, T. (2009). Design, Implementation and Evaluation of an Efficient Opportunistic Retransmission Protocol. In *Proceedings of IEEE/ACM Annual International Conference on Mobile Computing and Networking* (pp.73-84). IEEE. 10.1145/1614320.1614329

Lu, M., & Wu, J. (2009). Opportunistic Routing Algebra and its Applications. In *Proceedings IEEE Conference on Computer Communications (IEEE INFOCOM)* (pp. 2374-2382). Rio de Janeiro, Brazil: IEEE.

Mao, X., Tang, S., Xu, X., Li, X., & Ma, H. (2011). Energy Efficient Opportunistic Routing in Wireless Sensor Networks. *IEEE Transactions on Parallel and Distributed Systems*, *22*(11), 1934–1942. doi:10.1109/TPDS.2011.70

Mase, K. (2011). How to Deliver Your Message from/to a Disaster Area. *IEEE Communications Magazine*, *49*(1), 52–57. doi:10.1109/MCOM.2011.5681015

Mauve, M., Widmer, J., & Hartenstein, H. (2001). A Survey on Position-Based Routing in Mobile Ad Hoc Networks. *IEEE Network*, *15*(6), 30–39. doi:10.1109/65.967595

Menon, Prathap & Priya. (2016). Ensuring Reliable Communication in Disaster Recovery Operations with Reliable Routing Technique. Mobile Information Systems.

Menon, V. G., & Prathap, P. M. (2016). Routing in Highly Dynamic Ad Hoc Networks: Issues and Challenges. *International Journal on Computer Science and Engineering*, *8*(4), 112–116.

Menon V.G. & Prathap P. M. (2017). Vehicular Fog Computing: Challenges Applications and Future Directions. *International Journal of Vehicular Telematics and Infotainment Systems, 1*(2), 15-23.

Menon, V. G., & Prathap, P. M. J. (2016). Analysing the Behaviour and Performance of Opportunistic Routing Protocols in Highly Mobile Wireless Ad Hoc Networks. *IACSIT International Journal of Engineering and Technology*, *8*(5), 1916–1924. doi:10.21817/ijet/2016/v8i5/160805409

Menon, V. G., & Prathap, P. M. J. (2016). Comparative Analysis of Opportunistic Routing Protocols for Underwater Acoustic Sensor Networks. *Proceedings of the IEEE International Conference on Emerging Technological Trends*. 10.1109/ICETT.2016.7873733

Menon, V. G., Prathap, P. M. J., & Vijay, A. (2016). Eliminating Redundant Relaying of Data Packets for Efficient Opportunistic Routing in Dynamic Wireless Ad Hoc Networks. *Asian Journal of Information Technology*, *12*(17), 3991–3994.

Menon, V. G., Priya, P. M. J., & Prathap, P. M. J. (2013). Analyzing the behavior and performance of greedy perimeter stateless routing protocol in highly dynamic mobile ad hoc networks. *Life Science Journal*, *10*(2), 1601–1605.

Na, J., & Kim, C.-K. (2006). GLR: A Novel Geographic Routing Scheme for Large Wireless Ad Hoc Networks. *Computer Networks*, *50*(17), 3434–3448. doi:10.1016/j.comnet.2006.01.004

Nassr, M., Jun, J., Eidenbenz, S., Hansson, A., & Mielke, A. (2007). Scalable and Reliable Sensor Network Routing: Performance Study from Field Deployment. In *Proceedings of IEEE INFOCOM 2007 - 26th IEEE International Conference on Computer Communications* (pp. 670–678). Anchorage, AK: IEEE. 10.1109/INFCOM.2007.84

Navidi, W., & Camp, T. (2004). Stationary Distributions for the Random Waypoint Mobility Model. *IEEE Transactions on Mobile Computing*, *3*(1), 99–108. doi:10.1109/TMC.2004.1261820

Nelson, S., Bakht, M., Kravets, R., & Harris, A. F. III. (2009). Encounter-Based Routing in DTNs. *Mobile Computing and Communications Review*, *13*(1), 56–59. doi:10.1145/1558590.1558602

Park, V. D., & Corson, M. S. (1997). A Highly Adaptive Distributed Routing Algorithm for Mobile Wireless Networks. In *Proceedings of the Sixteenth Annual Joint Conference of the IEEE Computer and Communications Societies* (pp. 1405-1413). Kobe, Japan: IEEE. 10.1109/INFCOM.1997.631180

Perkins, C. E., & Bhagwat, P. (1994). Highly Dynamic Destination-Sequenced Distance-Vector Routing (DSDV) for Mobile Computers. In *Proceedings of the Conference on Communications Architectures, Protocols and Applications (SIGCOMM '94)* (pp. 234-244). New York: ACM. 10.1145/190314.190336

Perkins, C. E., & Royer, E. M. (1999). Ad-Hoc On-Demand Distance Vector Routing. In *Proceedings of Second IEEE Workshop on Mobile Computing Systems and Applications* (pp. 90-100). New Orleans, LA: IEEE. 10.1109/MCSA.1999.749281

Rosario, D., Zhao, Z., Braun, T., Cerqueira, E., Santos, A., & Alyafawi, I. (2014). Opportunistic Routing for Multi-Flow Video Dissemination Over Flying Ad hoc Networks. In *Proceeding of IEEE International Symposium on World of Wireless, Mobile and Multimedia Network* (pp. 1-6). Sydney, Australia: IEEE. 10.1109/WoWMoM.2014.6918947

Rozner, E., Seshadri, J., Mehta, Y., & Qiu, L. (2009). SOAR: Simple Opportunistic Adaptive Routing Protocol for Wireless Mesh Networks. *IEEE Transactions on Mobile Computing*, 8(1), 1622–1635. doi:10.1109/TMC.2009.82

Saha, B., Misra, S., & Obaidat, M. (2013). A Web-Based Integrated Environment for Simulation and Analysis with NS-2. *IEEE Wireless Communications*, 20(4), 109–115. doi:10.1109/MWC.2013.6590057

Shin, S. C. W. Y., & Lee, Y. (2013). Parallel Opportunistic Routing in Wireless Networks. *IEEE Transactions on Information Theory*, 59(10), 6290–6300. doi:10.1109/TIT.2013.2272884

Sozer, E. M., Stojanovic, M., & Proakis, J. G. (2000). Underwater acoustic networks. *IEEE Journal of Oceanic Engineering*, 25(1), 72–83. doi:10.1109/48.820738

Wang, Z., Chen, Y., & Li, C. (2012). CORMAN: A Novel Cooperative Opportunistic Routing Scheme in Mobile Ad Hoc Networks. *IEEE Journal on Selected Areas in Communications*, 30(2), 289–296. doi:10.1109/JSAC.2012.120207

Westphal, C. (2006). Opportunistic Routing in Dynamic Ad Hoc Networks: The OPRAH Protocol. In *Proceedings of the 2006 IEEE International Conference on Mobile Ad Hoc and Sensor Systems* (pp. 570-573). Vancouver, Canada: IEEE. 10.1109/MOBHOC.2006.278612

Witt, M., & Turau, V. (2005). BGR: Blind Geographic Routing for Sensor Networks. In *Proceedings of the Third International Workshop on Intelligent Solutions in Embedded Systems* (pp. 51-61). Academic Press. 10.1109/WISES.2005.1438712

Xiao, M., Wu, J., Liu, K., & Huang, L. (2013). Tour: Time-Sensitive Opportunistic Utility Based Routing in Delay Tolerant Networks. In Proceedings of IEEE INFOCOM (pp. 2085 2091). Turin, Italy: IEEE.

Yen, J. Y. (1970). An Algorithm for Finding Shortest Routes from All Source Devices to A Given Destination in General Networks. *Quarterly of Applied Mathematics*, *27*(1), 526–530. doi:10.1090/qam/253822

Yuan, Y., Yang, H., Wong, S., Lu, S., & Arbaugh, W. (2005). ROMER: Resilient Opportunistic Mesh Routing for Wireless Mesh Networks. In *Proceedings of the IEEE Workshop on Wireless Mesh Networks* (pp. 1-9). IEEE.

Zeng, K., Lou, W., Yang, J., & Brown, D. R. III. (2007). On Throughput Efficiency of Geographic Opportunistic Routing in Multihop Wireless Networks. *Mobile Networks and Applications*, *12*(5), 347–357. doi:10.100711036-008-0051-7

Zeng, K., Yang, Z., & Lou, W. (2009). Location-Aided Opportunistic Forwarding in Multirate and Multihop Wireless Networks. *IEEE Transactions on Vehicular Technology*, *58*(6), 3032–3040. doi:10.1109/TVT.2008.2011637

Zhang, X., & Li, B. (2009). Optimized Multipath Network Coding in Lossy Wireless Networks. *IEEE Journal on Selected Areas in Communications*, *27*(5), 622–634. doi:10.1109/JSAC.2009.090605

Zhao, Z., Rosario, D., Braun, T., & Cerqueira, E. (2014). Context-Aware Opportunistic Routing in Mobile Ad-Hoc Networks Incorporating Device Mobility. In *Proceedings of the IEEE Wireless Communications and Networking Conference* (pp. 2138–2143). Istanbul, Turkey: IEEE.

Zhao, Z., Rosario, D., Braun, T., Cerqueira, E., Xu, H., & Huang, L. (2013). Topology and Link Quality-Aware Geographical Opportunistic Routing in Wireless Ad-Hoc Networks. In *Proceedings of the IEEE International Wireless Communications and Mobile Computing Conference (IWCMC)* (pp. 1522-1527). Sardinia, Italy: IEEE. 10.1109/IWCMC.2013.6583782

Zhong, Z., Wang, J., Lu, G., & Nelakuditi, S. (2006). On Selection of Candidates for Opportunistic Any Path Forwarding. *Mobile Computing and Communications Review*, *10*(4), 1–2. doi:10.1145/1215976.1215978

Chapter 2
Dynamic Fog Computing:
Practical Processing at Mobile Edge Devices

Sander Soo
University of Tartu, Estonia

Chii Chang
University of Tartu, Estonia

Seng W. Loke
Deakin University, Australia

Satish Narayana Srirama
University of Tartu, Estonia

ABSTRACT

The emerging Internet of Things (IoT) systems enhance various mobile ubiquitous applications such as augmented reality, environmental analytics, etc. However, the common cloud-centric IoT systems face limitations on the agility needed for real-time applications. This motivates the Fog computing architecture, where IoT systems distribute their processes to the computational resources at the edge networks near data sources and end-users. Although fog computing is a promising solution, it also raises a challenge in mobility support for mobile ubiquitous applications. Lack of proper mobility support will increase the latency due to various factors such as package drop, re-assigning tasks to fog servers, etc. To address the challenge, this chapter proposes a dynamic and proactive fog computing approach, which improves the task distribution process in fog-assisted mobile ubiquitous applications and optimizes the task allocation based on runtime context information. The authors have implemented and validated a proof-of-concept prototype and the chapter discusses the findings.

DOI: 10.4018/978-1-5225-5693-0.ch002

INTRODUCTION

The information systems designed for integrating the Internet of Things (IoT) (Gubbi et al., 2013) are usually applying the global centralized model, in which the IoT devices rely on distant management systems. Such a model is considered to be a drawback in terms of agility (Bonomi et al., 2012). In many real-time ubiquitous applications such as augmented reality, environmental analytics, ambient assisted living, etc., mobile device users require rapid responses. However, the latency caused by the distant centralized model is too high, even though the mobile Internet speed has improved significantly during the last few years. To address this problem, Fog Computing (Fog) (Bonomi et al., 2012) introduces data pre-processing with the computers in the vicinity of the data sources and end-user applications located in the edge network of IoT systems.

In general, Fog computing resources, which are known as Fog nodes, are mediating devices that connect the edge network with the Internet. Some typical examples are industrial integrated routers (e.g., Cisco 829 Industrial Integrated Services Routers), home hubs or set-top boxes that are employed as wireless Internet access points together with embedded virtualization technologies (e.g., Virtual Machines) or containerization technologies (e.g., Docker containers (https://www.docker.com)), which allow clients to deploy software onto them. Compared to the traditional distant Cloud computing model, which requires sending all the data to the Distant Data Center (DDC) for the processing, Fog can provide much better agility.

Although Fog-driven IoT system provides explicit enhancement in performance, it also faces numerous challenges in terms of connectivity (Zhang et al., 2015), discoverability (Troung-Huu et al., 2014), efficient deployment (Ravi & Peddoju, 2014; Guo et al., 2016; Ceselli et al., 2017; Lin & Shen, 2017) and so on. While many of the previous works focused on Fog deployment for specific use cases, this chapter aims to address the mobility issue raised in the case of integrating Fog with ubiquitous mobile applications.

Imagine a mobile ubiquitous care application that needs to provide real-time environmental information to its user by continuously collecting and processing data derived from the surrounding environment while its user is moving in outdoor areas. For improving the efficiency, the mobile device (i.e. delegator) is distributing its computational tasks to vicinal Fog servers (i.e. workers). However, the delegator may need to repeatedly resend the tasks to different Fog nodes, due to the dynamic nature of the mobile environment, where the limited wireless signal coverage of the Fog nodes could cause failure in delivering results.

Consequently, it raises a question:

How can the system avoid the situation that requires the delegator to re-send tasks to the other workers due to the failed process result delivery?

In order to address the question, this chapter proposes a proactive task distribution framework for mobile Fog environments. The proposed framework consists of two core schemes:

- Proactive task distribution, which is an extension of the Work Stealing scheme (Loke et al., 2015) that provides the mechanism to hasten the speed of task distribution.
- Context-aware Work Stealing, which provides an optimal decision-making mechanism that helps workers (Fog nodes) to decide how they should participate in the distributed processes.

In essence, the contribution of the chapter is to study the potential of applying context-aware Work Stealing scheme in Fog computing towards improving the mobility-awareness, taking into account the characteristics of the workers and tasks in a dynamic and comprehensive manner. The study provides new insights about how distributed systems can achieve the high-performance process migration in the edge networks. Although the study is based on a specific ubiquitous application use case, the involved theoretical design still provides an important foundation for the discipline of mobile distributed computing.

Recent research trend of fog computing has considered distributing the processes to various gateway devices including the resource-constrained single-board computers (Petrolo et al., 2017; Verba et al., 2016; Pahl et al., 2016; Bellavista et al., 2017; Amento et al., 2016; Elkhatib et al., 2017; Krylovskiy et al., 2015; Hajji et al., 2016; Morabito et al., 2016; Kempen et al., 2017). In order to address the state-of-the-art of the Fog computing engineering, this chapter complements previous research (Soo et al., 2017) with additional discussions of recent related literature together with enhanced experiments to demonstrate the feasibility and validity of applying the proposed context-aware work-stealing model in different Fog computing environments.

This chapter is organized as follows. In the next section, the authors provide an overview and comparison of the related works. Afterwards, the details of the proposed system design are described. This is followed by the Evaluation section that provides detailed analysis of the performed experiments. Finally, this chapter is concluded along with future research directions.

BACKGROUND

Computation Offloading

Computation offloading is a common strategy to reduce the resource consumption and to improve the overall performance of ubiquitous mobile applications. Specifically, earlier works such as MAUI (Cuervo et al., 2010) or Cuckoo (Kemp et al., 2012) have introduced the schemes that assist the system in offloading the process from mobile devices to central surrogates such as the Cloud, where MAUI also takes the energy usage into account.

Considering the latency caused by the centralized offloading schemes, recent strategies have introduced the utilization of vicinal computational resources such as Virtual Machine (VM)-based Cloudlet (Satyanarayanan et al., 2009). In general, Cloudlet represents the VM-enabled server machines located on the same network as the mobile application nodes. For example, a local business may provide a Cloudlet machine to their customers to improve the Quality of Experience (QoE) of the mobile applications used by the customers.

In order to optimize the efficiency of the Cloudlet-based computation offloading, a number of researchers have proposed the machine learning algorithms to help the mobile applications' decision in whether or not to offload the tasks (Zhang et al., 2014, 2015; Truong-Huu et al., 2014). Truong-Huu et al. (2014) propose a Markov Decision Process (MDP) based approach for the opportunistic offloading from mobile to Cloudlet, taking into account the factors learned from past events. The work only considers a set of parallel uniform tasks as the work to be executed. Zhang et al. (2014) propose an MDP model that divides the application into multiple phases. The phases are then partitioned to run either on the mobile device or on the Cloudlet.

Machine learning algorithms are also included in other migration challenges. Wang et al. (2015) propose dynamic service migration in edge network based on user's movement using MDP, taking into account the distance between the user and the service in order to find the optimal policy.

Similarly, the offloading optimization is also an imperative research question in terms of balancing the workload between Cloud and Fog (Lin & Shen, 2017) and optimizing the task distribution in mobile ad hoc Clouds (Shi et al., 2015; Yousafzai et al., 2016).

The Mobile Complex Event Processing (MCEP) model (Ottenwälder et al., 2013) was proposed in the context of vehicle to infrastructure (V2I) in order to increase support for VM migration between infrastructural worker nodes for vehicular data sources. The focus of the proposed solution is on the continuous data stream processing for moving vehicles.

Although existing works described above have proposed numerous strategies for distributing the computational tasks from mobile devices to external resources, most of them have assumed the communication between the delegator and the workers is fairly stable, thus they did not fully address the challenge raised in this chapter.

Result Routing

An important aspect in mobile Fog is how to route the process result back to the delegator. In particular, the delegator may have moved out from the wireless network coverage of the Fog node, which has taken the computational tasks.

Instead of assuming the connectivity between the delegator and workers is fairly stable, a number of related research projects have proposed corresponding strategies. For example, Zhang et al. (2016) and Su et al. (2015) propose opportunistic collaborative caching with proximal peers, with the former also taking into account the social relationship of peers. Further, Fernando et al. (2013; 2016) and Shi et al. (2012) utilize the Time-To-Live (TTL) policy, which defines the work expiring time, i.e. the time before the delegator restarts its delegation process. However, these approaches can potentially cause extra latency. Hence, Ravi et al. (2014) have proposed the interconnected Cloudlet scheme in which the delegator can establish a data routing network among multiple Cloudlet machines on the move.

The proposed approach of this chapter addresses the issue of result routing by identifying the best approach for the results delivery, depending on the performance analysis of the approaches for the given context and scenario.

Load Balancing and Efficient Deployment

A lot of work (Ceselli et al., 2017; Lin & Shen, 2017; Guo et al., 2016; Hong et al., 2013) has introduced approaches for optimizing the workload or discussed the efficiency of deployment for applications that could be improved via Fog. Specifically, Ceselli et al. (2017) have proposed a scheme for the optimized placement of Virtual Machines (VM) that provides improved computation support. Hong et al. (2013) proposed a process placement algorithm based on utilizing the customized scaling policy acquired from the user. Arkian et al. (2017) proposed an optimized solution for cost-efficient task assignment and VM placement in the Fog environment, thereby providing an improvement for the deployment stage of Fog applications prior to runtime and creating opportunities, for example, for the support of crowd sensing applications in the context of IoT and Fog. Bitam et al. (2017) proposed Bees Life Algorithm (BLA) for Fog scheduling, taking into account CPU and memory requirements and distributing the tasks among all the Fog computing nodes. Ceselli et al. (2015) proposed a model for optimising the Cloudlet network design in terms of Cloudlet node placement and work assignments, with the model also including the optimisation in traffic routing. In comparison with Arkian et al. (2017), the latter is more focused on cost-efficiency, whereas Ceselli et al. (2015) focuses on mobility and further deals with runtime VM migration.

Existing works (Huerta-Canepa & Lee, 2010; Marinelli, 2009) did not consider the heterogeneous device capabilities of the worker nodes. Consequently, this raises the issue of assuming that heterogeneous and unknown devices have uniform capabilities. However, in the case of uniformly distributing the works, some nodes may be overloaded and cannot accept more works. Further, the weaker nodes may not be able to effectively complete the tasks they received in time and thereby result in the bottleneck issue.

Different to the previous works that were based on reactive strategies, the approach proposed in this chapter is a proactive task distribution scheme that combines the Work Stealing scheme with context-awareness.

Mobility-Aware Edge Computing

Prior works have considered the mobility-awareness of task distribution (Bittencourt et. al., 2017; Chamola et al., 2017; Li et al., 2013). Bittencourt et. al. (2017) propose policy-based task scheduling to improve the mobility-awareness of Cloudlet services. Chamola et al. (2017) propose a framework that allows mobile devices to offload computationally intensive tasks to Cloudlets, where the decision of which Cloudlet handles the task is made by a central Cloudlet manager. Li et al. (2013) propose a solution for finding the best and closest Cloudlet Wi-Fi Access Point based on predicting the user movement. Similar to Arkian et al. (2017), this work can be used in the plan-phase for the central server to determine the most optimal location to deploy the Fog server based on users' movement.

Existing works (Alam et al., 2016; Chamola et al., 2017; Lee & Shin, 2013) have also attempted to address the mobility-awareness of computation offloading. Specifically, Alam et al. (2016) proposed a mobility-aware extension to Fog, based on reinforcement learning. Lee & Shin (2013) proposed a mobile computation offloading scheme based on user mobility models, offering improved mobility support by predicting user movement and future network conditions.

Although previous research has considered mobility-awareness in various contexts, several of these works have only considered the strategies for task allocation in the context of a single computational node and have not focused on utilizing vicinal nodes. Furthermore, various assumptions, e.g., the offloadable application will be initially executed on the mobile device in order to analyze the execution beforehand, may not be applicable for real-world applications and tasks. Similarly, having a centralized manager in place to handle the decision of computation offloading or task allocation makes the system less adaptive and increases overall latency.

SYSTEM DESIGN

Overview

Figure 1 illustrates an overview of the proposed system based on a mobile sensing as a service scenario (Chang et al., 2015) in which the smartphone owner is participating in a system that applies smartphones

Figure 1. Overview of proactive fog computing in case of mobile sensing service. Fog servers actively 'steal' works from mobile host when the mobile host moved into their coverage.

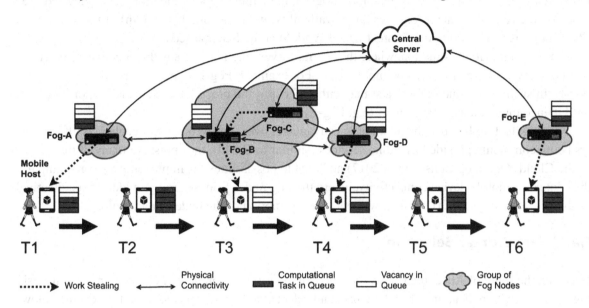

as servers (known as mobile hosts) and utilizes the sensors incorporated within smartphones (e.g., proximity sensor, camera, sound etc.) to collect environmental data and upload the data to the cloud-side central server after the mobile host pre-processes the raw data. In order to improve the life cycle of the mobile host, the central server has integrated its system with the local Fog service providers. Hence, mobile hosts can delegate their works of data pre-processing to the nearby Fog servers.

One assumption is that the route of the mobile host owner is predicable using the corresponding mechanism based on the historical records (Rahaman et al., 2017) or Google Maps API.

Another assumption is that the central server has configured the Fog servers before the mobile node starts moving.

Central server selects the candidate Fog servers based on the route of the user (see the Candidate Worker Selection section). Further, it also deploys the software to the Fog servers to trigger the proactive behavior.

In Figure 1, T1 to T6 represent the timestamps of the mobile host's route. In general, while the mobile host is moving, it (i.e., delegator) advertises its existence to Fog nodes (i.e., workers) with a host description message. Specifically, the message contains the information of the tasks in the delegator's queue, including the type of the computational tasks. For example, the task can be CPU-intensive, GPU-intensive, RAM-intensive and so on. This information is updated periodically via the Work Stealing requests.

As Figure 1 shows, in T1, a chosen candidate Fog-A actively 'steals' two work items from the delegator. While the mobile host is moving, it has generated more works (T2). At T3, the delegator has encountered a new worker Fog-B. Since there are two workers in the group where Fog-B belongs, both of them will assist the mobile host.

As can be seen from the figure, Fog-B and Fog-C have acquired a different number of tasks (Fog-C has stolen some tasks from the delegator via Fog-B). In summary, the number of works they acquire is based on the runtime context (e.g., resource availability, workload and bandwidth) of the Fog servers. The details are described in the Context-aware Work Stealing Scheme section.

There is an inevitable situation, where the central server does not find any direct connection between two Fog servers on the route of the mobile host. For example, Fog-D does not have the connection to Fog-E. In such a case, the mobile host may either perform the process by itself (at T5), or it can wait until it encounters the next Fog server (e.g. Fog-E).

Utilizing the Fog-based architecture offers advantages over various other approaches. With Fog servers being commonly provided as public services (similar to Cloud), they preserve many of the benefits of the Cloud (OpenFog Consortium, 2017) and enable possibilities for improvements, for example, in the context of mobile device connectivity, where the Fog nodes from several providers can collectively enable a greater degree of flexibility, without any assumption of the underlying topology, etc.

Candidate Worker Selection

Based on the route of Alice, the Cloud backend can identify the candidate Fog nodes needed in assisting Alice's AAL application. The Cloud backend selects the Fog servers based on the scheme below.

Let $E = \{E_1, \ldots, E_n\}$ be the set of all possible encounter Fog nodes on the end user's path. $E_k = \{e_i : 1 \leq i \leq N\}$ denotes the current encounter node(s) and E_{k+1} denotes the next encounter node(s) after E_k. E_1 are the closest encounter nodes to the starting point of the user.

Let E_x be one of the members of E. Then each of the $e_i \in E_x$ (denoted by e_i^x) would be included based on the following considerations:

- If e_i^x has a route through network infrastructure to e_i^{x+1} without utilizing the Cloud, then e_i^x is considered as a priority candidate.
- Let e_y^x to be one of the e_i^x. If e_y^x has a direct route to a priority candidate, it is also considered as a candidate.
- An isolated Fog node, with high computational capabilities, on the moving path of the user, where there are no alternatives, is also considered as a candidate.

Once the Cloud backend selected the candidate Fog servers, it will deploy the corresponding software to the Fog servers in order to trigger the proactive behavior.

The Worker Network

The central server forms a worker network for the mobile host based on configuring the candidate Fog nodes into non-overlapping subgroups beforehand, such that all the members of the group are aware of the other Fog nodes in their group. The Work Stealing process among Fog nodes only takes place within the boundaries of the group.

The members of the group receive updates on various characteristics of their peers, such as the current CPU usage, RAM usage, bandwidth usage, etc., delivered by a resource-efficient publish-subscribe protocol such as MQTT (Banks & Gupta, 2014). MQTT also offers a *Last Will and Testament* feature, to notify peers when a node unexpectedly goes offline. Such information is useful in the optimization process.

Upon receiving information from the delegator, the Fog node also notifies the group members about the number of work items available per type (e.g., CPU intensive, GPU intensive, RAM intensive, etc.).

Once the Fog node steals the task(s) from the delegator, the Fog node will also register to a topic of the delegator's events (e.g., an update of the current location). Thus, the Fog nodes would have the knowledge of the delegator's currently connected Fog node and could transmit the results to the delegator accordingly.

Transmitting the results back to the user via the network of Fog nodes could still cover as many Fog nodes as needed.

The grading value to partition the tasks indicates the expected number of tasks a Fog node will handle. This value can be used as an estimate for the initially stolen tasks. Any Fog node will effectively make the final decision on whether or not to take some tasks for processing.

Context-Aware Work Stealing Scheme

Basic Multi-Layered Work Stealing

Loke et. al. (2015) introduce an extension of the original Work Stealing approach for mobile ad-hoc Cloud computing. In addition to the workers and delegators (i.e. distributors of tasks), there exist intermediaries, who can act as workers for some nodes and delegators to others. This enables the delegator to distribute the works to the workers beyond the direct connection range, thereby increasing the resources available for handling tasks and creating a multi-layered view of the system. As the workers finish their tasks, they steal more work from each other and via the delegator.

Context-Awareness Extension

The context-aware extension of the Work Stealing scheme optimizes the multi-layered work item distribution within the worker network, considering the current capabilities of the workers. The scheduled work items (tasks) contain information about the primary hardware resource required for processing the tasks. Such information helps the workers to steal works based on their resource availability. For example, a worker will steal CPU-intensive tasks if it has available CPU resources. Fog nodes would also query for runnables to process the work, either directly from the delegator (e.g., jar files or offline Docker images) or from the Internet (e.g., Docker Hub).

To extend the basic possibility of stealing one task for each resource type, the proposed framework includes a rating for each of the primary resources of a Fog node. Further, the ratings will be fetched by each of the Fog nodes (e.g., at the start of every day from an external service) and would show the capability of the Fog node in the given context of the resource (e.g., CPU), thus enabling different Fog nodes to be mutually comparable based on these values. An example of such external service could be www.cpubenchmark.net for the ratings of CPUs.

The proposed framework aims to allow the Fog nodes to steal work until their resources are properly utilized. However, it is not sensible to steal much more work than what can be currently processed by the adjacent Fog nodes, due to the overhead of routing the computed results back to the continuously moving user.

A key aspect is thus to consider the estimation of the number of work items that should be stolen by a specific Fog node.

A set of context parameters considered in the performance measurement of Fog node can include values such as CPU capability, RAM capability, network speed capability and so on. As a basis, it requires a normalized value for each context element of each Fog node in a given group. The calculation for the case when a higher raw value is better is illustrated below.

$$v_l^x = \frac{raw_l^x \times uw_l^x}{\sum_{i=0}^{|O|} raw_l^i \times uw_l^i} \tag{1}$$

where:

- v_l^x denotes the normalized value of context element l of Fog node x.
- raw_l^x denotes the raw value of context element l of Fog node x. This is the rating value for the resource of the Fog node that denotes the capability of the Fog node, usually in regards to an execution of a common algorithm or a benchmark.
- uw_l^x denotes the utilization weight of context element l of Fog node x x. The weight can be used to account for the actual unutilized percentage of a resource on a Fog node. In this case, the weight would be equal to the idleness of the Fog node in the given context l (i.e. $1-U_l$, where U_l denotes the current load).
- O denotes the set of all Fog nodes, which belong in the same group.

For the case when a lower raw value is better (e.g., number of intermediate hops involved in delivering the result to the user), simply the formula $(1 - v_l^x)$ is used.

Based on the data from Equation 1, one can calculate the overall grade per resource context, in order to gain a preliminary estimate on the ratio of the work items to be taken by any given Fog node. This concept is illustrated as follows.

$$grade_l^x = \frac{\sum_{l=0}^{|C|} v_l^x \times cw_l}{\sum_{i=0}^{|O|} \sum_{l=0}^{|C|} v_l^i \times cw_l} \tag{2}$$

where:

- $grade_l^x$ denotes the preliminary estimate of the ratio of all work items to be taken by Fog node x that utilize the resource of context l.
- v_l^x denotes the normalized value of context element l of Fog node x.
- O denotes the set of all Fog nodes, which belong in the same group.
- C denotes the union set of the current context and common contexts. Common contexts are the ones that impact the performance of all the contexts (e.g. the network speed capability).
- cw_l denotes the context weight of context element l in the overall perspective. Not all contexts may be equally important, e.g. number of hops to the user may be considered less important than the actual CPU capability of the Fog node.

The estimate on the actual number of work items for fog_x to handle (denoted by $\#FW_l^x$) is thus deducible from the grading value and the number of work items. This concept is also illustrated as follow.

$$\#FW_l^x = \left\lceil grade_l^x \times |W_l| \right\rceil \tag{3}$$

where:

- W_l denotes the set of all work items with the given context as the primary resource.

The formulae have been validated experimentally and the results are reported in the Evaluation section.

Results Delivery

In general, there is a possibility that the delegator node has moved out from the range of the worker node before the worker node has completed the tasks and delivered the result to the delegator. Fundamentally, there are two basic approaches for handling the situation.

1. **Worker Network Routing:** As mentioned previously, central server has chosen the Fog nodes based on the priority of the connectivity (see Candidate Worker Selection). Hence, the workers can always attempt to route the process result to the node that is currently connected with the delegator.
2. **Central Server Assisted Routing:** In the case of missing routing path to the currently connected Fog node or due to heavy traffic among the nodes within the routing path, the worker can choose to route the process result to the delegator via the central server.

In order to identify the best approach for the process result delivery, the workers may need to continuously update the network status. Considering the status needs to be up-to-date, the workers will only keep the information in the vicinity (i.e. within the group). When a Fog node in the group should transmit data to the central server, it would also keep a record of the communication speed. Hence, if any node has a choice to possibly transmit data to the central server (or alternatively use the Fog node network), it would aggregate the communication speed data, and compute the weighted average of the times to the central server, where the most recent communications have the highest weight. For optimization reasons, some of the calculation results may be cached, in order to improve the performance of the system.

Another crucial aspect would be the distance of hops from the original worker, which handles the computation, to the current Fog node that the delegator is connected to.

The distance in hops could be statically calculated, assuming that each Fog node would know at least the Fog nodes in its vicinity (i.e. a subgraph of the vicinal network of Fog nodes) or a similar approach as for the central server context can be used. As the user is constantly moving, at each timestamp when the user connects with a Fog node, this node would send an MQTT message to the topic of the user's location. Only the Fog nodes that have some in-progress tasks from the delegator would subscribe to the topic.

EVALUATION

Reactive and Proactive Task Handling Performance

This section aims to evaluate the performance between the proposed proactive approach and a reactive approach. In the reactive approach, the delegator's work items are expired upon disconnection with one Fog node and retransmitted upon connection with a new Fog node. Conversely, in a proactive approach, the Fog nodes can transmit the results back to the user via the local network between Fog nodes.

The devices involved in the experiment were as follows:

- **Fog-1 and Fog-2:** HP Elitebook Folio 9470m (Intel i5-3437U, 8GB RAM).
- **Delegator:** Nexus 5 smartphone (LG-D821).

The experiment begins with the delegator transmitting a host description message to Fog-1. Fog-1 then steals work item(s) and also the runnable from the delegator. In our current experiment, only a single work item exists. As soon as Fog-1 begins the computation, the delegator disconnects from Fog-1 and connects with Fog-2.

The sizes of the runnable and result are constant values of 25MB. The computation time is a fixed value of 5 seconds on both of the Fog nodes. The work item data is a varying unit with a size of 25MB, 50MB, 75MB or 100MB.

Figure 2 illustrates the differences in time in the context of utilizing the Fog with either a reactive or a proactive approach.

In the reactive case, the delegator transmits the work item and runnable initially to Fog-1 and upon delegator disconnection from Fog-1, the same data is transmitted again to Fog-2. No data is transmitted between the Fog nodes and the computation is simply terminated by Fog-1.

In the proactive case, the work item and runnable data is transmitted once to Fog-1. When the delegator connects with Fog-2, there is nothing left for Fog-2 to steal since the only task has been taken by Fog-1, and the task has not yet expired. When Fog-1 finishes the processing, results are transmitted back to the delegator via Fog-2 (i.e. the Fog node where the delegator is currently connected).

A lot of data needs to be retransmitted when the system utilizes the reactive approach. Therefore, the proactive approach is shown to perform better under the circumstances.

Figure 2. Comparison of reactive and proactive approach of utilizing Fog

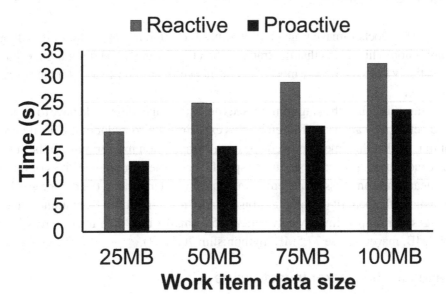

Docker Image Transfer Performance

This section aims to evaluate the performance of using Docker in the local scenario, where the user transmits the Docker image via smartphone using Wi-Fi, and also in the scenario, where the image is downloaded from Docker Hub, via image name and Docker provided API(s).

The experiments were conducted using Gigabit Ethernet connection for the Fog node to the Internet and 802.11n Wi-Fi network for smartphone communication.

The Docker images were chosen from the popular Docker Hub images listing, so that the file sizes (not compressed) would be near-linearly increasing (php:alpine 57.3MB, maven:alpine 115.7MB, python:slim 198.6MB).

The devices used in the experiments were as follows:

- **Fog-1:** HP Elitebook 840 (Intel i5-4200U, 12GB RAM).
- **Delegator:** Nexus 5 smartphone (LG-D821).

Local Docker Image Transfer

The process starts with downloading the image file from the delegator and ends with loading the image into the Docker infrastructure running on a Fog node.

The compression used in the experiment was 7-zip normal preset with the standard deflate compression method.

Docker Image File Transfer

Figure 3 illustrates the Docker image transmission times for different images for both with and without the use of compression. This shows that the transmission times can still be quite high, even in the local network. The relatively low speeds were most likely influenced by the Wi-Fi adapter hardware, especially that of the delegator.

In the case of transmission with using compressed files, the image files would have to be decompressed on the Fog node before they are used. Therefore, this experiment also includes the decompression time. It is important to note that the time improvement from compression may or may not be substantial, depending on the exact image. For example, in our specific case, the compressed sizes of the files did not turn out to be linear. The compressed maven:alpine image was larger than the compressed python:slim image, which reduced the benefit of compression to under a second for maven:alpine. The sizes of the compressed images were approximately between 30-65% smaller than their uncompressed counterparts (php:alpine 25.3MB, maven:alpine 77.5MB, python:slim 70.3MB).

Loading Image Into the Docker Infrastructure

Figure 4 illustrates the time taken to load the Docker image into the Docker infrastructure running on a Fog node. When this action completes, the Docker infrastructure will contain the new image that can be used to run a Docker container.

Figure 3. Docker image transmission time

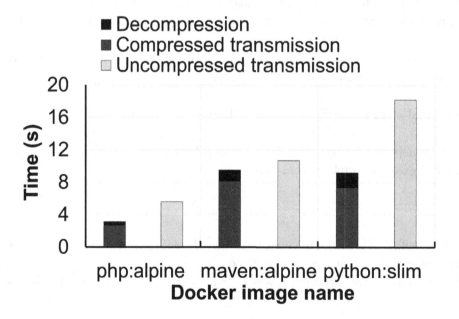

Figure 4. Loading time of Docker image to Docker infrastructure

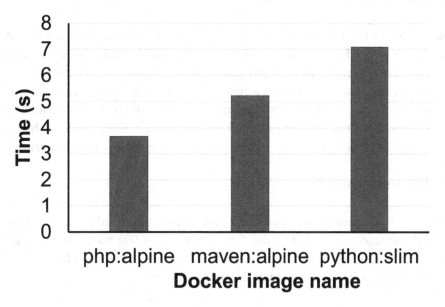

Total Latency of Local Docker Image Loading Into the Docker Infrastructure

Figure 5 illustrates the overall comparison of time spent utilizing the compressed and uncompressed approaches. This involves all the intermediary tasks required to migrate the Docker image from the delegator to the Fog node in a local network and is completed when the loaded image is ready to be deployed as a container on the Fog node.

Figure 5. Total latency of local Docker image loading into Docker infrastructure

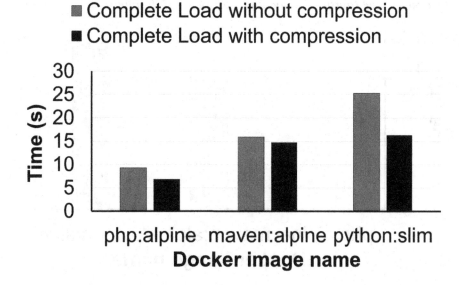

Docker Image Transfer via Docker Hub

Instead of transferring the runnable directly from the delegator to the Fog node, the system can choose to utilize a remote repository. For example, the delegator can specify the link of a Docker Hub image as the runnable. Such an option can reduce the file transmission overhead for the delegator.

Figure 6 illustrates the complete time to download the image with all its layers from Docker Hub and loading the image into Docker infrastructure. This is the over-the-Internet equivalent of Figure 5. Even though the local network has the reduced latency due to the devices being in close physical proximity, the physical hardware limitations become very relevant when the devices have constrained resources.

Task Execution Performance

This subsection aims to compare the performance of task execution by using direct node execution (tasks are not distributed, but solely processed by the directly connected Fog node), Work Stealing approach or a simple round-robin task assignment (even number of tasks distributed to all nodes).

The devices that were involved in this analysis were as follows:

- **Fog-1:** HP Elitebook 840 (Intel i5-4200U, 12GB RAM), where approximately 50% RAM and 50% CPU were utilized before the start of the experiment, in order to simulate a Fog node that is already busy with some other tasks beforehand.
- **Fog-2:** HP Elitebook Folio 9470m (Intel i5-3437U, 8GB RAM), where the OS utilized approximately 10-15% of RAM by default.
- **Fog-3:** Lenovo V570 (Intel i7-2670QM, 16GB RAM), where the OS utilized approximately 10-15% of RAM by default.
- **Fog-4:** Raspberry Pi 3 (Broadcom BCM2837, 1GB RAM), where OS utilized approximately 10% of RAM by default.

Figure 6. Total Docker image transfer and loading via Docker Hub (Internet)

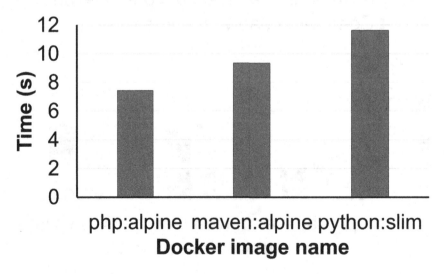

- **Fog-5:** Raspberry Pi 2 (ARM Cortex-A7, 1GB RAM), where OS utilized approximately 10% of RAM by default.
- **Delegator:** Nexus 5 smartphone (LG-D821).

The work items were either CPU-intensive or RAM-intensive tasks, where each category utilized mainly the CPU or RAM resources respectively.

CPU-intensive tasks were further subcategorized as small-cpu and large-cpu tasks, where the time to process small-cpu task was approximately equivalent to half of the large-cpu task. The tasks were designed such that the CPU would be kept utilized at about 70-80% usage level by one task on average, in the case of Fog nodes 1 to 3.

RAM-intensive tasks were further subcategorized as small-ram and large-ram tasks, where the time to process small-ram task was approximately equivalent to half of the large-ram task. The tasks were designed such that approximately 3-3.5GB of RAM would be utilized by one task on average, in the case of Fog nodes 1 to 3.

Since the Raspberry Pi does not have sufficient and comparable CPU and RAM capabilities, the Pi would simply be utilized for a correspondingly longer period of time, until the task is finished.

The size of the inputs for the tasks was 5MB for the small tasks and 10MB for the large-tasks. The size of the result data was 1MB.

The evaluation was conducted in the scenario of completing 20, 40, 60, 80 and 100 tasks. In the case of having 20 tasks in total, 12 tasks were CPU-intensive (6 small-cpu and 6 large-cpu tasks) and 8 were RAM-intensive (4 small-ram and 4 large-ram tasks). The cases of 40, 60, 80 and 100 tasks in total followed the same overview as the 20 tasks variant, except there were correspondingly exactly twice, three times, four times and five times the number of each type of task. The delegator of the work was connected to Fog-1. Since there were tasks with CPU type and RAM type, then the formula used for estimated distribution of works also contained these main context values.

Task Execution Time

Figure 7 illustrates the differences of task execution times for the three approaches. Firstly, the round-robin approach statically assigns equivalent number of tasks to workers, which produced the worst overall results. This was to be expected, because the different Fog nodes have vastly different capabilities (different by an order of magnitude). Therefore, assigning the tasks in such a manner creates the scenario, where the strongest node has long completed its assigned computation and the weakest may have not even reached half way.

Secondly, the direct node execution approach relies on one node to handle all the tasks. Although it is already partially utilized beforehand, it is still a more powerful node than the weakest ones and thus finishes the computation faster than the round robin variant.

Finally, the Work Stealing approach, which considered the runtime context factors, has outperformed the other two approaches.

The experimental results indicate the importance of considering the heterogeneous specification and the runtime context factors of the Fog nodes in the mobile Fog computing.

Figure 7. Task execution time

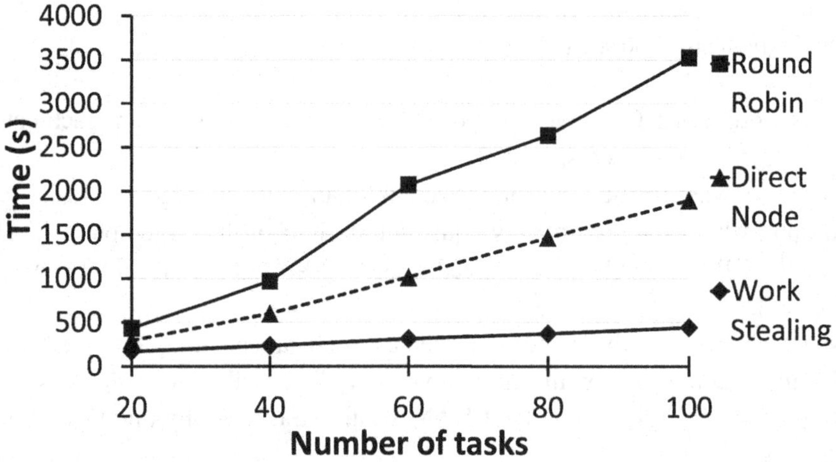

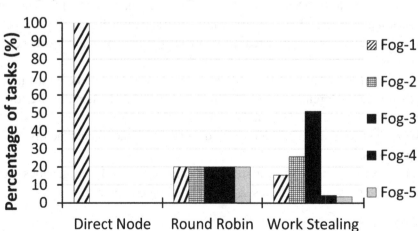

Partition of Tasks Allocated to Fog Nodes

Figure 8 shows the partitioning of all the tasks that were executed at any Fog Node, for all the task distribution methods.

The direct node execution approach relies on one single Fog node to handle all the tasks, and the Fog node does not distribute the works further. Hence, it has taken all the tasks.

Since the round-robin approach does not consider the different capabilities of the Fog nodes, the works are distributed uniformly over all the Fog nodes.

The Work Stealing approach, on the other hand, considers the heterogeneous capabilities of the workers. Therefore, the Fog node with the highest characteristics has handled the greatest number of work items and the one with the lowest current capabilities has handled the least work items.

Figure 8. Partitioning of tasks distributed to Fog nodes

Task Distribution

Figure 9 illustrates the partitioning of types of work between Fog nodes, e.g., how were the small-cpu type of tasks partitioned between Fog nodes.

Since Fog-3 has a much more capable CPU than the other Fog nodes, it has taken the majority of both the small-cpu and large-cpu typed tasks. Since Fog-1 is already considerably utilized both in the CPU and RAM categories, it has taken a lesser number of work items. Fog-4 and Fog-5 are the nodes with the least capabilities, therefore they have taken a rather small number of work items. The context of the RAM follows a similar approach. The reason for the partitions to not differ as greatly in this context is most likely because CPU-intensive tasks also use up a part of the RAM. Therefore, if a Fog node is busy with many CPU-intensive work items, then it directly affects the amount of available RAM and thus influences the amount of RAM tasks to be processed. The CPU usage by the RAM-intensive work items is comparably smaller, thus not producing the opposite effect in the other context. This is yet another aspect that cannot be easily taken into account by the static task distribution methods.

A similar figure regarding the round-robin approach was omitted, due to the fact that it would show a uniform distribution of all types of work items between all Fog nodes. Similarly, the direct node execution approach would show everything executing on Fog-1.

FUTURE RESEARCH DIRECTIONS

A future research direction is to address additional optimizations, where work items that share a runnable would preferably be distributed to Fog nodes that already possess the runnable to further improve efficiency. In addition, it would be interesting to research the pre-scheduling or reservation of Fog nodes in a given area, where a higher priority of execution and a more aggressive variant of freeing up resources (or executing fewer tasks) for the reservation would be used. Additionally, analysis of the energy efficiency of the proposed approach, along with an exploration of network fluctuations of mobile scenarios and the susceptibility to errors and a more in-depth evaluation in the context of more complex IoT systems are also considered as future research directions.

Figure 9. Work stealing task distribution by task type

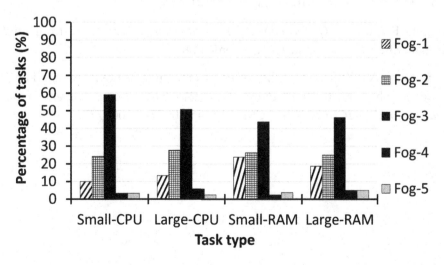

CONCLUSION

This chapter introduced a mobility-aware framework for proactive Fog service provisioning. In contrast to the previous works that assumed the stable connectivity between the delegator and worker nodes, or required prior analysis of historical data in order to provide an efficient offloading approach; the proposed scheme of this chapter enables mobile ubiquitous applications to perform computation offloading with Fog computing servers with minimal prerequisite analysis. Further, the approach provides for an adaptable environment in which stable connectivity need not be guaranteed. Specifically, the Fog nodes can deliver the computational results to the mobile delegator node either through the local worker network or via the assistance of the central server, depending on the performance analysis of the approaches. Moreover, the proposed system utilizes an extended Work Stealing paradigm with worker groups, which considers heterogeneous capabilities of Fog nodes and the heterogeneous nature of incoming tasks to be distributed amongst these Fog nodes. The elements of the framework were evaluated on real devices.

ACKNOWLEDGMENT

This research was supported by Study IT in Estonia.

REFERENCES

Alam, M. G. R., Tun, Y. K., & Hong, C. S. (2016). Multi-agent and reinforcement learning based code offloading in mobile fog. *International Conference on Information Networking (ICOIN)*, 285–290. 10.1109/ICOIN.2016.7427078

Amento, B., Balasubramanian, B., Hall, R. J., Joshi, K., Jung, G., & Purdy, K. H. (2016). FocusStack: Orchestrating Edge Clouds Using Location-Based Focus of Attention. *2016 IEEE/ACM Symposium on Edge Computing (SEC)*, 179-191. 10.1109/SEC.2016.22

Arkian, H. R., Diyanat, A., & Pourkhalili, A. (2017). MIST: Fog-based data analytics scheme with cost-efficient resource provisioning for IoT crowdsensing applications. *Journal of Network and Computer Applications*, *82*, 152–165. doi:10.1016/j.jnca.2017.01.012

Banks, A., & Gupta, R. (2014). *MQTT Version 3.1.1. OASIS standard*. Retrieved from http://docs.oasis-open.org/mqtt/mqtt/v3.1.1/csprd02/mqtt-v3.1.1-csprd02.html

Bellavista, P., & Zanni, A. (2017). Feasibility of Fog Computing Deployment based on Docker Containerization over RaspberryPi. In *Proceedings of the 18th International Conference on Distributed Computing and Networking (ICDCN '17)*. ACM. 10.1145/3007748.3007777

Bitam, S., Zeadally, S., & Mellouk, A. (2017). Fog computing job scheduling optimization based on bees swarm. *Enterprise Information Systems*, 1–25.

Bittencourt, L. F., Diaz-Montes, J., Buyya, R., Rana, O. F., & Parashar, M. (2017). Mobility-aware application scheduling in fog computing. *IEEE Cloud Computing*, *4*(2), 26–35. doi:10.1109/MCC.2017.27

Bonomi, F., Milito, R., Zhu, J., & Addepalli, S. (2012). Fog Computing and Its Role in the Internet of Things. In *Proceedings of the First Edition of the MCC Workshop on Mobile Cloud Computing* (pp. 13-16). New York: ACM. 10.1145/2342509.2342513

Ceselli, A., Premoli, M., & Secci, S. (2015). *Cloudlet network design optimization IFIP Networking Conference* (pp. 1–9). Toulouse: IFIP Networking.

Ceselli, A., Premoli, M., & Secci, S. (2017). Mobile Edge Cloud Network Design Optimization. *IEEE/ACM Transactions on Networking,* 1-14.

Chamola, V., Tham, C. K., & Chalapathi, G. S. S. (2017). Latency aware mobile task assignment and load balancing for edge cloudlets. In *IEEE International Conference on Pervasive Computing and Communications Workshops (PerCom Workshops).* (pp. 587–592). IEEE. 10.1109/PERCOMW.2017.7917628

Chang, C., Srirama, S. N., & Liyanage, M. (2015). A Service-Oriented Mobile Cloud Middleware Framework for Provisioning Mobile Sensing as a Service. *Proceedings of the 21st IEEE International Conference on Parallel and Distributed Systems (ICPADS 2015),* 124-131. 10.1109/ICPADS.2015.24

Cuervo, E., Balasubramanian, A., Cho, D.-k., Wolman, A., Saroiu, S., Chandra, R., & Bahl, P. (2010). Maui: Making smartphones last longer with code offload. In *Proceedings of the 8th International Conference on Mobile Systems, Applications, and Services.* ACM.

Elkhatib, Y., Porter, B., Ribeiro, H. B., Zhani, M. F., Qadir, J., & Riviere, E. (2017, March). On Using Micro-Clouds to Deliver the Fog. *IEEE Internet Computing*, *21*(2), 8–15. doi:10.1109/MIC.2017.35

Fernando, N., Loke, S. W., & Rahayu, W. (2013). Honeybee: A Programming Framework for Mobile Crowd Computing. In *Proceedings of the 2012 International Conference on Mobile and Ubiquitous Systems: Computing, Networking, and Services* (pp. 224–236). Springer Berlin Heidelberg. 10.1007/978-3-642-40238-8_19

Fernando, N., Loke, S. W., & Rahayu, W. (2016). Computing with nearby mobile devices: a work sharing algorithm for mobile edge-clouds. *IEEE Transactions on Cloud Computing*, 1–1.

Gubbi, J., Buyya, R., Marusic, S., & Palaniswami, M. (2013). Internet of things (IoT): A vision, architectural elements, and future directions. *Future Generation Computer Systems*, *29*(7), 1645–1660. doi:10.1016/j.future.2013.01.010

Guo, P., Lin, B., Li, X., He, R., & Li, S. (2016). Optimal deployment and dimensioning of fog computing supported vehicular network. In *Proceedings of the 2016 IEEE Trustcom/BigDataSE/ISPA*, (pp. 2058–2062). IEEE. 10.1109/TrustCom.2016.0315

Hajji, W., & Tso, F. P. (2016). Understanding the Performance of Low Power Raspberry Pi Cloud for Big Data. *Electronics (Basel)*, *5*(2), 29. doi:10.3390/electronics5020029

Hong, K., Lillethun, D., Ramachandran, U., Ottenwalder, B., & Koldehofe, B. (2013). Mobile fog: A programming model for large-scale applications on the Internet of things. In *Proceedings of the Second ACM SIGCOMM Workshop on Mobile Cloud Computing* (pp. 15–20). New York: ACM. 10.1145/2491266.2491270

Huerta-Canepa, G., & Lee, D. (2010). A virtual cloud computing provider for mobile devices. In *Proceedings of the 1st ACM Workshop on Mobile Cloud Computing & Services: Social Networks and Beyon* (pp. 6:1–6:5). New York: ACM. 10.1145/1810931.1810937

Kemp, R., Palmer, N., Kielmann, T., & Bal, H. (2012). Cuckoo: A Computation Offloading Framework for Smartphones. In *Proceedings of the Second International ICST Conference on Mobile Computing, Applications, and Services* (pp. 59–79). Springer Berlin Heidelberg. 10.1007/978-3-642-29336-8_4

Kempen, A., Crivat, T., Trubert, B., Roy, D., & Pierre, G. (2017). MEC-ConPaaS: An Experimental Single-Board Based Mobile Edge Cloud. *2017 5th IEEE International Conference on Mobile Cloud Computing, Services, and Engineering (MobileCloud)*, 17-24.

Krylovskiy, A. (2015). Internet of Things gateways meet Linux containers: Performance evaluation and discussion. In *Proceedings of the 2015 IEEE 2nd World Forum on Internet of Things (WF-IoT) (WF-IOT '15)*. IEEE Computer Society.

Lee, K., & Shin, I. (2013). User mobility-aware decision making for mobile computation offloading. *IEEE 1st International Conference on Cyber-Physical Systems, Networks, and Applications (CPSNA)*, 116–119.

Li, J., Bu, K., Liu, X., & Xiao, B. (2013). ENDA: embracing network inconsistency for dynamic application offloading in mobile cloud computing. In *Proceedings of the second ACM SIGCOMM workshop on Mobile cloud computing (MCC '13)*. ACM. 10.1145/2491266.2491274

Lin, Y., & Shen, H. (2017). Cloudfog: Leveraging fog to extend cloud gaming for thin-client mmog with high quality of service. *IEEE Transactions on Parallel and Distributed Systems*, *28*(2), 431–445. doi:10.1109/TPDS.2016.2563428

Loke, S. W., Napier, K., Alali, A., Fernando, N., & Rahayu, W. (2015). Mobile computations with surrounding devices: Proximity sensing and multilayered work stealing. *ACM Transactions on Embedded Computing Systems, 14*(2), 22:1–22:25.

Marinelli, E. E. (2009). Hyrax: Cloud Computing on Mobile Devices using MapReduce. *Science, 0389*(September), 1–123.

Morabito, R., & Beijar, N. (2016) Enabling Data Processing at the Network Edge through Lightweight Virtualization Technologies. *2016 IEEE International Conference on Sensing, Communication and Networking (SECON Workshops)*, 1-6. 10.1109/SECONW.2016.7746807

OpenFog Consortium. (2017). OpenFog Reference Architecture for Fog Computing. *OPFRA001, 20817*, 162.

Ottenwälder, B., Koldehofe, B., Rothermel, K., & Ramachandran, U. (2013). MigCEP: operator migration for mobility driven distributed complex event processing. In *Proceedings of the 7th ACM international conference on Distributed event-based systems (DEBS '13)*. ACM. 10.1145/2488222.2488265

Pahl, C., Helmer, S., Miori, L., Sanin, J., & Lee, B. (2016). A Container-Based Edge Cloud PaaS Architecture Based on Raspberry Pi Clusters. *2016 IEEE 4th International Conference on Future Internet of Things and Cloud Workshops (FiCloudW)*, 117-124.

Petrolo, R., Morabito, R., Loscrì, V., & Mitton, N. (2017). Article. *Annales des Télécommunications*, 1–10.

Rahaman, M. S., Mei, Y., Hamilton, M., & Salim, F. D. (2017). Capra: A contour-based accessible path routing algorithm. *Information Sciences, 385*, 157–173. doi:10.1016/j.ins.2016.12.041

Ravi, A., & Peddoju, S. K. (2014). Mobility managed energy efficient Android mobile devices using cloudlet. In *2014 IEEE Students' Technology Symposium (TechSym)* (pp. 402-407). Kharagpur, India: IEEE.

Satyanarayanan, M., Bahl, P., Caceres, R., & Davies, N. (2009). The case for VM-based cloudlets in mobile computing. *IEEE Pervasive Computing, 8*(4), 14–23. doi:10.1109/MPRV.2009.82

Shi, C., Lakafosis, V., Ammar, M. H., & Zegura, E. W. (2012). Serendipity: Enabling remote computing among intermittently connected mobile devices. In *Proceedings of the Thirteenth ACM International Symposium on Mobile Ad Hoc Networking and Computing* (pp. 145–154). New York: ACM. 10.1145/2248371.2248394

Shi, H., Chen, N., & Deters, R. (2015). Combining mobile and fog computing: Using CoAP to link mobile device clouds with fog computing. In *Proceedings of the 2015 IEEE International Conference on Data Science and Data Intensive Systems* (pp. 564–571). IEEE. 10.1109/DSDIS.2015.115

Soo, S., Chang, C., Loke, S., & Srirama, S. N. (2017). Proactive Mobile Fog Computing using Work Stealing: Data Processing at the Edge. *International Journal of Mobile Computing and Multimedia Communications*, 8(4), 1–19. doi:10.4018/IJMCMC.2017100101

Su, J., Lin, F., Zhou, X., & Lu, X. (2015). Steiner tree based optimal resource caching scheme in fog computing. *China Communications*, 12(8), 161–168. doi:10.1109/CC.2015.7224698

Truong-Huu, T., Tham, C. K., & Niyato, D. (2014). To Offload or to Wait: An Opportunistic Offloading Algorithm for Parallel Tasks in a Mobile Cloud. In *2014 IEEE 6th International Conference on Cloud Computing Technology and Science* (pp. 182-189). Singapore: IEEE.

Verba, N., Chao, K.-M., James, A., Goldsmith, D., Fei, X., & Stan, S.-D. (2017). Platform as a service gateway for the Fog of Things. *Advanced Engineering Informatics*, 33, 243–257. doi:10.1016/j.aei.2016.11.003

Wang, S., Urgaonkar, R., Zafer, M., Chan, T., He, K., & Leung, K. K. (2015). Dynamic service migration in mobile edge-clouds. *IFIP Networking Conference (IFIP Networking)*, 1-9.

Yousafzai, A., Chang, V., Gani, A., & Noor, R. M. (2016). Directory-based incentive management services for ad-hoc mobile clouds. *International Journal of Information Management*, 36(6, Part A), 900–906. doi:10.1016/j.ijinfomgt.2016.05.019

Zhang, C., Sun, Y., Mo, Y., Zhang, Y., & Bu, S. (2016). Social-aware content downloading for fog radio access networks supported device-to-device communications. In *Proceedings of the 2016 IEEE International Conference on Ubiquitous Wireless Broadband* (pp. 1–4). IEEE. 10.1109/ICUWB.2016.7790392

Zhang, Y., Niyato, D., & Wang, P. (2015). Offloading in Mobile Cloudlet Systems with Intermittent Connectivity. *IEEE Transactions on Mobile Computing*, 14(12), 2516–2529. doi:10.1109/TMC.2015.2405539

Zhang, Y., Niyato, D., Wang, P., & Tham, C. K. (2014). Dynamic offloading algorithm in intermittently connected mobile cloudlet systems. *IEEE International Conference on Communications (ICC)*, 4190-4195. 10.1109/ICC.2014.6883978

KEY TERMS AND DEFINITIONS

Context-Aware Computing: A method of computing augmentation, where the computational systems take factors of their environment into account and adapt their behavior accordingly.

Fog Computing: A method of optimizing computational data processing by distributing the computational data processing tasks to the edge network nodes located near data sources and end-users to reduce the latency.

Mobility-Aware Computing: A method of computing augmentation, where the computational systems take physical location into account and adapt their behavior accordingly.

Proactive Computing: A concept of the future of computing, where computers proactively anticipate the user's needs and take actions on the user's behalf and/or for the benefit of the user.

Ubiquitous Computing: A concept of computing, which outlines that computing can occur using any device, in any location, using any format, etc.

Work Stealing: A strategy of task distribution, where the computational nodes take tasks from the delegator node(s) to be processed, instead of having tasks assigned to them.

Chapter 3
Predictive Methods of Always Best-Connected Networks in Heterogeneous Environment

Bhuvaneswari Mariappan
University Chennai, India

ABSTRACT

Heterogeneous networks are comprised of dense deployments of pico (small cell) base stations (BSs) overlaid with traditional macro BSs, thus allowing them to communicate with each other. The internet itself is an example of a heterogeneous network. Presently, the emergence of 4G and 5G heterogeneous network has attracted most of the user-centric applications like video chatting, online mobile interactive classroom, and voice services. To facilitate such bandwidth-hungry multimedia applications and to ensure QoS (quality of service), always best-connected (ABC) network is to be selected among available heterogeneous network. The selection of the ABC network is based on certain design parameters such as cost factor, bandwidth utilization, packet delivery ratio, security, throughput, delay, packet loss ratio, and call blocking probability. In this chapter, all the above-mentioned design parameters are considered to evaluate the performance of always best-connected network under heterogeneous environment for mobile users.

INTRODUCTION

The advent of telecommunication has greatly reduced the communication distance between users. However, the emergence of cellular generations from 1G to present 5G has attracted a large number of users to enjoy many value added services like mobile multimedia applications, videoconferencing, online classroom besides voice services. The main challenge is to provide seamless and always best connected services to the demanding mobile user in hostile environments (Marques V., et al., 2003).However, the trade- off between bandwidth and delay of Always Best Connected network for bandwidth hungry applications in the vicinity of heterogeneous network is still a research issue.

DOI: 10.4018/978-1-5225-5693-0.ch003

In a wireless network, mobile multimedia users face challenges when the user is handed over to an entirely different wireless network from the recently served network. This is due to the incompatibility of the architectural and technical specification of different networks. For instance, consider a mobile user roaming between networks such as 3G cellular and WLAN during an ongoing video streaming session. Here, the most critical parameters of interest for seamless service to the user are minimum delay, efficient bandwidth utilization factor, low call blocking probability and minimum cost factor.

Background

In any wireless network, minimum delay for real time applications is obtained by using a higher data rate interface, merging of network elements to reduce the hop distance and reserving the resources along the shortest path. However as in Xiao, Chen, and Wang (2000) by carefully rearranging the scheduling access of non-real time applications and also by bandwidth up gradation and degradation methods, bandwidth could be efficiently utilized in wireless networks. A network can have low call blocking probability by using load balancing concept. Lastly, the cost of the service could be reduced by proper network planning and management process. In the proposed approach, the behavior of network is predicted based on novel parameter namely recent call history-'rch'.

The estimation of 'rch' reflects the call blocking probability of networks. This anticipation of wireless network status are more pronounceable during handover situations besides conventional channel borrowing strategy. So, by estimating the 'rch' of a network, the future behavior of the network is predicted and the prediction is utilized for further planning to achieve the performance evaluation parameters of Always Best Connected networks. As far as heterogeneous network is concerned, the optimal network selection is itself is a research issue (Sehgal & Agrawal, 2010).The authors Sehgal and Agrawal (2010), Lahby, Mohammedia, Morocco and Adib (2013), Shen and Zeng (2008) selected the optimal network based on user preference and QoS parameters such as cost, bandwidth, distance and security but, however the network is not studied overlong term/average period similar to Malanchini,Cesana and Gatti (2012) and Jiang, Li, Hou, etal. (2013).

The popular research work in QoS identified from literature of Bari F & LeungVCM(2007) are European Research Funded Projects, Two IST STREP and 4 MORE-MCCDMA Multiple- Antenna System On Chip for Radio Enhancements. Its main objective is to emphasize on research, develop, integrate and validate a cost effective low power system on chip solution for multi antenna multi carrier CDMA Mobile terminals based on joint optimization of layer 1 and layer 2.

To develop an end to end optimized wireless communication link C. Verikoukis, L. Alonso and T.Giamalis (2005) proposed PHOENIX, a scheme offering the possibility to let the application world (source coding, ciphering) and the transmission world(channel coding, Modulation) talk to each other over an IPV6 protocol stack. Verikoukis, Alonso and Giamalis (2005) also proposed NEWCOM,IST-507325, NEWCOM Project E of IST Network of Excellence on Wireless Communication proposal to identify existing gaps in European Knowledge in cross-layer and also investigated the potential benefits of cross-layer in wireless network design in relation to the methodology of separate layer design. It also considered the coupling of higher layers with physical layer and elaborated the information to be exploited from the physical layer to optimize the network performance.

However, some possible approaches of QoS Enhancement in Multimedia Mobile Networks identified from literature includes IMS [IP Multimedia subsystem] to support multimedia traffic with QoS along with Software Defined Radio (SDR) to provide access to network independent services with IPV6forming common platform for 4G networks.

To achieve higher data rate, modification of radio and core network could be done to enable new as well as emerging networks for seamless connectivity to the general framework. Cross-Layer coordination to enhance QoS parameter among different layers is facilitated through well defined message interfaces such as Application Programming Interface(API), Inter Signaling Pipe (ISI) and ICMP. A common control/signaling mechanism could be utilized to assist access network discovery, location management and vertical handoff by periodically computing and broadcasting a list of available Radio Access Network (RANs). List of surrounding BS IDs, their associated AR IDs and network alternatives and their QoS parameters are analyzed in Rubaiyat Kibria and Abbas Jamalipour (2007).However, the common signaling problem could be resolved using an overlay structure on existing BS / APS (as proposed by MIRAI). For resource management, a distributed bandwidth broker (BB) is proposed inside each router domain, along with backup facility where BB includes SLA, SLS, ACS, A&A, PDP (BaderAl-Manthari, Hossam Hassanein, & NidalNasser, 2007).

In 4G heterogeneous networks in order to provide connectivity at anytime along with horizontal hand-off, forced vertical hand-off was also proposed to upgrade the QoS of application against dropped sessions. To be specific, periodical vertical handoff algorithm could also be used. This algorithm can be embedded within terminal architecture for Mobile Terminal controlled against Mobile Terminal assisted handoffs as in 3G (Ren, Koutsopoulos and Lean dros Tassiulus, 2002). As 4G deals with heterogonous networks, so for access network selection, the proposed method will be Analytic Hierarchy Process (AHP) and Grey Relational Analysis (GRA) taking throughput, delay, jitter, reliability, BER, burst error, average retransmissions, security, cost & power consumption (Zhikui Chen, 2007).

In this chapter, the behavior of networks is studied over some average time instants such as busy/peak hour, business hour and festival hours. Any network is considered as best if and only if it has least call blocking(Cbp) and call dropping probability apart from other QoS parameters. The estimation of Cbp can be best judged by analyzing the behavior of networks in terms of their traffic handling capacity and the type of users and their services. This analysis of networks over some average period of time reduces unnecessary handoff of calls in the network thereby enhancing the system efficiency.

OVERVIEW OF DESIGN PARAMETERS

The growing popularity of multimedia based value added services in ABC networks has urged the telecom operators to indulge in some viable solutions to meet the objectives. The acronym of4G networks - Always Best Connected network could be achieved in terms of various design parameters for heterogeneous applications. As in Sehgal A & Agrawal R.(2010) a network is termed as best network based on their service for different type of applications. For instance, for real time multimedia applications, the network having high throughput with minimal delay and low call blocking probability is termed as best network, whereas for applications such as FTP, web browsing like, the network with minimum packet loss, maximum packet delivery ratio and less cost factor is termed as best network. So, measures must be devised to select ABC network amidst various available networks pertaining to the application request of the user. Some of the design parameters of interest Xiao, Chen and Wang (2000), Lahby, M., Mohammedia, Morocco and Adib (2013) are discussed in the following sub-sections.

ISSUES, CONTROVERSIES, PROBLEMS

The growing popularity of multimedia based value added services in ABC networks has urged the telecom operators to indulge in some viable solutions to meet the objectives. The acronym of 4G networks - Always Best Connected network could be achieved in terms of various design parameters for heterogeneous applications.

As in Sehgal A & Agrawal R.(2010) a network is termed as best network based on their service for different type of applications. For instance, for real time multimedia applications, the network having high throughput with minimal delay and low call blocking probability is termed as best network, whereas for applications such as FTP, web browsing like, the network with minimum packet loss, maximum packet delivery ratio and less cost factor is termed as best network. So, measures must be devised to select ABC network amidst various available networks pertaining to the application request of the user. Some of the design parameters of interest Xiao, Chen and Wang (2000), Lahby, M., Mohammedia, Morocco and Adib (2013) are discussed in the following sub-sections.

Packet Delivery Ratio (PDR)

The calculation of Packet Delivery Ratio (PDR) is based on the received and generated packets as recorded in the trace file. The ratio of total packets received to that transmitted is termed as packet delivery ratio. Higher the value of PDR, better is the network status. PDR is also a measure of good channel condition and compromising values could be achieved at low load network condition:

$$PDR = \sum_i \frac{\text{Total packets sent}}{\text{Total packets received}} \tag{1}$$

Bandwidth Utilization Factor(BUF)

Bandwidth utilization is one of the most basic and one of the most critical statistics available in a network. It shows the current traffic levels on the segment or link, compared to the theoretical maximum.

The ratio of utilized bandwidth to that of offered one is measured as bandwidth utilization factor. Higher the value of the BUF, better will be the efficiency of the network. However, effective bandwidth utilization is achieved by rescheduling the non-real time applications followed by bandwidth up gradation/degradation methods. This adaptive bandwidth scheme also enhances the throughput of the system:

$$BUR = \sum_i \frac{\text{Total bandwidth sent}}{\text{Total offered bandwidth}} \tag{2}$$

$$Delay = \sum_i (\text{packet arrival time} - \text{Packet start time}) \tag{3}$$

Delay/Latency

The transmission time taken for requested service delivery to recipient is termed as delay. Delay is a very important parameter for real time services like voice communication, in which a smaller deviation in its value causes very annoying effect to the users. For example, 80ms and 150ms is tolerable delay for voice and video streaming respectively.

Delay is mainly caused due to network congestion and non-availability of channels for traffic communication. However, end to end delay could be minimized by optimal route selection, merging the network elements based on their functionality similar to 3G networks where the RNC –radio network controller does the function of BTS and BSC of 2G networks. This merging of network elements along with the functionality reduces the hop distance taken between the network elements.

For instance, as in Figure 1 a typical 3G network comprising of Node B,RNC-Radio Network Controller, SGSN-Serving GPRS Support Node, GGSN/NGME and Server/router has a total end to end delay of 180ms for a multimedia(optional) user .The typical hop time taken between any two network element also vary between 2 to 120ms according to Xiao, Chen, & Wang, (2000).So, by merging the network elements the total end to end delay could be minimized from 180-200ms to a lesser value up to 50ms based on the network status. The merging of network elements reduces the total end to end delay at the expense of control overhead of the network elements.

Call Blocking Probability

Blocking in telecommunication systems occurs when a circuit group is fully occupied and unable to accept further calls. It is also referred to as congestion. Due to blocking in telecommunications systems, calls are either queued or are lost.

In wireless networks, the randomly varying channel condition makes difficult for error free reception of applications. The randomly varying channel condition and non-availability of channels causes call blocking probability in networks (Shen & Zeng, 2008). However, by adapting the allocation of channels based on traffic class such as real time and non-real time, call blocking probability is minimized. Hence, call blocking probability is defined as:

Figure 1. Schematic diagram of typical 3G network exhibiting network element latency details

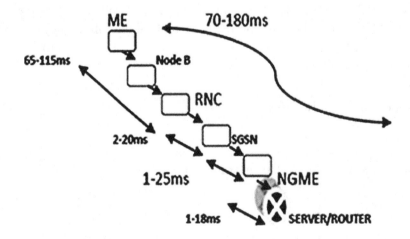

$$\text{Call blocking probability} = \frac{\text{number of calls serviced}}{\text{number of calls arrived}}$$

By minimizing call blocking probability, user satisfaction factor is improved. So, an adaptive approach based on recent call history 'rch' for the design parameter- call blocking probability is proposed to enhance the effective bandwidth utilization of a cellular network under heavy load condition. The novel parameter 'rch' estimates status of cellular region over an average period of time from which the call blocking probability of the cellular region is determined. However, as in Xiao, Chen and Wang (2000) an adaptive approach is proposed in which available channels are rescheduled based on the traffic class to effectively utilize the available bandwidth thereby minimizing the call blocking probability of the wireless networks.

ESTIMATION OF PROPOSED 'RCH' PARAMETER FOR CALL BLOCKING PROBABILITY

ABC network is a network which provides always best connection with desired quality of service. It could be estimated by predicting rch-recent call history parameter of networks. The estimated rch parameter reflects the average call blocking probability of networks measured over average instant of peak/busy, festival and business hour.

A Bayesian network model is assumed for estimation of rch is detailed in Figure 2 in which the status of network N at any time instant depends on previous state of network(whose rch is to be calculated) and the previous status of all its neighboring networks. Bayesian network is ideal for combining prior knowledge to predict future events. So, the neighboring network status is considered anticipating handover in those networks. Likewise, the rch of all networks are periodically estimated and updated in the QoS mapper. The estimated rch is then averaged and the minimum value is always chosen as it reflects the minimal call blocking probability of the network.

The architecture considered in our analysis is as in Marques, Xavier, Aguiar, Marco Liebsch and Manuel Oliveira Duarte (2005) comprising of a heterogeneous wireless network domain with AAAC server, QoS broker + PDP network element equipped with AP (AR) +PEP to serve the users. The users after authentication by AAAC server are assigned channels based on the available quality of the service from QoS broker of the traffic class.

A QoS mapper module is proposed as shown in Figure 3 in QoS broker to estimate and map the available channel with QoS and to reschedule the available resources to be distributed for requesting services. From the authentication information of the users, their recent call history-rch over a average period of time is also estimated jointly from AAAC server, QoS broker and QoS mapper. The estimated rch is reflected in terms of Cbp of a network.

The prediction of rch parameter in a network avoids unnecessary handoff by estimation of call blocking probability-Cbp of a network and the prediction reveals a Always Best Connected networks among the available networks, thus enhancing the QoS of the network. In a scenario shown in Figure 4 there are four heterogeneous network domains viz., N1,N2, N3,N4 as 2G(1),3G,WLAN and 2G(2) networks respectively with their corresponding BSC/AP along with PEP(Policy Enforcement Point) attached to

QoS broker and PDP(Policy Decision Point) (Marques, Xavier, Aguiar, Liebsch, & Manuel Oliveira Duarte, 2005).

A QoS mapper is proposed in QoS broker to map the traffic class with available quality of service along with the recent call history of the network domain. Of all the available networks, their recent call history is estimated over an average period of time from the information provided by NME –Network Monitoring Entity present at various points in a network. The networks are then arranged along with other design parameters such as cost factor, bandwidth utilization, packet delivery ratio, delay and call blocking probability .

The call blocking probability of a network is measured over an average period based on the rch of the available networks. This estimation improves the performance of Always Best Connected networks pertaining to the various traffic classes. Let N1, N2, N3,N4 be 2G(1),3G,WLAN and 2G(2) networks respectively have users demanding for voice, video, file transfer and interactive sessions. Assume all four networks i.e. a 3G network, two 2G networks and a WLAN networks are available to all users. Now, the optimum selection of Always Best available networks is not only based on the above mentioned design parameters but mainly depends on the estimation of call blocking probability. The network status of each network is estimated in terms of available TCH(traffic channels),busy TCH, average call blocking rate, call drop rate, call set up success rate and total traffic in each network as detailed in Table 1.The parameters are measured over different time instants at peak hour, festival hour and business hour and their average is taken for simulation.

The measured parameters as mentioned in Table 1 about various networks corresponds to the measurement taken at various parts of Tiruchirappalli district, Tamilnadu, India. From these measurements three types of case studies are performed to appreciate the effectiveness of rch parameter.

Table 1. Measured network status in and around Tiruchirappalli district over an average period of time

Cell Type	Available TCH (max)	Busy TCH (max)	TCH Blocking Rate	TCH Usage (ms)	Call Setup Success Rate	DropRate	OutgoingHandover Success Rate	Outgoing Handover Failure Rate	Incoming Handover Success Rate	Total Traffic (Erlangs)	%Utilization	Total Data Volume DL (KB)	Total Data Volume UL (KB)
3G	29	38	0.13	841046135	99.14	0.18	97.77	0.08	98.15	233.76	46.294976	41516.31	18378.2
2G(1)	20	19	0.08	255251910	99.46	0.54	97.43	0.18	98.77	70.8	22.379002	17783.69	5334.57
2G(2)	29	10	0.37	137723130	98.86	1.23	96.29	0.27	96.54	38.4	21.039	3442.58	1105.96
WLAN	11	10	2.28	27760130	98.09	1.89	88.94	0.51	91.09	12.96	9.244201	6764.84	894.63

Figure 2. Bayesian network modeling of heterogeneous networks

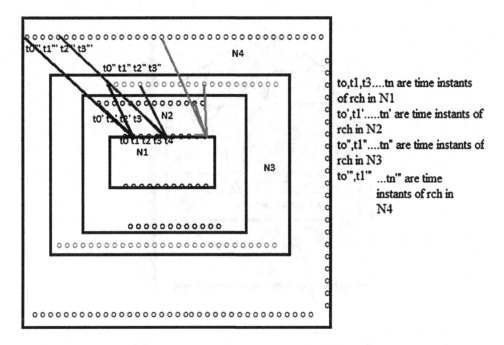

Case Study 1: Selection of ABC Network for Voice Traffic

Consider all four networks as detailed in Table 1{2G(1),2G(2),3G and WLAN} are readily available to a user requesting for voice service. If the user selects 3G network, then it causes inefficient resource usage, as 3G attracts multimedia users rather than voice users. It is also clearly emphasized in the total traffic generation column of the Table 1.The next option is two 2G networks, of which2G(1) could be selected since it has less call blocking probability and call drop rate compared to 2G(2).But, it is interesting and contradictory if available and busy TCH of 2G(1) and 2G(2) are investigated over an average period.2G(2) network has less busy TCH than 2G(1) at the expense of higher call blocking probability. The geographical terrain, channel conditions and type of frequently handled traffic by the network are studied at various average period such as busy/peak, festival and business hours and their average values are measured in terms of the rch parameter of that particular network. The network with lowest rch is always chosen. On the other hand, if WLAN is chosen, then, it will eventually drop the call during call progress phase, as WLAN (from measured average value) has highest call drop rate. So, ABC network for voice traffic is 2G(1).On the other hand, the selection of 2G(2) leads to unnecessary handoff for the call completion.

Case Study 2: Selection of ABC Network for Video Traffic

The availability feature of networks is best estimated during session initiation which fits to voice and file transfer and e-mail like applications. However, this is not suitable for bandwidth hungry applications as they require adequate channels for session progressing till the entire downloading/streaming process. This is facilitated, only if the network status from recent history is estimated. Even, if all four

Figure 3. Schematic diagram of QoS mapper

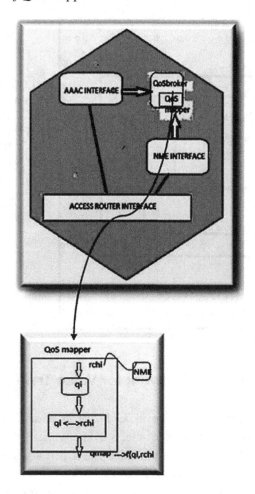

networks are available, best connected network is selected based on the estimation of the call blocking probability of the network.

Consider α_{ci}, β_{cj}, γ_{ck} be the traffic class of i voice users, j multimedia users and k non-real time applications users with 'c' number of occupied channels respectively. Let q_i be the quality of channels available in QoS broker module and a_i be the authenticity of the ith user from AAAC server. rch(A,t) denotes the recent call history of networks over a period of time, where ' A' is the type of network and 't' is the time period taken for simulation.

The function of QoS mapper sub- module proposed in QoS broker module maps the corresponding allotted channels to the available networks based on the estimated call blocking probability. To calculate the call blocking probability, the novel parameter namely recent call history 'rch' over an average period of time pertaining to networks is estimated. The estimated 'rch' parameter implies the call blocking probability of 3G network is the least of all available network. Based on the status of call blocking probability, Always Best Connected network is predicted and assigned to the users for enhanced performance. For analysis, consider three networks A,B and C such as WLAN,3G and 2G wireless networks. Each network has voice services(α), bandwidth hungry services(β) and other non-real time services(γ) with

Figure 4. Proposed heterogeneous network domain

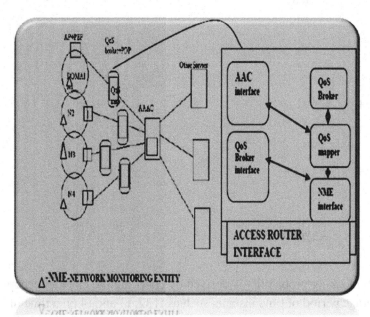

their respective allotted channels. At time period t1, the summation of all services occupation in each network is estimated over allotted channels as follows:

- Network A:

$$At\ t1 : \Sigma\ \alpha_{ci} + \beta_{cj} + \gamma_{ck} = At1 \rightarrow \tag{5}$$

$$At\ t2 : \Sigma\ \alpha_{ci} + \beta_{cj} + \gamma_{ck} = At2 \rightarrow \tag{6}$$

$$At\ t3 : \Sigma\ \alpha_{ci} + \beta_{cj} + \gamma_{ck} = At3 \rightarrow \tag{7}$$

- Network B:

$$At\ t1 : \Sigma\ \alpha_{ci} + \beta_{cj} + \gamma_{ck} = Bt1 \rightarrow \tag{8}$$

$$At\ t2 : \Sigma\ \alpha_{ci} + \beta_{cj} + \gamma_{ck} = Bt2 \rightarrow \tag{9}$$

$$At\ t3 : \Sigma\ \alpha_{ci} + \beta_{cj} + \gamma_{ck} = At3 \rightarrow \tag{10}$$

- Network C:

$$At\ t1 : \Sigma\ \alpha_{ci} + \beta_{cj} + \gamma_{ck} = Ct1 \rightarrow \tag{11}$$

$$At\ t2 : \Sigma\ \alpha_{ci} + \beta_{cj} + \gamma_{ck} = Ct2 \rightarrow \tag{12}$$

$$At\ t3 : \Sigma\ \alpha_{ci} + \beta_{cj} + \gamma_{ck} = Ct3 \rightarrow \tag{13}$$

The status of call blocking probability is derived by checking the channel occupancy status of each network over an average period of time instants such as t1,t2, t3 and t4 on a time unit:

$$If\ At1 > C_A \rightarrow high\ C_{bp} \rightarrow low\ rch_A$$
(least selected network) \hfill (14)

$$If\ At1 < C_A \rightarrow low\ C_{bp} \rightarrow high\ rch_A$$
(most selected network) \hfill (15)

$$If\ At1 = C_A \rightarrow medium\ C_{bp} \rightarrow medium\ rch_A$$
(optional selected network) \hfill (16)

A similar procedure is followed for network B and C. However, by bandwidth adaptation, call blocking probability of the network is improved from least selected status to optional selected status, thus improving the throughput of the network as presented in Table 2.

Case Study 3: Selection of ABC Network for Non- Real Time Data Traffic

The optimum ABC network for data traffic is WLAN, as it provides best resource utilization when compared to other networks. The QoS broker module is responsible for the channel allocation with permissible/available quality of service to the users, admission control, management of network resources and load balancing(Song, Zhuang, & Cheng, 2007).However, the non-availability of channels is delivered as blocked calls in the network. The proposed QoS mapper module limits the call blocking probability by predicting the network status in terms of recent call history, 'rch' parameter over an average unit time. Thus, by adapting the time instants, weight is calculated based on the estimation of recent call history by which the call blocking probability of the networks is predicted. Weights are normalized after 0-1 scale. Higher the weights, lower is the call blocking probability and better is the status of the network. The simulation details for network selection based on 'rch' at various instants of time is given in Table 3.

EXPERIMENTAL CLASSIFICATION RESULTS AND ANALYSIS

For analysis, three regions namely A,B and C are considered to comprise three network domain namely X,Y and Z. Here X is a WLAN (11Mbps),Y is a 3G (1.5Mbps) and Z is a 2G (256Kbps) network. Region A has 5 mobile nodes in addition to other stationary nodes, Region B is simulated with 4mobile nodes and many immobile nodes, Region C with 5 mobile nodes apart from other nodes. Various applications like voice, video/multimedia and e-transaction are taken for simulation. The network status is estimated from NME-Network Management Entity present at various points in the network. QoS Broker module estimates the available channel with the requested QoS. The QoS mapper maps the estimated network status which is observed over a average period of time instant as 'rch' parameter. The 'rch' parameter is utilized for calculating the probability of call blocking. This anticipated network status lowers the unnecessary handovers among the available networks thereby improving the performance of always best connected network. Besides the estimation of distance, cost factor, security issues and bandwidth of the available networks, prediction of 'rch' parameter avoids unnecessary handover. The simulated results clearly exhibit the selection of ABC networks. This estimation shall be useful particularly for multimedia applications. Even though, all the three networks are available, the selection of best/optimum network based on the parameters such as cost, bandwidth, distance and security, shall be effective for low bandwidth applications only. However, for bandwidth hungry applications such as multimedia streaming, mere availability of networks does not suffice the progress of the application in the same network except initialization phase. But, at the same time, the applications need to face unnecessary handover if the recent call history ('rch') parameter of the available networks is not estimated. Therefore, by predicting the behavior of the network in terms of 'rch' parameter, the performance of always best connected network could be improved.

Table 2. Weight assignment based on 'rch' at instants of time (numerical example)

Time Instant	Networks			Weights
	A	B	C	
t1	G	G	M	0.8
t2	G	M	M	0.7
t3	B	M	M	0.4
t4	G	B	G	0.7
Weights	1	0.8	0.9	G-Good M-Medium B-Bad

Table 3. Selection of network based on 'rch' at instants of time

Time Instant	Networks		
	A	**B**	**C**
t1			x
t2		x	x
t3	x		
t4		x	
Network Selected over average period 't' time units		x	x

SELECTION OF BEST

Network for Applications Considering 'rch' Parameter

In real time applications, user can select any type of application like as video calling, voice calling, internet and e-transaction and so on. An application selection module for setting the specific application with some fixed properties preference is given in Table 4 (Xiao, Chen, & Wang, 2000). From the simulated experimental results of Figure 5, it is very clear that at the time instant t1, of all three available networks, Always Best Connected network is net A/B which has least call blocking probability. Here, time instant is and t4.A buffer size of 80 for a channel capacity of 340 channels are assumed, which produces a worst case delay of 235ms.This assumption of 235ms is small enough for streaming applications. However, at t2, net A is Always Best Connected network, at t3 either B/C is selected and at t4, net A/C is best network with least call blocking probability.

Thus, by predicting 'rch' of networks over a period of time instants, Always Best Connected network could be selected among the available best networks for low call blocking probability.

FUTURE RESEARCH DIRECTIONS

The upcoming 5G systems are expected to be dense and irregular heterogeneous networks (HetNets), where the user should be able to access the system through different points of access. In this context, it is crucial to develop predictive techniques that can efficiently leverage the available radio resources across different spectral bands using multi radio access technology (RAT).The predictive methods discussed in this chapter is very limited and in future, research towards hybrid predictive methods involving both network and user preferences shall be considered. However the behavior and performance evaluation of wireless channels and noise in heterogeneous networks is still a open research problem.

Table 4. Design parameters for various applications

Application	Cost	Distance	Security	Bandwidth
Voice	Low	Medium	Low	Low
Video	Medium	Medium	Medium	High
e-transaction	Low	Low	High	Low

Figure 5. Simulated results depicting availability of ABC networks (Y-axis in seconds) at various time instants (X-axis)

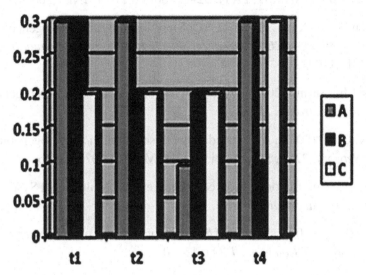

CONCLUSION

The prediction of the proposed 'rch'-Recent Call History parameter in QoS mapper over a time period estimates call blocking probability and lowers unnecessary handoff among the available networks. In existing methods, selection of best network is carried out by estimating distance, cost, security and bandwidth. However, in the existing method, the availability of optimal network is estimated but, in the proposed method, the 'rch' parameter prediction of networks translates the network status from best connected to Always Best Connected network, thereby enhancing the network performance and user satisfaction. Thus this chapter discussed on various predicting parameters required for estimation of always best connected networks in a heterogeneous environment.

ACKNOWLEDGMENT

This research received no specific grant from any funding agency in the public, commercial, or not-for-profit sectors. This research work was supported by Anna University Chennai.

REFERENCES

Abdulkafi, Kiong, Sileh, Chieng, & Ghaleb. (2016). *A Survey of Energy Efficiency Optimization in Heterogeneous Cellular Networks*. Academic Press.

Al-Manthari, B., Hassanein, H., & Nasser, N. (2007). Packet scheduling in 3.5G high –speed downlink packet access networks. *IEEE Network*, *21*(1), 52–57. doi:10.1109/MNET.2007.314537

Alshamrani, A., Shen, X., & Xie, L. L. (2011). QoS provisioning for heterogeneous services in cooperative cognitive radio networks. *IEEE Journal on Selected Areas in Communications*, *29*(4), 819–830. doi:10.1109/JSAC.2011.110413

Awad, Mohamed, & Chiasserini. (2016). Dynamic Network Selection in Heterogeneous Wireless Networks. *IEEE Consumer Electronics Magazine*.

Bari, F., & Leung, V. C. M. (2007). Automated network selection in a heterogeneous wireless network environment. *IEEE Network*, *21*(1), 34–40. doi:10.1109/MNET.2007.314536

Devasia. (2017). Iterative Control for Networked Heterogeneous Multi Agent Systems With Uncertainties. *IEEE Transactions on Automatic Control*.

Fan, Tian, Zhang, & Zhang. (2017). Virtual MAC concept and its protocol design in virtualised heterogeneous wireless network. *IET Commun.*, *11*(1), 53–60.

Ghosh, A., & Misra, I. S. (2016). An analytical model for a resource constrained QoS guaranteed SINR based CAC scheme for LTE BWA Het-Nets. *International Conference on Advances in Computing, Communications and Informatics (ICACCI)*. 10.1109/ICACCI.2016.7732376

Jiang, J., Li, J. D., & Hou, R. (2013). Network selection policy based on effective capacity in heterogeneous wireless communication systems. *Science China. Information Sciences*, 56.

Kang, Z., & Wang, X. (n.d.). Load balancing algorithm in heterogeneous wireless networks oriented to smart distribution grid. *12th International Conference on Natural Computation, Fuzzy Systems and Knowledge Discovery (ICNC-FSKD)*. DOI: 10.1109/FSKD.2016.7603496

Kappler, C., Poyhonen, P., Johnsson, M., & Schmid, S. (2007). Dynamic network composition for beyond 3G networks: A 3GPP viewpoint. *IEEE Network*, *21*(1), 74–77. doi:10.1109/MNET.2007.314538

Khandekar, Bhushan, Tingfang, & Vanghi. (2010). *LTE-Advanced: Heterogeneous Networks*. Qualcomm Inc.

Kibria, M. R., & Jamalipour, A. (2007). On designing issues of the next generation mobile network. *IEEE Network*, *21*(1), 6–13. doi:10.1109/MNET.2007.314532

Lahby, M., Mohammedia, M., & Adib, A. (2013). Network selection mechanism by using MAHP/ GRA for heterogeneous networks. *Proceedings of the 2013 6th Joint IFIP IEEE Wireless and Mobile Networking,Conference (WMNC)*, 1-3.

Lei, Z., & Gang, C. (2016). QoS-aware user association for load balancing in heterogeneous cellular network with dual connectivity. *Computer and Communications (ICCC), 2016 2nd IEEE International Conference*. DOI: 10.1109/CompComm.2016.7925224

Li, W., & Chao, X. (2007). Call admission control for an adaptive heterogeneous multimedia mobile network. *IEEE Transactions on Wireless Communications*, 6(2), 515–525. doi:10.1109/TWC.2006.05192

Liu & Lau. (2017). Joint BS-User Association, Power Allocation, and User-Side Interference Cancellation in Cell-free Heterogeneous Networks. *IEEE Transactions on Signal Processing*, 65(2).

Malanchini, I., Cesana, M., & Gatti, N. (2012). *Network selection and resource allocation games for wireless access networks. IEEE Transactions on Mobile Computing.*

Marques, V., Aguiar, R. L., Garcia, C., Moreno, J. I., Beaujean, C., Melin, E., & Liebsch, M. (2003). An IP-based QoS architecture for 4G operator scenarios. *IEEE Transactions on Wireless Communications*, 10(3), 54–62. doi:10.1109/MWC.2003.1209596

Marques, V., Xavier, P. C., Aguiar, R. L., Liebsch, M., & Duarte, M. O. (2005). Evaluation of a mobile IPv6-based architecture supporting user mobility QoS and AAAC in heterogeneous networks. *IEEE Journal on Selected Areas in Communications*, 23(11), 2138–2151. doi:10.1109/JSAC.2005.856825

Marques, V., Xavier, P. C., Aguiar, R. L., Liebsch, M., & Manueloliveira, D. (2005). Evaluation of a mobile IPv6-based architecture supporting user mobility QoS and AAAC in heterogeneous networks. *IEEE Journal on Selected Areas in Communications*, 23(11), 2138–2151. doi:10.1109/JSAC.2005.856825

Mohammad, G., & Khoshkholgh, V. C. M. (2017). Analyzing Coverage Probability of Multi-tier Heterogeneous Networks Under Quantized Multi-User ZF Beam forming. *IEEE Transactions on Vehicular Technology*. DOI: 10.1109/TVT.2017.2780519

Ren, T., Koutsopoulos, I., & Tassiulus, L. (2002). QoS provisioning for real time traffic in wireless packet networks. *Proceedings of the IEEE GLOBECOM*. 10.1109/GLOCOM.2002.1188482

Sehgal, A., & Agrawal, R. (2010). QoS based network selection for 4G systems. *IEEE Transactions on Consumer Electronics*, 56(2), 560–565. doi:10.1109/TCE.2010.5505970

Shen, W., & Zeng, Q. A. (2008). Cost-function-based network selection strategy in integrated wireless and mobile networks. *IEEE Transactions on Vehicular Technology*, 57(6), 3778–3788. doi:10.1109/TVT.2008.917257

Shi, C., Li, Y., Zhang, J., & Sun, Y. (2017). A Survey of Heterogeneous Information Network Analysis. IEEE Transactions on Knowledge and Data Engineering, 29(1).

Song, W., Zhuang, W., & Cheng, Y. (2007). Load balancing for cellular/WLAN integrated networks. *IEEE Network*, 21(1), 27–33. doi:10.1109/MNET.2007.314535

Verikoukis, C., Alonso, L., & Giamalis, T. (2005). *Cross-layer optimization for wireless systems: A European research key challenge. Global Communications Newsletter.*

Verma. (2017). Pheromone and Path Length Factor-Based Trustworthiness Estimations in Heterogeneous Wireless Sensor Networks. *IEEE Sensors Journal, 17*(1).

Wu, Yuen, & Cheng. (2016). Energy-Minimized Multipath Video Transport to Mobile Devices in Heterogeneous Wireless Networks. *IEEE Journal on Selected Areas in Communications, 34*(5).

Xiao, Y., Chen, C. L. P., & Wang, B. (2000). *Quality of service provisioning framework for multimedia traffic in wireless/mobile networks.* IEEE IC Computer Communication Networks.

Yiping, C., & Yuhang, Y. (2007). A new 4G architecture providing multimode terminals always best connected services. *IEEE Wireless Communications, 14*(2), 33–39. doi:10.1109/MWC.2007.358962

Zhang, D., Chen, Z., Ren, J., & Zhang, N. (2015). *Energy Harvesting-Aided Spectrum Sensing and Data Transmission in Heterogeneous. Cognitive Radio Sensor Network.* IEEE.

Chapter 4
Classification of Channel Allocation Schemes in Wireless Mesh Network

Abira Banik
Tripura University, India

Abhishek Majumder
Tripura University, India

ABSTRACT

Wireless mesh network (WMN) is a widely accepted network topology due to its implementation convenience, low cost nature, and immense adaptability in real-time scenarios. The components of the network are gateways, mesh routers, access points, and end users. The components in mesh topology have a dedicated line of communication with a half-duplex radio. The wireless mesh network is basically implemented in IEEE 802.11 standard, and it is typically ad-hoc in nature. The advantageous nature of WMN leads to its extensive use in today's world. WMN's overall performance has been increased by incorporating the concept of multi-channel multi-radio. This gives rise to the problem of channel assignment for maximum utilization of the available bandwidth. In this chapter, the factors affecting the channel assignment process have been presented. Categorizations of the channel assignment techniques are also illustrated. Channel assignment techniques have also been compared.

INTRODUCTION

Wireless mesh networks (WMN) is a kind of network topology where every node is connected to the other node as that in mesh network topology. There is a dedicated line of communication of half duplex nature between the nodes. Here the nodes use wireless channel. The nodes are basically access points or mesh routers or both incorporated in a single node, which provide connection to the end users (Saini et al., 2016). The wireless mesh networks are widely used in deploying WLAN's. The advantageous nature of the deployment of WLAN and its working principle leads to extensive use of it in today's world. Basically IEEE 802.11 standard has three major protocols IEEE 802.11 a, b, g (Low et al., 2002)

DOI: 10.4018/978-1-5225-5693-0.ch004

which has different bandwidth and modulation. It is a typical network showing ad-hoc nature. The IEEE 802.11 has incorporated the ability of using multi channel and multi radio. Introduction of multiple radio interfaces on the device causes improvement in network capacities, latency and fault tolerance. But, this leads to the problem of assigning channels from the available channel bandwidth so that they perform the best irrespective of the design issues. An example of WMN is shown in Figure 1. Introduction of different channels over multiple radios on single mesh node compels to retrospect different issues such as interference, channel diversity, and channel switching (Islam et al., 2016). Table 1 shows the frequency, bandwidth and modulation used by the above mentioned protocols.

The IEEE802.11 WLAN standards allow multiple orthogonal frequency channels to be used simultaneously to have maximum frequency utilization. In the ad-hoc mode the nodes are connected in the network region in peer-to-peer basis. There is no as such infrastructure maintained for the communication. All the components of the network are immobile. The bandwidth aggregation is rarely used in the context of multi-hop 802.11-based LAN's that operate in the ad-hoc mode (Skalli et al., 2007). The chapter mainly focus on the multi radio multi channel (MRMC) model. In MRMC model the nodes are incorporated with multiple network cards or NIC's to operate in different radio channel. Channel assignment (Si et al., 2010) is the main tool for efficient utilization of MRMC. Channel assignment is the process of assigning separate orthogonal channel or partially over-lapped orthogonal channels to all the nodes in the communication range.

This chapter is regarding the various factors affecting the channel assignment process, the categorization of the channel assignment techniques. And the illustrative study of the channel assignment algorithm under the different heads of the category of the channel assignment techniques. The various factors that affect the channel assignment processes are: connectivity, interference, throughput, load balancing, dynamicity, distributiveness, stability, fault tolerance, convergence rate and fairness. The channel assignment strategies can be classified based on two schemes namely, point of decision and rigidness of the decision. Again they can further be subdivided into sub parts. Based on point of decision the schemes can be classified as centralized and distributed scheme. Again based on the rigidness of the decision the schemes can be sub divided into static, semi dynamic and dynamic schemes.

The entire chapter is organized as follows. The first section discusses the wireless mesh networks, channel assignment process, its need and the strategies in brief. The second section presents the factors affecting the channel assignment process. This section contains the various basic issues that should be dealt with and considered while designing the channel assignment technique. The following section is the preliminaries which discusses a few basic terms used in this chapter. The next section presents the classification of channel assignment strategies in wireless mesh network where the channel assignment strategies are divided depending upon two factors. The first factor is based on the point of decision which considers the node that takes the decision regarding the channel assignment. It has two subparts namely, centralized and distributed. On the other hand, the second factor is, rigidness of the channel

Table 1. Frequency, bandwidth and modulation used by different protocols of IEEE 802.11 standard

	IEEE 802.11 protocol	Frequency (in GHz)	Bandwidth (in MHz)	Modulation
1.	A	5/ 3.7	20	OFDM
2.	B	2.4	22	DSSS
3.	G	2.4	20	OFDM

Figure 1. Overview of wireless mesh networks

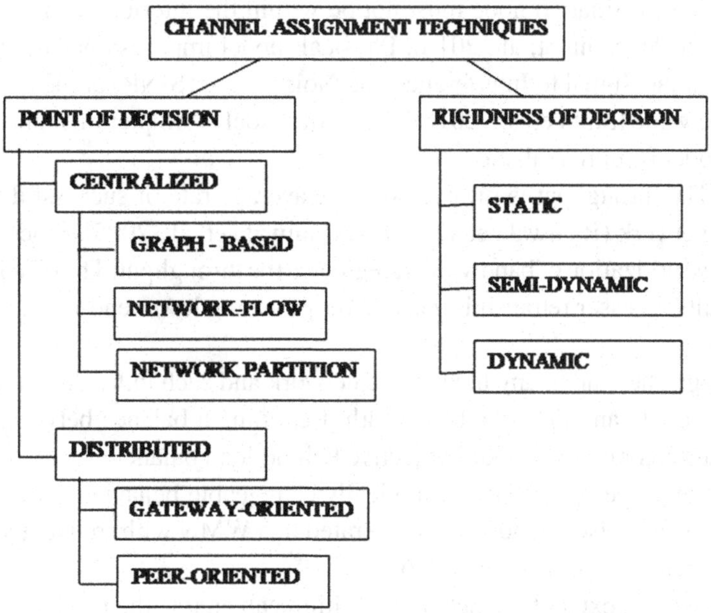

assignment decision which considers the behavior of the channel assignment technique and its stability. It is further sub-divided into-static, semi-dynamic and dynamic schemes. The next section presents a comparison table of the different channel assignment techniques. Finally, the chapter has been concluded in the last section.

FACTORS AFFECTING THE CHANNEL ASSIGNMENT PROCESS

The issues that affect the decisions in the channel assignment process are listed below –

1. **Connectivity:** In the channel assignment process it should be ensured that the connectivity is maintained throughout the topology so that the each node is reachable from the other node. It leads to efficient transmission from each group of nodes to the other increasing the total allowable transmission (Islam et al., 2016).

2. **Interference:** If two nodes try to simultaneously transmit data on the same channel and transmission of both can be sensed from a common position, the two signals cause interference. Interference is basically of three types- intra-flow, inter-flow and external interference (Guerin et. al., 2007). If different simultaneous transmissions of data interfere with each other, it causes intra-flow inter-ference. If simultaneous transmissions of different data flows interfere with each other, it causes inter-flow interference. If transmissions from any device outside the WMN interfere with trans-mission from any node inside the WMN, it is known as external interference (Islam et. al., 2016). Interference basically uses two models namely, Protocol model and Physical model (Gupta et. al., 2000). Protocol model assumes two different ranges for transmission and interference. It considers interference longer than transmission range. The two constraints for successful data transmission

are: the immediate destination node must be within the transmission range of the source range and the immediate destination node must not be within the interference range of any node other than the source node (Saini et. al., 2016). Physical model imposes constraint for successful data transmission that the Signal to Interference and Noise Ratio (SINR) at the receiver must be larger than a threshold value (Saini et. al., 2016). Protocol model is simpler than physical model whereas the physical model is more realistic.

3. **Throughput:** The throughput is measured as the average rate of successful transmission of bits over the entire network (Raniwala et. al., 2004; Raniwala et. al., 2005). Since channel assignment controls effective utilization of bandwidth it regulates the throughput. The efficient utilization of the bandwidth results in lesser retransmissions thereby causing lower end-to-end delay for successful transmissions.

4. **Load Balancing:** There are many links in the network and each individual link will have different data transmission rate and different bandwidth. Retaining a balance between the bandwidth and the data transmission rate is the main objective behind load balancing.

5. **Dynamicity:** The property to adapt dynamically changeable behaviour of the network is termed as the dynamicity. It is also intended to be operated in a WMN with frequently changing topology, traffic, environment etc (Islam et al., 2016).

6. **Distributiveness:** The extent to which an algorithm can enable the mesh nodes to take their own decisions indicates the distributive nature of the algorithm. But if the individual decisions change frequently, the WMN may face high overhead to propagate those (Islam et al., 2016).

7. **Stability:** The traffic interruption and delay in transmission affects the stability of the network. The two parameters that affect the stability are: oscillation and ripple effect. Oscillation occurs due to the frequent changes in the channel assignment decisions i.e. it should not converge and change back and forth among several choices. To overcome oscillation a threshold is used to determine whether the channel assignment decision should be taken or not. Ripple effect mainly occurs when a change in decision requires a number of propagations through the network. Changing the decision in only one node imposes incremental changes on the other two nodes. Both the oscillation and ripple effect reduce data transmission efficiency of the network. And frequently switching the channels will incur switching delay and traffic interruption thereby impairing the network performance (Islam et. al., 2016).

8. **Fault Tolerance:** The nodes in WMN can fail because of software or hardware problems. Even the wireless links can also fail due to external interference and temporary obstacles. Channel assignment algorithm should quickly adapt to the transmission faults and maintain the connectivity (Islam et al., 2016).

9. **Convergence Rate:** It refers to the time required by the algorithm to converge to a final decision. High convergence rate of a channel assignment algorithm indicates a small time requirement for the finding the ultimate channel assignment decision which is essential for the applicability of the algorithm (Islam et al., 2016).

10. **Fairness:** It is considered for designing channel assignment algorithms in WMN's. It implies maintenance of balance among usages of different available channels by the mesh nodes. To achieve a high degree of fairness, channel assignment algorithms must ensure synchronization of selection of channels for the WMN's links (Islam et al., 2016).

PRELIMINARIES

In this section basic concepts used in channel assignment have been presented.

1. **Unit Disk Graph:** A unit-disk graph is the intersection graph of unit circles (Clark et. al., 1990).A vertex for each circle is taken and two vertices are connected by an edge whenever the corresponding unit radius circles crossover each other (Islam et al., 2016). In this graph model, each node is mapped into vertex and each pair of nodes in the transmission range of each other is mapped as an edge. An example of unit disk graph is shown in Figure 2.
2. **Connectivity Graph:** Connectivity graph is a uni-directed graph in which each vertex denotes a mesh node and an edge between two vertices denote the existence of an active transmission link between the corresponding nodes (Islam et al., 2016; Marina et. al., 2010). An example of unit disk graph is shown in Figure 3.
3. **Conflict Graph:** Conflict graph (Ramachandran et. al., 2006) is uni-directed graph used to represent the potential interferences among the mesh nodes. Each edge in the connectivity graph is mapped to a vertex in a conflict graph. Edge of the conflict graph is drawn if and only if there are two links corresponding to the vertices in the interference range of each other (Islam et al., 2016). An example of unit disk graph is shown in Figure 4.
4. **Multi Conflict Graph:** Multi-radio conflict graph (Siek et. al., 2001) is also a uni-directed graph used to represent the potential interference present among the radios of WMN. It is more complicated in comparison to the conflict graph as the multi radio conflict graph takes one vertex per radio pair and each edge represents the potential difference in the interference between two end vertices (Islam et al., 2016). An example of unit disk graph is shown in Figure 5.
5. **Graph Coloring:** Graph coloring (Marina et. al., 2010) is a type of graph labeling algorithm where the components of graph (vertex and edges) either of them are colored based on some constraints. One of the simplest constraints is coloring the vertices of a graph in such a way that the adjacent vertexes of the graph do not have the same color (Marina et al., 2010).
6. **BFS:** Breadth First Search (BFS) (IEEE Std 802.11, 2012) is a tree traversal algorithm where it starts visiting the root node followed by the neighboring nodes horizontally and then to the next level nodes (Ramachandran, 2006).
7. **Graph Decomposition:** Graph decomposition (Clausen et. al., 2003)is a collection of disjoint sub graphs of the original graph such that every edge belongs to only one disjoint graph (Wang et al., 2007).
8. **MST:** Minimum Spanning Tree (MST) is a spanning tree which has the value of the summation of the weights of the edges as minimal as possible (Raniwala et al., 2005).

The examples of unit disk graph, connectivity graph, conflict graph and multi radio connectivity graph are shown in Figure 2, Figure 3, Figure 4 and Figure 5. All the figures are connected with the initial conditions of having five (5) nodes namely A, B, C, D and E. The available channel set is<C1, C2>.

CLASSIFICATION OF CHANNEL ASSIGNMENT STRATEGIES

The channel assignment strategies can be categorised depending upon the few factors.

Figure 2. Unit disk graph

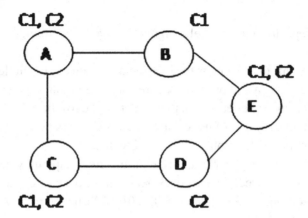

Figure 3. Connectivity graph

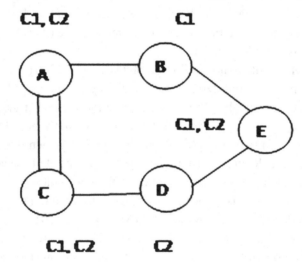

Figure 4. Conflict graph

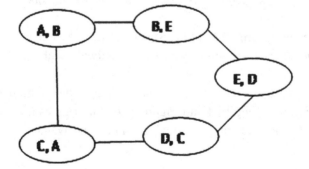

Figure 5. Multi radio connectivity graph

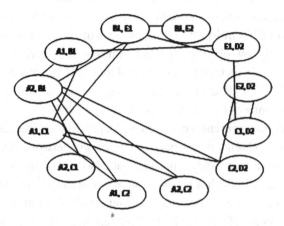

1. **Point of Decision:** The channel assignment to the nodes and the other decisions regarding the channel assignment can be taken either by a single centralised node or can be cumulative decision from the nodes. This leads to classification of the techniques in two broad categories:

   a. **Centralised Techniques:** In centralised technique decisions are taken by only one point or one node called central node or central entity. The central entity takes all the decisions regarding the channel assignment and contains all information related to the channel assignment (traffic load, band width, data rate, priority of the nodes if applicable etc). Under this centralised scheme the central entity can take decision of channel assignment based on the following methods:

      i. **Graph Based:** The central node can visualise the nodes in the multi radio WMN as a graph consisting a set of vertices and edges. Few techniques that fall under this category are: Connected Low Interference Channel Assignment (CLICA) (Marina et al., 2010), Multi-Radio Breadth First Search Channel Assignment (MRBFS) (Ramachandran et al., 2006), Routing, Channel assignment and Link scheduling (RCL) (Alicherry et al., 2005).

      ii. **Network Flow Based:** In the network flow based techniques all the links in the same flow are assigned the same channel. They are dependent on the flow of the link. Examples of this category are: Balanced Static Channel Assignment (BSCA) (Kodialam et al., 2005), Packing Dynamic Channel Assignment (PDCA) (Kodialam et al., 2005), Routing, Channel assignment and Link scheduling (RCL) (Alicherry et al., 2005).

      iii. **Network Partition Based:** In network partitioning based techniques the entire wireless mesh network is partitioned into partitions or groups of nodes. An example of this category is Maxflow-based Centralized Channel Assignment Algorithm (MCCA) (Avallone et al., 2008).

   b. **Distributed Techniques:** In distributed techniques each node is capable of taking decision regarding the channel assignment. Depending on the choice of the local node for the channel assignment schemes the techniques can be divided into two categories:

      i. **Gateway-Oriented Technique:** In this technique a certain number of nodes in the entire mesh which connect the internal nodes of the network to the external network or the internet called mesh gateways take the sole responsibility of the decisions regarding the channel assignment in the mesh network. For example: Hyacinth (Raniwala et al., 2005).

ii.   **Peer-Oriented Technique:** In this approach there is no assumption on the traffic pattern and the actual flow of the traffic is taken into consideration. Example of the schemes belonging to this technique are: Joint Optimal Channel Assignment and Congestion Control (JOCAC) (Rad et al., 2006), Cluster based Multipath Topology control and Channel assignment (CoMTaC) (Naveed et al., 2007).

2.   **Rigidness of Channel Assignment Techniques:** Depending on the behaviour of the channel assignment technique upon change in the network topology, network flow, addition or omission of any node etc. the channel assignment techniques can also be divided into some categories namely:

a.   **Static Techniques:** In this technique once the channels are assigned they are never changed again for long stretch of time. Few examples are Connected Low Interference Channel Assignment (CLICA)(Marina et al., 2010), Multi-Radio Breadth First Search (MRBFS) (Ramachandran et al., 2006), and Hyacinth (Raniwala et al., 2005).

b.   **Semi-Dynamic Technique:** In this technique the flexibility of channel assignment decision is more in comparison to the static technique. The entire working period of the WMN is divided into short intervals of time (Saini et al., 2016). And channel assignment process is carried out before each of the intervals. Example of this category is – Balanced Static Channel Assignment (BSCA) (Kodialam et al., 2005), Multi-Radio Breadth First Search Channel Assignment (MRBFS-CA) (Ramachandran et al., 2006).

c.   **Dynamic Techniques:** The dynamic techniques involves techniques which have complete dynamic behaviour of decisions regarding assignment of channels to the links of the WMN's. Since the assignment of channels is done continuously on the requirement of the action and the process is absolutely flexible. Few examples of this category are–Packing Dynamic Channel Assignment (PDCA) (Kodialam et al., 2005), Joint Optimal Channel Assignment and Congestion Control (JOCAC) (Rad et al., 2006), Cluster based Multipath Topology control and Channel assignment scheme (CoMTaC) (Naveed et al., 2007).

The classification is shown in Figure 6.

Point of Decision

The channel assignment to the nodes and the other decisions regarding the channel assignment can be taken either by a single centralised node or by the cumulative decision from the nodes. This leads to the categorization of the techniques in two broad categories: centralized and distributed.

Centralised Techniques

In centralised techniques decisions are taken at only one point or one node called central node or central entity. The central entity takes all the decisions regarding the channel assignment and contains all information related to the channel assignment (traffic load, band width, data rate, priority of the nodes if applicable etc.). This technique has the main advantage as they are more prone to have a highly optimised technique which considers all the factors. It is more hassle free since the decision is taken from a single end thereby no collision of decision or relative decision. But if the central entity fails the entire

Figure 6. Categorization of channel assignment techniques

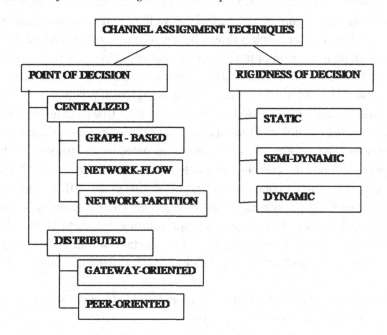

system fails leading to exhibit single point of failure. There is more possibility of the central entity to face bottleneck since all the information of the network resides with the central entity. More load on the central node since it has to face a heavy load of control message in order to collect the information about the node in order to assign channel. Under this centralised scheme the central entity can take decision of channel assignment based on the three methods namely, Graph based, Network flow based and Network partition based.

Graph Based

The central node can visualise the nodes in the multi radio WMN as a graph consisting a set of vertices and edges, where the vertices can be the mesh nodes and the edges can be links between the nodes or vice versa. This approach can be further extended in four different approaches – unit disk graph, conflict graph, multi radio conflict graph and connectivity graph. Irrespective of the graph approaches the techniques to traverse through the graph can also be subdivided as depth-first-search, breadth-first-search, graph-colouring and graph-partitioning (Saini et al., 2016).

Connected Low Interference Channel Assignment (CLICA)

CLICA (Marina et al., 2010) is a greedy trial-and-error method for channel assignment to find connected low interface topologies by implementing multiple channels. It is also a heuristic approach which finds optimal solution from the available solutions. The main idea is to assign channels to links to obtain an initial, well-connected topology which is traffic-independent so that neighboring nodes have a common default channel for communication, named the base channel. The base channel assignment facilitates for different adaptation modes depending on the channel switching delays and relative number of available channels and radio interfaces per node. After the base channel is allocated the mesh nodes use their default channel for further reassignment of channels as well as to register the channel assignment for future communication. The conflict graph is drawn between radios and channels using the link conflict weight. Link conflict graph is the sum of the weights of the edges incident to the vertex in the conflict graph corresponding to the network link. The graph coloring technique is applied in order to maintain all the links in the network topology after all the channel assignment process.

The graph coloring technique is processed just to get a conflict free channel assignment. The central idea is to get a wide number of choices in number of channels (or colors) in determining the order of future coloring decisions. The coloring decisions are taken based on the priority of the nodes. The set of coloring decisions at a node i include choosing colors for radios at i and its adjacent nodes in order to color all links incident to i in the connectivity graph (Marina et al., 2010). The priority is assigned based on the distance from the gateway node in decreasing order and the gateway is assigned the highest priority. These priorities determine the ordering of graph coloring decisions. Before the algorithm initiates, for each node, a set called NodeSet is defined which contains all the nodes neighboring the node or reachable from the node and another set called ColorSet. The algorithm flows as below:

- If the link between the vertices (nodes) is uncolored and both ends of the vertex has a common color initially then color the link with a common color (radio).
  - If there is any uncolored path between the concerned node (suppose v) to any node in the NodeSet (suppose x) via any other node (suppose w) and satisfies the constrain as each intermediate node on the link v-w has only one uncolored radio then assign the link with any color from the ColorSet.
  - If the node (suppose v) has a link with another node (suppose w) then assigning the link a color greedily from the ColorSet based on the constraint as if all radios at node v are colored then choose any color from the already assigned color or choose any color from the unassigned color from the ColorSet. And if not all radios colored at v then assign any color from ColorSet on both ends of the link from v-w.
- If all radios at the node w from the NodeSet (v) are colored then the priority of the node w is increased.

After execution of each step in the algorithm the conflict graph is updated. The assignment of the channel through the graph coloring manner is done using a token. And the token is passed from the gateway node to the other nodes according to priority. Depth First Search is used in this procedure.

Multi-Radio Breadth First Search Based Channel Assignment (MRBFS-CA)

MRBFS-CA (Ramachandran et al., 2006) technique presents a centralized, interference aware channel assignment algorithm and an assignment protocol which can be used to improve the capacity of WMN. The algorithm is designed to intelligently select the channels from the available channels in order to minimize interference within the mesh network and between the mesh network and co-located network. Interference is estimated using a metric WCETT for each mesh node. WCETT is a metric which assigns weights to individual links based on the expected transmission time of each packet (Draves et al., 2004).

$ETT = f$ (loss rate and bandwidth).

This individual link weights are combined into weighted cumulative ETT (Draves et al., 2004)that keeps track of interference among links that use the same channel. The above mentioned technique uses OLSR (Optimized Link State Routing) (Clausen et. al., 2003) and weighted Cumulative Expected Transmission Time (WCETT) for route selection.

Given,

p_f =Packet loss probability for forward transmission.

p_r = Packet loss probability for reverse transmission.

p = Probability of packet transmission not successful

$$p = 1 - (1-p_f) * (1-p_r) \tag{1}$$

If the packet transfer is not successful then it retransmits required the packet. Probability of successful delivery of packet after k-attempts is,

$$s\ (k) = p^{k-1} * (1-p) \tag{2}$$

The value of ETX can be calculated as,

$$\text{ETX} = \sum_{k=1}^{\alpha} k * s\left(k\right) = \frac{1}{1-p} \tag{3}$$

The path metric is the sum of the ETX values for each link in the path. Protocol selects the path with minimum path metric. The ETX is independent of the size, identically distributed and bidirectional.

The routing protocol Link Quality Source Routing (LQSR) is a source routed link state protocol which is derived from Dynamic Source Routing (DSR). From this protocol few assumptions are made:

- All nodes are immobile.
- Each node is equipped with more than one NIC's.
- Each node is equipped with more than one radio. These radios will be tuned to different non interfering channels.

$$\text{WCETT} = \sum_{i=1}^{n} ETT_i \tag{4}$$

Considering a n-hop path and a system containing k channels. Therefore, sum of transmission times of the hops on channel j, X_j can be calculated as,

$$X_j = \sum ETT_j \tag{5}$$

i.e. considering hop i is on channel j.

The total path throughput will be influenced by the bottleneck channel, which has the maximum X_j.

Therefore, $\text{WCETT} = \max_{l \leq j \leq k} X_j$ (6)

considering eq. 4 and6.

$$\text{WCETT} = (1 - \beta) * \sum_{i=1}^{n} ETT + \beta * \max_{l \leq j \leq k} X_j \tag{7}$$

where β is tunable parameter $0 \leq \beta \leq 1$

The MRBFS-CA uses multi-radio conflict graph. The algorithm also adopts Breadth First Search Channel Assignment. It uses the BFS to assign channels to mesh radios. The search begins with links emanating from the gateway nodes and then in decreasing levels of priority to links fanning outward towards the edge of the network.

The gateway node chooses a default radio. The default channel chooses a default radio. The default channel is chosen such that its use in mesh network minimizes interference between the mesh network and collocated wireless networks. The CAS then creates the MCG for the non-default radios in the mesh. MCG created using neighbor information is sent by each MR to the CAS. Then BFS-CA is used by CAS to select the non-default radios.

$$R_c = \frac{\sum_{i=1}^{n} Rank_c^i}{n} \tag{8}$$

where,

n = number of routers

$Rank_c^i$ = rank of channel c at router i

The least R_c value is the least interfered channel.

After the rank calculation, non-default channel selection starts.

Neighbor information sent by a router contains the information of the neighbors like delay, interference estimation for all channels supported by the router's radios etc. The distance of a radio is obtained from beacon frames initiated from the gateway. A beacon frame is a gateway advertisement broadcasted hop-by-hop throughout the mesh. There is a special operating mode called RFM on mode by capturing packets from the medium. The captured packets are then used to measure the number of interfering radios and the bandwidth consumed by these radios. The radios with RMN on mode cannot transmit data packets for the duration of sniffing. Therefore, there will be disruption of the data flow. To prevent the disruption of flow the link redirection is implemented.

Link redirection is achieved when a flow intended for a router's non default radio is directed to the router's default radio instead. The link redirection process is adopted only in two situations-

- When a router's intended transmitter is incapable of delivering packets.
- When the intended receiver on the neighboring router is incapable of receiving packets.

It invokes link redirection, then initiates configuration of radios after assignment. Neighboring radio is configured to be active and use ping command to confirm link is activated.

The algorithm runs as follows:

- The algorithm is started from the least hop count node or vertex in the graph. A queue (Q) is maintained which contains all the vertices in increasing order of the hop count.
- Initially the least hop count vertex is taken and traversed. A set is maintained which contains vertex with which the current node has a link. And a highest ranked channel is assigned (A highest rank channel is assigned to the least hop count vertex because the least hop count vertex is generally the nodes near to the gateway which faces the bottleneck condition). If there is a conflict in assigning the highest ranked channel to the links of vertex then assign any random channel to the current node.
- If all the radios of the current node are assigned some channel yet another link is left out then the algorithm temporarily assigns a channel to the link of the current node (this is usually done to satisfy the constraint that only one channel is assigned to each radio).
- Maintain another queue (named tail) which contains all the active links from the neighboring node or vertex. Sort the queue and add the elements of the queue to the initial queue (Q). Repeat the steps from 1.

Traffic-Aware Routing-Independent Channel Assignment (TARICA) Algorithm

TARICA (Wang et al., 2007) assumes to mitigate signal interference, increases network capacity in multi-channel multi-radio wireless mesh network. This technique has few assumptions. It measures traffic load and channel utilization on each network link. There is no frequent channel switching since it assumes long term stable network topology. There exist a tight coupling of channel assignment and routing which result in sub-optimal performance when radio fails.

The channel assignment process assumes the graph decomposition problem which divides the entire graph into almost k-sub graphs where the value of k is number of available channels. It follows the following three ideas:

- Each node distributes its links to the sub-graphs based on the utilities of the links, so that the sum of utilities of the links in each sub-graph is almost equal.
- The connected components in each sub-graph should be as small as possible. The interference degree of the neighboring links is always one because the links are half duplex.
- Different connected components in a sub graph should be as distant as possible to avoid interference between the links.

The algorithm is divided into two main components:
Initial channel setting:
The graph is traversed using Breadth First Search algorithm.

1.   While any of the links in the graph new_link remains unassigned with channels
2.   Take up a start node N (which has maximum work time first and least vacant radio first).
3.   Select a least used channel around N.
4.   Node N is put in the queue.
5.   While queue is not empty, first element is dequeued. Select the lines whose end points are already in the queue. With the given channel assign the link. The endpoints of the links are enqueued.
6.   After assigning the channel to the new_link, add the end point of the link to a queue.
7.   End while

Iterative improvement:
This process checks for the bottleneck links and removes the channel assigned to the link with the least channel used from the surrounding.

1.   Old_MCU = maximum channel utility of current settings and l_i = the link with max CU
2.   If the end point of l_i has vacant radio, update the channel of l_i with another channel. New_MCU = maximum channel utility of resulting setting.
3.   If the end point of l_i has another common channel, update the channel of l_i with another channel. New_MCU = maximum channel utility of resulting setting.
4.   Else, Channel_2 = Least used channel around l_1. Update the channel of l_1 with Channel_2. Update the channel of other links if necessary. New_MCU = maximum channel utility of resulting setting.
5.   Iterate steps (1) to (4) till New_MCU<Old_MCU.
6.   Restore the channel setting of last round.

Network Flow Based

In the network flow based techniques all the links in the same flow are assigned to the same channel. They are dependent on the flow of the link. They do not experience channel switching in between the flow thereby no channel switching delay. And before channel switching the central node or the entity will collect the information about all the nodes and then take the decision of channel assignment or channel reassignment (Saini et al., 2016).

Balanced Static Channel Assignment (BSCA)

This technique jointly optimizes the routing, assigns channel to the links and schedules the flow in order to obtain upper and lower bounds for the capacity region under an objective function. It uses protocol model of interference and provides a valid feasible solution for linear programming. Scheduling of multi hop multi radio is more complex. So, in this technique single hop multi radio is considered. BSCA (Kodialam et al., 2005) performs static link channel assignment followed by greedy coloring for conflict resolution.
The main deciding factor for this technique is the utilization of channel-

Utilization of channel i over link e $= f_i(e) / c_i(e)$.

where

$f_i(e)$ is the flow on data link e using channel i

$c_i(e)$ is the capacity of the channel i.

At the beginning of each time slot every node has to make two decisions

- To which node it is going to communicate.
- The channel on which the communication will take place.

The decision can be taken in per time slot basis or will remain constant for life time. If the mapping of links to the channels is fixed, it takes up the static link channel assignment.

The scheme performs link channel assignment along with scheduling based on greedy coloring to remove potential conflicts and obtain a feasible schedule. The equation mentioned below has three necessary and sufficient constraints to check the feasibility of a link schedule in MC-MR network graph G.

$$\sum_{y_i}^{t} (e') \leq 1, \ \forall_i \in OC, \ \forall_e \in E \cup E\psi, \ \forall \, l \, e' \in E(t(e)) \cup E(h(e)) \tag{9}$$

$y_i^t(e) = 1$ if link e is active on channel i on time slot t $\tag{10}$

0 otherwise

$$y_i^t(e) = 0, \ \forall_i \in OC, \ \forall_e \in E \cup E\psi$$

i.e. interference links that do not carry data.

Over a period of time [0,T] the fraction of time link e is active on channel I is given by,

$$\left(\sum_{y \leq T} y_i^t(e) \right) / T$$

$\therefore$ Mean flow on channel i over link e

$$f_i(e) = c_i(e) . \left(\sum_{y \leq T} y_i^t(e) \right) / T \tag{11}$$

Any solution satisfying the necessary condition will be an upper bound of the optimal solution.

First, a greedy approach is used in solving the static channel assignment problem and then time-slots are assigned to each channel using coloring algorithm. The assignments of the time slots are almost independent. The information related to the channels is exchanged only during the assignment of the time slots to resolve node-radio constraint at each node in the network. The node- radio constraint is

$$\sum_{e \in E(v)} \sum_{i \in OC} y_i^t\left(e\right) < k(v), \forall v \in V \forall t$$

i.e. node can use at most k(v) radios in a given time slot for transmission or reception.

The algorithm is as follows:

1. Assigning a channel to link e such that each channel is an independent orthogonal channel. The algorithm calculates the maximum load on any constraint set (link, channel) pair (e, i).
2. Link e is assigned to the channel c that has the minimum value of m(e, i) over all channels. The flow on the set T (e, j) is incremented by the summation of value of all flow on channel i of link e as all the flow on link e is now assigned to channel j. The mapping of links to the channels is max-min allocation in order to ensure that the total load on the constraint set is distributed among the given channels.
3. At the end of the static link channel assignment, a set of flows assigned to links that have been assigned to a particular channel is obtained. Time slot is also assigned to each channel separately while taking node-radio constraint through greedy coloring algorithm.

Packing Dynamic Channel Assignment (PDCA)

In dynamic link channel assignment the channel is switched for every link once every T_d time slots ($T_d \geq 1$). For this coordination of nodes is required. Coordination across different links is also required to perform link channel assignment at the beginning of each time period. PDCA (Kodialam et al., 2005) performs link channel assignment and scheduling simultaneously. The algorithm works as follows:

1. First all the link flows on different channel are assigned to a given link into a single scaled flow on the link.
2. Then the flow is packed for each time slot. The amount of unassigned flow is also packed on link e.
3. At the beginning of each time period the links are sort in descending order of the unassigned flows.
4. The first link e is assigned the channel j which has the highest channel capacity value.

Routing, Channel Assignment and Link Scheduling (RCL)

RCL (Alicherry et al., 2005) technique implements the mathematical formulation of joint channel assignment and routing problem, taking interference constraints, number of channels and number of radios. The assumptions made in this scheme are: Mesh router u, has the wireless interface I(u) each operating on single channel. A set of orthogonal channels (*F*). There is an ordered pair *F*(u) which contains mappings of ith interface of node (u) operating on i^{th} channel. Aggregate traffic on u is *l*(u) which is only outgoing not incoming (for simplicity). Communication possible if there is a common channel between F(u) and F(v). Channel to radio mappings are maintained as long as there is no change in the traffic demand. Using protocol model of interference the transmission range of the node is considered as R_T.

Following steps are executed in the algorithm:

- The flow on the flow graph is solved by linear programming to ensure minimum interference in the channels in order to yield the lower bound of the demand for channel and routing of the nodes.
- Channel assignment process is used to adjust the flow on the flow graph but maintaining minimum interference.
- Post processing and flow scaling process in carried out to ensure that the flow on the flow graph is readjusted so that the interference is minimized thus yielding a feasible routing
- Interference free link scheduling process is followed.

The algorithm attempts to uniformly divide the interference among the links so that within each channel the maximum interference is bounded. In addition the algorithm also tries to minimize the interference within connected components of each channel. The connected components are formed by the edges that have positive flows on the given channel.

Network Partition Based

In the network partitioning technique the entire wireless mesh network is partitioned into partitions or groups of nodes depending upon the network flow, or interference in the links, packet transfer delay etc.

Maxflow-Based Centralized Channel Assignment Algorithm (MCCA)

In MCCA (Avallone et al., 2008) the most critical link in context of carrying data traffic is identified and is replaced with another radio on the link in order to limit the interference. The algorithm visits all the links in decreasing order of the expected link load and selects the channel which minimizes the sum of the expected load from all the links in the interference region that are assigned the same radio channel.

A channel assignment A assigns a set A (u) of channels to each node u $\in$ V. A induces a new graph model G = (V, E) where two nodes are connected iff they are in transmission range of each other and share at least one common channel. If this constraint is written in linear programming then,

$$A(u) \cap A(v) \cap A(x) \cap A(\mathrm{y}) = \varnothing,$$

this depicts that $u \leftrightarrow v \in E$ and $x \leftrightarrow y \in E$ may interfere with each other if node x or node y are in one of the interferences ranges of u and v and the communication along the two links take place in the same channel. The criticality of the link is calculated as the amount of units of flow it must carry.

The algorithm goes as follows:

Phase I: Links are grouped (Link group) based on the flows they carry. A group may contain links from many different nodes. For each node, the first stage assures that the number of different groups assigned to its links does not exceed the number of radio interfaces.

Phase II: The second stage selects a channel for each group and assigns the selected channel to all links of group. This process is knows as group channel assignment. The attempt is to assign different channels to groups containing interfering links.

Link group:

- The idea is to repeatedly merge a pair of group until the number of groups equals the number of radios.
- At every iteration, the groups chosen to be merged are the two with the least total flow. The total flow F_{tot} of a group is the sum of the flows associated with its entire links. A merging reassigns all the links belonging to group j' to group j''.
- An attempt is made to give the links with the target flow preference, in the sense that they are assigned the same group as the links with the smallest flow.
- The 'n' number of links is sorted in descending order of the flow they carry and denote the neighbor of node (u) associated with the (i^{th}) link by (u_i). The strategy is to assign a group to the elements until the local flow of the group exceeds its share of the flow passing or there are no other links left.

Group channel assignment:

Group channel algorithm performs this task with the objective of protecting critical links. This is achieved by sorting the groups in decreasing order of the maximum flow associated with any of the links of the group and visiting them one-by-one in such an order. The algorithm is to maximize the capacity of the induced graph to enhance the links those are most critical to carry traffic and protect those links against interference. Interference is the major cause of throughput decrease and accordingly channels are assigned in a way such that the most critical links experience the least possible interference. The term criticality depends upon the placement of mesh routers and on the transmission capability of their radio interference. Mesh routing devices will collect user traffic and do not have to forward each packet to a specific mesh gateway, but can direct it to any of the mesh gateways. To identify critical links maximum throughput flowing from aggregation devices to the gateways on the initial graph is computed.

Distributed Techniques

In distributed approaches there is no central control among the nodes and each node runs its own copy of the algorithm to assign the channels. The algorithm can have two approaches either it can be gateway-oriented where the main network traffic is associated with the gateways or it can be peer-oriented where the network traffic occurs between any pair of nodes without any definite pattern of network traffic (Saini et al., 2016). The load for channel assignment is equally distributed among the nodes. All the nodes have to circulate their information to the neighbouring nodes. Locally optimising and collecting the message leads to low control message overhead. Single point of failure is not exhibited here. Since the channel is assigned from the local information high degree of optimisation is not possible also. It may some time lead to degradation of performance (Si et al., 2010). Since every node can make decision, this sometime may lead to interoperability problem. The choice of the nodes for local optima is important for taking the local decisions for channel assignment. The difference in the local optima can also change the total channel assignment of the nodes. Depending on the choice of the local optima the techniques can further be divided.

Gateway-Oriented Technique

In this approach it is assumed that a certain number of nodes in the entire mesh network functions as the mesh gateway which connect the internal nodes of the network to the external network or the internet. The main advantage of this technique is that, the gateway nodes can also be used in the distributed technique there by simplifying the assignment process and reducing the load of the mesh routers. But the gateways cannot accommodate the different kinds of traffic patterns (Saini et al., 2016).

Hyacinth

The distributed algorithms utilize only local traffic load information to dynamically assign channels and to route packets. Hyacinth (Raniwala et al., 2005) is designed to work directly with 802.11 based interfaces and requires only systems software modification.

The channel assignment algorithm is divided into 2 sub problems:

- Neighbour to interface binding: It determines through which interface a node communicates with each of its neighbours and with whom it intends to establish a virtual link. Each node typically uses one interface to communicate with multiple of its neighbour.
- Interface to channel binding: It determines which radio channel a network interface should use.

The channel assignment depends on the load of each virtual link, which in turn depends on routing. The routing process aims at determining routes between each traffic aggregation device and the wired network. Load balancing helps in avoiding the bottleneck links and increases the network resource utilization efficiency and the network good-put. Mathematically the cross-section good-put of network can be expressed as,

$$X = \sum_a \min(\sum_i C(a, g_i), B(a)) \tag{12}$$

The distributive nature of the routing and channel assignment algorithm utilizes only local topology and local traffic load information to perform channel assignment and route computation. Information is collected from a (k+1) hop neighbourhood, where k is the ratio between the interface and communication range is (Skalli et. al., 2007; Si et. al. 2010). The routing algorithm gives the output a tree like structure where each wired gateway node is the root of a spanning tree and each WMN node attempts to participate in one or multiple such spanning tree. It may distribute its load among the trees and use them as alternative routes when nodes or links fail. The routes are reliability ensured as they are built as a part of the neighbour discovery protocol.

Each node periodically exchanges individual channel usage information. The total load of a channel is a sum of weighted combination of the aggregated traffic load and the number of nodes which uses the channel. In channel load balancing phase a WMN node evaluates its current channel assignment based on the channel usage information it receives from neighbouring nodes as soon as the node finds a relatively less loaded channel after accounting for priority and its own usage. It also provides failure recovery. When a node fails, node in its sub tree loses their connectivity to the wired network. Hyacinth reorganizes the network to by-pass the failed node and restores the connectivity.

Joint Optimal Channel Assignment and Congestion Control (JOCAC)

JOCAC (Rad et al., 2006)is a decentralised utility maximisation scheme. It solves the problem that arises from the interference of the neighbouring transmission. This scheme works on the non-overlapped channel as well as partially overlapped channel. Interference between neighbouring links can potentially cause network congestion. The JOCAC scheme depends on number of channels, number of allocated NIC's, link's congestion price, transmission power, wireless path loss information, channel frequency response.

If a particular link is congested, its transmission rate can be increased by either increasing the SINR or reducing the interference level. The objective is achieved by not allocating the same channel used by the congested link to other link within the neighbourhood.

The throughput is further improved if the system uses partially overlapping channels. The filter used here is a low pass filter. For pulse shaping, raised cosine filter is used. Non-Zero portion of the frequently spectrum of its simplest form is a cosine function above the horizontal axis.

The channel weighting matrix is W. For any two arbitrary channels m & n, value of W_{mn} is W.

$$W_{mn} = W_{nm} = \frac{A_o}{A_o + A_{no}} \qquad (13)$$

A_{no} is power spectral density of non-overlapping area.

A_o is power spectral density of overlapping area.

In distributed scenario, each node n assigns optimal channels to some links $L_n \subset L$.

Periodically individual channel usage X_l and congestion price on link l, λ_l will be exchanged. Here, $l \in L_n$.

$$\text{Traffic load for TCP} = \frac{cwnd}{rtt} \qquad (14)$$

where,

cwnd = TCP congestion window size

rtt = packets round trip time.

Cluster Based Multipath Topology Control and Channel Assignment Scheme (CoMTaC)

CoMTaC (Naveed et al., 2007) creates a separation between the channel assignment and topology control functions for minimizing flow disruptions. It is a cluster based approach as it ensures basic network connectivity. Intrinsic support for broadcasting with minimal overheads is provided. It uses the concept of developing a spanner of the network graph instead of Minimum Spanning Tree (MST) since MST does not provides advantage of the inherent multiple paths that exist in WMN. Spanner is nothing but a graph in which each pair of vertices is connected by a short path. This scheme also provides a novel interference estimation mechanism based on the average link layer queue length within the interference domain. It is an approach that explicitly creates a separation between the topology construction and channel assignment functions thus minimizing flow disruptions. The primary objective of CoMTaC is maximizing the network capacity while minimizing the interference and taking advantage of multiple paths in the underlying network topology. It constituents of two phases-

Phase I: The nodes are grouped into cluster of small nodes of small radii. Within each cluster, a common channel is used by all member nodes on one of their interfaces. The default channel provides an efficient broadcasting facility that incurs significantly low overheads. The nodes bordering the clusters have their second interfaces tuned to the default channel of highest priority cluster resulting in inter cluster connectivity. The network model assumed here is an undirected unplanar multigraph $G(V, E)$ representing the Wireless mesh network. $V \rightarrow R^d$. Where $|V| = n$ and $d \geq 1$. $E \rightarrow$ set of edges representing the wireless links between WMN nodes. Each node has k radio interfaces among which one is default interface. $G_T(V, E')$ represents the graph topology control scheme, $G_A(V, E'')$ induced by channel assignment scheme. $G_A' \subseteq G_T \subseteq G$.

Phase II: The second step of the topology control scheme aims at identifying multiple feasible paths, enhancing the initial base nodes connectivity. The interference experienced by the non-default channel is calculated using the average link layer queue length within the interference domain.

Transmission and interference model used in this scheme is given as:

R_t, R_i be the transmission range and interference range. $R_i > R_t$. dist (u,v) = Euclidean distance between 2 nodes. $u, v \in V$. For two nodes $u, v \in V$, direct communication is only possible if the dist $(u, v) \leq R_t$ and at least one of the interfaces of the two nodes operates on a common channel. Two links $e1 = (u1, v1)$ and $e2 = (u2, v2)$ interfaces with each other if both edges operates on a common channel and any of the distances $d(u1,u2)$ $d(u1, v2)$ $d(v1, v2)$ $d(v1, v2) \leq R_i$. Interferences $I(e)$ on link is $e \in E$.

$I(e) = \sum load_i$, sum of the traffic load on all the interfering links.

$I_T(c)$ = number of interfering links.

$\therefore$ Interferences in path $(u, v) \in V = I_T(path(u, v)) = \sum I_T(e) \forall e$ on path (u, v). (15)

Low time complexity of O (kn) uniform sized clusters. Clusters radius determining parameter r = 2. For realistic approach since interference domain of a node typically covers 2 – hop neighbouring. The cluster head is gateway and all nodes connected to this gateway are parts of clusters assuming that the nodes are capable of retaining the information about their distance from the gateways.

The two cost metrics are link and path interferences. A connected graph G_s (V, E') is a spanner of the graph G (V, E). If E' $\subseteq$ E and for any two vertices u, v $\in$ V, cost G_s (u, v) = t* cost$_G$ (u, v) for t $\geq$ 1. For every edge (u, v) $\in$ E, if a path with lesser path interference exists between u and v as compared to t times the interference on this edge. Then the path is added into the resulting topology, otherwise the edge itself is added. Same thing is repeated for each node ensuring its distributed in nature.

For selection of default channel within each cluster; the interference from external source is taken into account. Accurately measure of external interference is by periodic passive monitoring of traffic load on the channel. The interference estimation process for the default channel of a cluster is done collaboratively by constituent nodes.

Passive monitoring mechanism has parameters-

- Channel utilization
- Channel quality

One of the non-default interfaces of the nodes is configured periodically to the packet capture mode for specific interval of time on each channel. But the captured traffic may not be representing the actual load on the channel therefore channel quality is based on the number of lower layer metrics such as bit or frame error rate, received signal strength etc. This is expensive due to channel switching delay to overcome disruption of traffic flows. All flows that use the non-default interface are redirected to the default interface. Cluster- head computes best channel.

U_{ij} = Channel utilization by node i on channel j.
C_{ij} = Channel quality by node i on channel j.

For default channels

$$\text{Cost}_j = (1 - \alpha)\sum_{i=1}^{S} U_{ij} + \alpha . \frac{1}{\sum_{i=1}^{S} C.Q_{ij}} \quad \forall j \in C \tag{16}$$

The channel with least cost is selected as default channel. If it is different from current default channel then cluster head informs the nodes and nodes configure their default interfaces to this channel.

For non-default channel each node periodically (every T_A units of time) transmits the information about the channel and queue length for all interfaces to the cluster head. Border nodes transmit information to its own cluster head and its neighbouring cluster head.

Let C = set of available channels.
For node *v,*

$$A_v(\text{matrix})_{mXn} = a_{ij} = \begin{cases} 1, & if \text{ int}erface \ j \ operates \ on \ channel \ i \\ 0, & otherwise \end{cases} \tag{17}$$

Q_{max} =Maximum possible queue length.

$$q_i = \frac{Average \ queue \ length \ of \ \text{int}erface \ i}{Q_{max}} \tag{18}$$

$Q_{(non\text{-}default)} = C_i$ = column matrix m = |c|

$$C_i = \sum_{j=1}^{n} a_{ij} * q_j = A*Q \ [\text{for non-overlapping channels}] \tag{19}$$

[Overlapping channels]

$$C_i = \sum_{k=1}^{m} x_{ik} * \left(\sum_{j=1}^{n} a_{ij} * q_j \right) \tag{20}$$

$X = x_{ik}$ = matrix of order m x n

$X_{ij} = I \ (i, j) = 1$ if i = j

= 0 non-overlapping channels

Based on the channel cost, the cluster head computes the new channel assignment. First priority is the border nodes. If the interface i of the border node v is connected to a higher priority cluster, then this interface is assigned to the cluster head by its neighbouring nodes from that cluster. If the best channel calculated differs from currently used channel, the node and neighbours of node bound to this interface are informed to update.

Rigidness of Channel Assignment Techniques

Channel assignment schemes can be classified based on the behaviour of the channel assignment technique upon change in the network topology, network flow, addition of omission of any node etc. The channel assignment techniques can be divided into three categories:

Static Techniques

In this technique once the channels are assigned before the transmission in the network initiates, they are never changed again for long stretch of time. It can be implemented in networks where the total information of the network is known in advance and is never going to change. But in real life scenarios it is hard to implement because the topology or the flow characteristics change continuously and prior information about the characteristics is not present (Islam et al., 2016). The main advantage of this type of channel assignment is it avoids channel switching delay and thereby no control message overhead. This technique fails to adapt any client mobility, has no fault tolerance of the system and does not consider the external interference. Few schemes under this categorization are Connected Low Interference Channel Assignment (CLICA) (Marina et al., 2010), Multi-Radio Breadth First Search Channel Assignment (MRBFS-CA) (Ramachandran et al., 2006), Hyacinth (Raniwala et al., 2005). The schemes are elaborately in the previous sections.

Semi-Dynamic Technique

In this technique the flexibility of channel assignment decision increases in comparison to the static technique. The entire working period of the WMN's is divided into short intervals of time (Islam et al., 2016). Channel assignment process is carried out before each of the intervals. It is more practical and there is no need for having any prior knowledge of the network or estimating the characteristics of the WMN's. There is a bit of fault tolerance adopted in this technique by keeping track of the external interference, environmental effects and client mobility. The channel assignment process is carried out before the start of each interval. Consideration of the external factors leads to inclusion of the control messages. Schemes lying under this category are: Balanced Static Channel Assignment (BSCA) (Kodialam et al., 2005) and Multi-Radio Breadth First Search Channel Assignment (MRBFS-CA) (Ramachandran et al., 2006). The schemes have been presented in the previous sections.

Dynamic Techniques

The dynamic techniques have complete dynamic behaviour of assigning channels to the links of the WMN's. Since they assign channels continuously on the requirement of the action, the channel assignment technique is absolutely flexible. They need no prior knowledge of the characteristics of the WMN but require continuous monitoring of the scenario of the network in order to implement total dynamicity (Islam et al., 2016). Since the channel assignment technique is on the fly, there is a heavy control message overhead. The main challenge for the dynamic and adaptive channel requirement is the multi hop coordination requirement that arises because of channel dependency among nodes. Such coordination becomes more complex when efficient and flexible support for different communication patterns is desirable. Few examples of the techniques are: Packing Dynamic Channel Assignment (PDCA) (Kodialam et al., 2005), Joint Optimal Channel Assignment and Congestion Control (JOCAC) (Rad et al., 2006) and Cluster based Multipath Topology control and Channel assignment scheme (CoMTaC) (Naveed et al., 2007).

COMPARISON OF THE DIFFERENT CHANNEL ASSIGNMENT TECHNIQUES

The comparison of the schemes is shown in Table 2.

Table 2. Comparison table

Category		Algorithm	Deciding Parameter	Technique Used	Advantage	Disadvantage
Centralized	a. Static	(i) CLICA (Marina et al., 2010)	Link conflict weight	Graph-colouring problem	1. Better throughput performance. 2. Mitigation of ripple effects. 3. Despite of traffic patterns no switching required.	1. Continuous visit to node increase network overhead. 2. Greedy and heuristic approach. 3. High channel switching delay.
		(ii) TARICA (Wang et al., 2007)	Traffic load and channel utilization.	Graph decomposition problem.	1. Ensures connectivity. 2. Reduction of control message overhead. 3. Helps in load balancing and increase in overall throughput.	1. Does not consider external interferences. 2. Does not provide locality of information. 3. Cannot be implemented in frequently changing environments.
		(iii) MCCA (Avallone et al., 2008)	Network capacity and link load.	Linear programming	1. Independent of any particular traffic pattern. 2. Maximizing throughput. 3. Maintains connectivity in the network.	1. It is not fault tolerant.
		(iv) RCL (Alicherry et al., 2005)	Interference	Linear programming.	1. Considers intra- flow and intra-flow interferences in the network. 2. Ensures connectivity in the network. 3. Increases overall throughput of the network.	1. Does not consider the convergence rate. 2. Does not provide load balancing.
	b. Semi-Dynamic	(i) MRBFS-CA (Ramachandran et al., 2006)	WCETT (Draves et al., 2004)	Breadth First Search algorithm.	1. No disruption of flow due to link redirection. 2. Better performance due to frequency diversified alternative.	1. Requires changes in MAC layer due to per-packet channel switching. 2. Cannot be implemented in frequently changing topology.
		(ii) BSCA (Kodialam et al., 2005)	Channel utilization.	Graph greedy colouring problem.	1. Considers all interferences. 2. Ensuring load balancing and load distribution of channels.	1. NP - hard problem.
	c. Dynamic	(i) PDCA (Kodialam et al., 2005)	Channel utilization.	Graph greedy colouring problem.	1. Considers all interferences. 2. Ensures load balancing and load distribution of channels.	1. NP – hard problem.
Distributed	a. Static	(i) Hyacinth (Raniwala et al., 2005)	Network traffic.	Multiple spanning tree based load balancing routing algorithm.	1. Fault tolerant. 2. It also supports load balancing among the channels. 3. It is distributive in nature.	1. Does not incorporate locality of information.
	b. Semi-Dynamic	(i) JOCAC (Rad et al., 2006)	Transmission power and link capacity.	Decentralized utility maximization problem	1. Considers locality of information. 2. It also incorporates convergence rate and maximizes throughput. 3. It takes into consideration the interflow and intra flow interference. 4. It mitigates transmission delay.	1. It does not consider external interference. 2. It is not fault tolerant.
	c. Dynamic	(i) CoMTaC (Naveed et al., 2007)	Channel utilization and channel quality.	Cluster-based approach.	1. Low time complexity. 2. Maximizes throughput. 3. Ensures connectivity in the network.	1. Choice of default and non-default interface regulates the entire assignment. 2. High switching delay.

CONCLUSION

The main purpose of the investigation is to study the channel assignment techniques in wireless mesh networks. The factors affecting the channel assignment process such as, interference, connectivity, fault tolerance, load balancing, fairness etc. have been discussed. These factors affect the channel assignment process and should be considered while designing any channel assignment technique. The channel assignment techniques have been categorised based on two characteristics: point of decision and rigidness of decision. Based on point of decision the schemes are classified into two categories: centralized and distributed. Centralized schemes can further be subdivided into three categories: graph based, network flow based and network partition based. Again distributed schemes have been classified into two sub-categories: gateway oriented and peer oriented. On the other hand, based on rigidness of decision, the channel assignment schemes have been classified into three categories: static, semi-dynamic and dynamic. The existing channel assignment algorithms have been presented in brief. Finally the deciding parameters, techniques, advantages and disadvantages of the schemes have been compared in tabular format.

ACKNOWLEDGMENT

The authors are thankful to Mobile Computing Laboratory, Department of Computer Science & Engineering, Tripura University for providing necessary infrastructure.

REFERENCES

Alicherry, M., Bhatia, R., & Li, L. E. (2005, August). Joint channel assignment and routing for throughput optimization in multi-radio wireless mesh networks. In *Proceedings of the 11th annual international conference on Mobile computing and networking* (pp. 58-72). ACM. 10.1145/1080829.1080836

Avallone, S., & Akyildiz, I. F. (2008). A channel assignment algorithm for multi-radio wireless mesh networks. *Computer Communications*, *31*(7), 1343–1353. doi:10.1016/j.comcom.2008.01.031

Clark, B. N., Colbourn, C. J., & Johnson, D. S. (1990). Unit disk graphs. *Discrete Mathematics*, *86*(1-3), 165–177. doi:10.1016/0012-365X(90)90358-O

Clausen, T., & Jacquet, P. (2003). *Optimized link state routing protocol (OLSR)* (No. RFC 3626).

Draves, R., Padhye, J., & Zill, B. (2004, September). Routing in multi-radio, multi-hop wireless mesh networks. In *Proceedings of the 10th annual international conference on Mobile computing and networking* (pp. 114-128). ACM. 10.1145/1023720.1023732

Guerin, J., Portmann, M., & Pirzada, A. (2007, December). Routing metrics for multi-radio wireless mesh networks. In *Proceedings of Australasian Telecommunication Networks and Applications Conference* (pp. 343-348). IEEE. 10.1109/ATNAC.2007.4665270

Gupta, P., & Kumar, P. R. (2000). The capacity of wireless networks. *IEEE Transactions on Information Theory*, *46*(2), 388–404. doi:10.1109/18.825799

IEEE Std 802.11-2012. (2012). Information Technology—Telecommunications and information exchange between systems—Local and metropolitan area networks—Specific requirements—Part 11: wireless LAN medium access control (MAC) and physical layer (PHY) specifications.

Islam, A., Islam, A. A., M. J., Nurain, N., & Raghunathan, V. (2016). Channel assignment techniques for multi-radio wireless mesh networks: A survey. *IEEE Communications Surveys & Tutorials, 18*(2), 988-1017.

Kodialam, M., & Nandagopal, T. (2005, August). Characterizing the capacity region in multi-radio multi-channel wireless mesh networks. In *Proceedings of the 11th annual international conference on Mobile computing and networking* (pp. 73-87). ACM. 10.1145/1080829.1080837

Low, S. H., Paganini, F., & Doyle, J. C. (2002). Internet congestion control. *IEEE Control Systems, 22*(1), 28-43.

Marina, M. K., Das, S. R., & Subramanian, A. P. (2010). A topology control approach for utilizing multiple channels in multi-radio wireless mesh networks. *Computer Networks, 54*(2), 241–256. doi:10.1016/j.comnet.2009.05.015

Naveed, A., Kanhere, S. S., & Jha, S. K. (2007, October). Topology control and channel assignment in multi-radio multi-channel wireless mesh networks. In *Mobile Proceedings of IEEE International Conference on Adhoc and Sensor Systems, 2007*(pp. 1-9). IEEE.

Rad, A. H. M., & Wong, V. W. (2006, June). Joint optimal channel assignment and congestion control for multi-channel wireless mesh networks. *Proceedings of IEEE International Conference on Communications, 5*, 1984-1989.

Ramachandran, K. N., Belding, E. M., Almeroth, K. C., & Buddhikot, M. M. (2006, April). Interference-aware channel assignment in multi-radio wireless mesh networks. *Proceedings of 25th IEEE International Conference on Computer Communications*, 1-12. 10.1109/INFOCOM.2006.177

Raniwala, A., & Chiueh, T. C. (2005, March). Architecture and algorithms for an IEEE 802.11-based multi-channel wireless mesh network. *Proceedings of 24th Annual Joint Conference of the IEEE Computer and Communications Societies, 3*, 2223-2234.

Raniwala, A., Gopalan, K., & Chiueh, T. C. (2004). Centralized channel assignment and routing algorithms for multi-channel wireless mesh networks. *Mobile Computing and Communications Review, 8*(2), 50–65. doi:10.1145/997122.997130

Saini, J. S., & Sohi, B. S. (2016). A Survey on Channel Assignment Techniques of Multi-Radio Multichannel Wireless Mesh Network. *Indian Journal of Science and Technology, 9*(42).

Si, W., Selvakennedy, S., & Zomaya, A. Y. (2010). An overview of channel assignment methods for multi-radio multi-channel wireless mesh networks. *Journal of Parallel and Distributed Computing, 70*(5), 505–524. doi:10.1016/j.jpdc.2009.09.011

Siek, J. G., Lee, L. Q., & Lumsdaine, A. (2001). *The Boost Graph Library: User Guide and Reference Manual, Portable Documents*. Pearson Education.

Skalli, H., Ghosh, S., Das, S. K., & Lenzini, L. (2007). Channel assignment strategies for multiradio wireless mesh networks: Issues and solutions. *IEEE Communications Magazine, 45*(11), 86–95. doi:10.1109/MCOM.2007.4378326

Wang, J., Wang, Z., Xia, Y., & Wang, H. (2007, September). A practical approach for channel assignment in multi-channel multi-radio wireless mesh networks. *Proceedings of Fourth International Conference on Broadband Communications, Networks and Systems*, 317-319.

Chapter 5
SMARC:
Seamless Mobility Across RAN Carriers Using SDN

Walaa F. Elsadek
American University in Cairo, Egypt

Mikhail N. Mikhail
American University in Cairo, Egypt

ABSTRACT

Next-generation network promises to integrate cross-domain carriers; thus, infrastructure can be provided as a service. 5G-PPP's vision is directed toward solving existing 4G LTE mobility challenges that congest core networks, disrupt multimedia and data transfer in high mobility situations such as trains or cars. This research adopts 5G methodology by using software-defined networking (SDN) to propose a novel mobile IP framework that facilitates seamless handover, ensures session continuity in standard and wide area coverage, and extends residential/enterprise indoor services across carriers under service level agreement while ensuring effective offload mechanism to avoid core network congestion. Performance excels existing protocols in setup and handover delays such as eliminating out-band signaling in bearer setup/release and isolating users' packets in virtual paths. Handover across cities in wide area motion becomes feasible with lower latency than LTE handover inside city. Extending indoor services across carriers becomes equivalent to LTE bearer setup inside a single carrier's PDN.

INTRODUCTION

Next generation cellular system is expected to be a transformational shift in telecommunication. 5G aims to mobilize service cross-domain networks so carriers can provide their infrastructure on a need-for-service basis (5G Americas, 2016). Consistent user quality of experience across heterogeneous topologies becomes a real challenge hindering the applicability of wide range of Internet of Things (IoT) real time services (Abdulhussein et al., 2015). Tremendous innovation in smart homes, offices, remote health care, and multimedia streaming services raise an urgent need for continuous connectivity to real

DOI: 10.4018/978-1-5225-5693-0.ch005

time indoor services. Effective solutions are required with optimal cost structures through the benefits gained by leveraging automation. This is to ensure seamless coverage with millisecond latencies and wire rate of transfer during motion in high speed vehicle crossing wide geographically separated areas with large population density under different administrative domains. 4G Long Term Evolution (LTE) exhausted several trials to provide an effective solution that guarantees session continuity in wide area motion with uninterruptable access to residential, enterprise, and internet services. Unluckily, existing solutions suffer dramatically from inefficient data forwarding that leads to the Evolved Packet Core (EPC) congestion and induces high latency in the services offered. LIMONET trials in 3GPP releases from 9 to 12 as well as the IETF Multipath TCP, discussed in the background section, highlight these problems (ETSI, 2016; Gupta & Rastogi, 2012; Hampel, Rana, & Klein, 2013; Wang, 2015).

This research proposes a novel network based Mobile IP (MIP) framework using SDN, called SMARC, which guarantees uninterruptible accessibility to indoor services with an effective offload mechanism inside and across carriers under service level agreement (SLA). SMARC is a successive research for SRMIP: Software-Defined RAN Mobile IP Framework for Real Time Applications in Wide Area Motion (Elsadek & Mikhail, 2016). SRMIP ensures session continuity in normal and challenging situations within a single carrier RAN. These researches target solving most mobility challenges that hinder services applicability in the next hyper interconnected IoT world (Elsadek, 2016). SMARC prototype is established to assess the feasibility of extending residential/enterprise indoor services inside and across carriers under SLA while ensuring seamless handover and wire speed forwarding of Mobile Node's (MN) packets without congesting core networks. Experimental results show that cross-carriers mobility setup delay is equivalent to existing LTE mobility setup delay inside single carrier's PDN. Strong improvements are achieved in mobility setup delay for standard and wide area motion inside single carrier over LTE total bearer setup time in Proxy Mobile IP (PMIP). Handover delay inside city becomes equivalent to L2 handover in Software-Defined Wireless Networks (SDWN) that is highly better than LTE. SMARC ensures session continuity during handover across cities that is currently unfeasible in LTE as of EPC congestion problem. The clue is replacing LTE bearer in PMIP and GPRS Tunneling Protocol (GTP) with OpenFlow virtual paths. Furthermore, LTE out-band signaling has been eliminated as of replacing the control messages associating bearer setup/release during MN's join/handover with a recursive procedure occurring in-line during the allocation of MN's IP address.

The research is organized as follows; the second section gives an overview on existing mobility protocols and highlights the performance degradation associating tunneling. In addition, it describes the inefficient data forwarding plan of existing mobility protocols that induces core congestion problem and limits session continuity in real deployments of wide area motion. The third section illustrates an overview over SDN principles, OpenFlow SDN-based architecture, the advantages gained from adopting such technology in LTE, and related research work. The fourth section presents SMARC framework, illustrates the key design concepts and states how the described problems are solved. The fifth section analyzes SMARC experimental results and compares them to SDWN and PMIP. The sixth and seventh sections provides the research summary and highlights the contributions respectively. The last two sections briefly conclude the research and present future research directions.

MOBILE IP BACKGROUND

Overview

Mobility management is divided into two main categories; host-based and network-based architectures (Al-Surmi, Othman, & Ali, 2010). The former enforces MN's involvement in mobility signaling while the later carries signaling through the network transparent from MN. Mobile IP was defined in RFC 3344 as a way for enabling mobile user to keep the same IP address while traveling to different networks for continuous communication (Perkins, 2002). The initial proposed MIP was host based. Its' deployment required MN's kernel modification and introduced inefficient data forwarding, called triangle problem. These problems hindered its applicability in large deployments (Esmat, 2000; Mikhail, 2001; Weyland, 2002). Later, MIPv6 was introduced in RFC 3375 as host-based mobility protocol for supporting MNs' global mobility (Johnson, 2004). It solved several issues in MIP; however, it suffered from high latency, complex signaling overheads, and security issues (Al-Surmi et al., 2010). Several researches tried to decrease handover latency as MIPv6 Fast Handover followed by proposals to decrease signaling overhead through hierarchical layout finally that ended with PMIPv6 introduction (Koodli, 2009; Schmidt, 2015; Soliman, Malki, & Bellier, 2005).

Proxy Mobile IP

PMIPv6 is IETF standard network-based mobility protocol as specified in RFC 5213 (Gundavelli, 2008). Initially designed to support IPv6 latter IPv4 is incorporated in RFC 5844 (Wakikawa & Gundavelli, 2010).

Main Functional Entities

1. **Mobile Node (MN):** Is a registered PMIP subscriber attached to foreign network. MN can be an IPv4-only node, an IPv6-only node, or a dual-stack node.
2. **Corresponding Node (CN):** Is a node outside MN's home with established sessions to the roaming MN.
3. **Mobile Access Gateways (MAGs):** Intercepts MN's communications and related signaling. It proxies MN's authentication, relays MN's IP allocation messages to LMA or external DHCP, tunnels MN's packets to LMA, and facilitates MN's handover across MAGs managing other Access Points (AP).
4. **Local Mobility Anchors (LMAs):** Is MN's home agent managing MIPv6 bindings as stated in RFC 3775 and PMIPv6 (Gundavelli, 2008; Johnson, 2004). LMA and MAG tunnel can be GRE or IP-in-IP.

Modes of Deployments

PMIP has two main modes of deployments; Flat Domain Model and Domain Chaining Model. In Flat Domain Model, MAG tunnels MN's packets to LMA home IP address or domain name. The major drawback of this approach is that every office/home registered for residential/enterprise service must state a Public IP address or a known Private IP address to which LTE/WiFi service provider can terminate the tunnel. This solution is almost inapplicable and violates the security of involved enterprises. Domain

Chaining Model adopts a hierarchical deployment of Flat Domain Model. Foreign MAG tunnels MN's packets to local LMA. The session chaining mode transfers local LMA to MAG for tunneling MN's packets to home LMA. Then, home LMA performs another session chaining to tunnel MN's packets to office/home registered LMA. This means that at least three tunnels are created for full inter-domain mobility. PMIP Domain Chaining solution adoption in real environment induces severe latency associated with several tunnels establishments.

Mobility in 4G LTE

LTE was designed to support only packet-switched services for LTE and WiFi Interworking as shown in figure 1. LTE ultimate goal is seamless IP connectivity between User Equipment (UE) and Packet Data Network (PDN), without any disruption to end users' applications during mobility while ensuring higher data rate, lower latency, flexible bandwidth deployments, and great quality of service (QoS). Evolved Packet System (EPS) adopts IP packet flow concept with defined QoS to route UE's traffic (Alcatel-Lucent, 2016).

Main Functional Entities

1. **Home Subscriber Server (HSS):** Holds centralized database of subscription data, PDNs information, and UE's dynamic records of attachment. HSS may carry Authentication Center (AUC) functionality.
2. **Evolved Base Station (eNodeB or e-NB):** Provides UE and EPC wireless connection that ensures both control/data integrity and security through datapath encryption.
3. **Serving Gateway (S-GW):** Is responsible for forwarding packets between e-NB and P-GW. It serves as MAG that buffers downlink data during bearers' re-establishment in inter-eNB/inter-3GPP handovers.
4. **PDN Gateway (P-GW):** Is the sole UE's exit to outside world. P-GW allocates UE's IP from the attached PDN, anchors inter S-GW handover, and acts as LMA when interworking residential/enterprise networks.
5. **Mobility Management Entity (MME):** Is the EPS brain in HSS signal for UE's authentication and location.
6. **Wireless Access Gateway (WAG):** Seamlessly integrates UE and P-GW in trusted 3GPP WiFi authentication.
7. **ePDG (Evolved Packet Data Gateway):** Terminates UEs' IPsec tunnels in untrusted non-3GPP access.

Tunneling Protocols Overheads

IP-session continuity is supported in EPC using MIP, GTP, and PMIP. Due to complicated signaling and overheads introduced by host based MIP, majority of vendors overlooked MIP interoperability (Ruckus, 2013). The main differences between GTP and PMIP are highlighted in figure 2, 3 and table 1.

Figure 1. LTE and WiFi Interworking Architecture

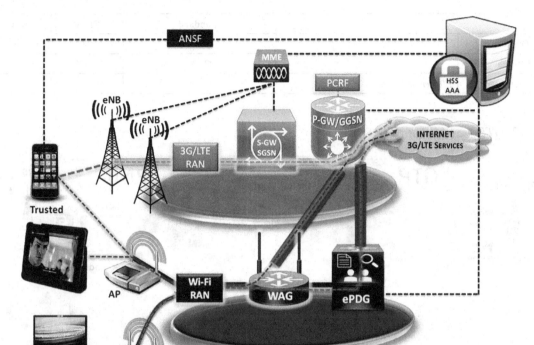

Minimum Tunnel Overheads

Calculations in table 1 highlight that minimum theoretical improvement in L2/L3 standard network over GTP/PMIP is ~2% as of tunnel headers. In real environment, PMIP tunnel induces at least 10% overhead in addition to the latency associating re-encapsulation at LTE-Uu, S1-U, S5/8, and SGi interfaces shown in figure 3 (Savic, 2011).

Maximum Transfer Unit (MTU) in 100BASE-TX Ethernet = 1500 octet payload

Minimum Packet Size = 1500 (payload) + 8 (preamble) + 14 (header) + 4 (trailer) + 12 (min. gap) = 1538 octets

GRE Payload = 1500 (payload) – 20 (min IPv4) – 4 (GTP) = 1476 octets

GTP Payload = 1500 (payload) – 20 (min IPv4) – 8 (UDP) – 8 (GTP) = 1464 octets

Existing Challenges Hindering Mobility in Wide Area Motion

4G LTE solution in challenging mobility situations within a geographical location, like city, is through a centralized P-GW to which all UE's packets are tunneled. P-GW serves as the LMA allocating UE

Table 1. Tunnel Protocols Overheads

		Standard	**PMIP**	**GTP**
Overhead		2.47%	4.03%	4.81%
Efficiency		97.53%	95.97%	95.19%
Throughput	*Efficiency/Net Bit*	97.53%	95.97%	95.19%

Figure 2. GTP vs. PMIP in LTE

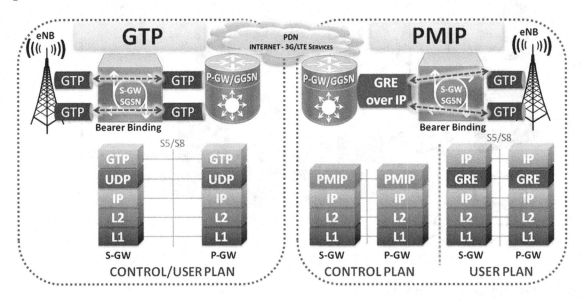

Figure 3. LTE PMIP User Protocol Stack

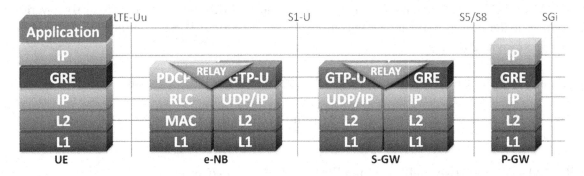

an IP address and the gateway to internet and indoor services. Few P-GWs serve carriers for country coverage. The centralized approach ensures session continuity during motion within a region but not for wide area motion as train or car crossing cities' boundaries. Handover across P-GWs at different cities PDNs is currently not supported in LTE as the second city P-GW allocates UE a new private IP mapped to new public IP for internet access. Thus, all UE's active sessions are disconnected. To avoid session disconnects, HSS instructs S-GW, acting as MAG, to tunnel all UE's packets to previous P-GW regardless of path optimality or core congestion as shown in figure 4. In residential/enterprise mobility, UE is registered with a predefined private IP assigned from closest P-GW to the registration location (NMC, 2015). Whenever UE with indoor service registration joins, HSS directs all UEs' packets to home P-GW regardless of core congestion problem.

The centralized approach for solving both problems; indoor service accessibility and session continuity in wide area motion introduces two main problems. The first is long latency as of non-optimum path selection and the second is EPC congestion. These problems occur as Transmission Control Protocol (TCP) and User Datagram Protocol (UDP) packets need to cross EPC to initial P-GW assigned during registration or to P-GW that handled initial UE join to avoid interrupting active internet and indoor TCP sessions. PMIP dynamic nature advances it over GTP to make selective breakouts for minimizing EPC congestion problem. 3GPP releases from 9 to 12 introduce some traffic offloading mechanisms named LIMONET; Local IP Access (LIPA) Mobility and Selected Internet IP Traffic Offload (SIPTO) for efficient utilization of core networks (Gupta & Rastogi, 2012; Wang, 2015). LIPA mobility focuses on residential/enterprise with indoor femtocells and picocells, while SIPTO is designed for internet access in both femtocell and macrocell setups (Gupta & Rastogi, 2012). Despite the exhausted efforts, almost all offloading mechanisms lack an accurate traffic classification mechanism to ensure minimum latency during handover. Offered solutions put carriers in a tie of either providing session continuity or making efficient usage of their core networks.

Figure 4 illustrates both inefficient data forwarding and EPC congestion problems associating centralized P-GW. LIPA packets are correctly forwarded to home networks and SIPTO feature facilitates for

Figure 4. 4G LTE Inefficient Data Forwarding

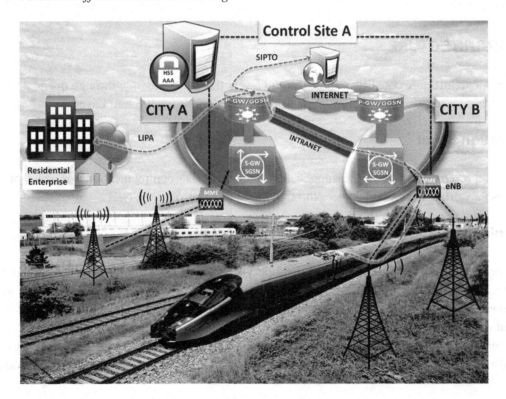

non-LIPA packets finding an internet breakout in City-A P-GW at initial point of registration without enforcing full traffic redirect to home network. However, the offered solutions overlook that best breakout to UDP and new TCP sessions are City-B P-GW.

Later, several IETF researches were developed to isolate TCP from UDP. TCP is reliable protocol requiring session continuity while this doesn't matter for UDP. Bell Labs presented a lightweight MultiPath TCP (MPTCP) proxy that is inserted in datapath for header rewrite without packet buffering or stream assembly. MPTCP facilitates seamless session end-point migration across multi-provider network (Hampel, 2013). Cisco tried to extend PMIP to support MPTCP as an initiative toward path selection on flow basis to facilitate MAG registrations of multiple transport end-points with LMA (Cisco, 2016; Seite, Yegin, & Gundavelli, 2016). The goal is finding an optimal solution for filtering indoor packets and active TCP sessions to cross the EPC only while providing instant breakout from UE's current point of attachment for both UDP and new TCP sessions.

SDN PRINCIPLES

Open Networking Foundation (ONF) introduces a novel approach called Software-Defined networking (SDN) that decouples network control plane from the underlying data forwarding plane. The open programmable North Bound Interface (NBI) facilitates development of wide range of off-the-shelf and custom network applications. SDN logically centralized control plane is considered the network brain that dramatically improves network agility and automation with significant reduction in operations' cost through the South Bound Interface (SBI) by complete manipulation of users' packets and underlying resources as switches and router switches (Bailey et al., 2013; ONF, 2012).

SDN ARCHITECTURE

ONF's SDN architecture consists of three distinct layers that are accessible through open APIs. Figure 5 shows these layers; Application Layer, Controller Layer, and Infrastructure Layer.

Application Layer

This layer represents end-user business applications providing SDN communications services that directly programs network behavior and devices through the SDN Controller via NBI.

Control Layer

This layer consolidates the control functionality supervising network forwarding behavior through an open interface. SDN Controller is a logically centralized entity that transfers the requirements of Application layer down to Infrastructure Layer through NBI and provides an abstract view of the network up to the Applications layer. SDN Control-Data-Plane Interface (CDPI) is the interface defined between SDN Controller and Datapath, which is responsible for network capabilities advertisement, statistics reporting, and events notifications. Physically, SDN controller has a hierarchical layout of sub-controllers and communication interfaces with virtualized framework called Network Function Virtualization (NFV) for slicing network resources. This physical layout facilitates the federation of multiple controllers.

Infrastructure Layer

This layer represents Network Elements (NE) and devices providing packet switching and forwarding functions. SDN Datapath is a logical network device, which exposes visibility and exceptional control over its advertised forwarding and data processing capabilities through a CDPI agent and a set of traffic forwarding engines and traffic processing functions. These engines and functions may include simple forwarding between datapath's external interfaces or internal traffic processing or termination functions (Bailey et al., 2013).

OPENFLOW SDN-BASED TECHNOLOGY

Overview

OpenFlow protocol is key enabler for SDN and the first standard SDN protocol in SBI to facilitate relay of information and packets between control and forwarding planes through complete manipulation of

Figure 5. SDN Architecture

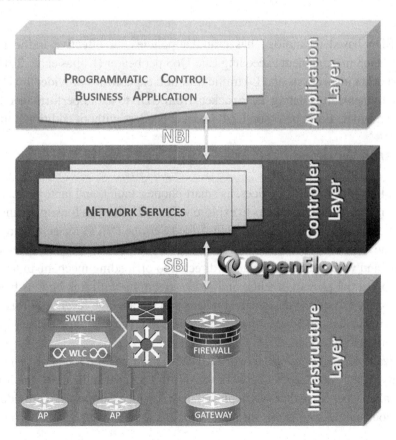

network devices' forwarding plane such as switches and routers. OpenFlow adopts the concept of flows to identify network traffic based on pre-defined match rules that can be statically or dynamically programmed by the SDN controller software. For enterprises and carriers, OpenFlow SDN-based technology facilitates unprecedented programmability gain capable to trail the network behavior in a matter of hours through virtualization of network infrastructure and abstraction from individual network services. Traffic policy can be defined based on usage patterns, applications, and cloud resources. Such unprecedented granular control facilitates for network instant response to real-time changes at application, user, and session levels unlike the monolithic, closed, and mainframe-like today's networking (Kolias et al., 2013; Nygren, 2013; ONF, 2012).

Motivation for Using OpenFlow Based SDN Technology in RAN

Similarities

- Both EPS and OpenFlow adopt the flow concept. The former uses it to route packets between UE and PDN while the later identifies traffic based on pre-defined match rules programmed by the controller.
- Both technologies adopt centralized decision architectures. In EPS, HSS holds the centralized database used by MME to process UE--HSS signaling while in SDN, the controller holds all information with complete manipulation of UEs' packets and network resources.

Advantages

- OpenFlow SDN-based technology has granular policies for effective traffic isolation, service chaining, and QoS management exceeding EPS QoS per bearer (Kolias et al., 2013; Zhang, 2014).
- L2/L7 deep packet inspections and L4 traffic filtration capabilities provided by OpenFlow replace MPTCP deployments in isolating TCP packets from UDP to avoid performance degradation.
- OpenFlow can isolate flows based on L2/L4 packet headers without additional tunnel header.

Related Research Work

The enormous explosion of wireless devices, as smart phones, tablet, and laptop, increasingly captures attentions of carriers and service providers to explore variety of solutions for supporting higher volumes of traffic while offering new sophisticated Value-Added Service (VAS) for generating more revenues. The ONF formed the Wireless & Mobile Working Group (WMWG) to assess new investigatory solutions for mobile carriers and vendors. Strong focus is directed to offloading mechanism for accommodating the subsequent exponential growth in audio and video streaming with tremendous bandwidth requirements that dramatically exceed the designated budgets and the average revenue per user (ARPU). Such investigations are vital to make efficient usage of the scarce RAN capacity though service discrimination to maximize revenues (Kolias et al., 2013, ONF, 2012).

Recently, software-defined wireless network (SDWN) becomes the focus of several researchers to increase the agility of the radio access network (RAN) through virtualization. Such initiative started with a prototype called OpenRoad that was deployed in the college campuses to provide interconnection among heterogeneous wireless networks through virtualization (Yap et al, 2010). The objective of this research is showing how MNs can effectively use available coverage for session continuity, effective offload, and load balance, regardless of being different wireless technologies; as WiFi and WiMAX. This is a turn point for interconnecting heterogenous networks. In real deployments, different wireless technologies are almost owned by different carriers. "OpenRoad" can be expanded to include mobility across PDNs managed by single controller for wide area coverage as well across PDN of different carriers.

"Odin" is a prototype built on light virtual AP abstraction to simplify client management without any modification to client side. The research uses SDN to implement enterprise WLAN services as network applications providing seamless mobility, load balancing and hidden terminal mitigation (Suresh et al., 2012). "Odin" is an early initiative to replace vendor specific Wireless LAN Controller (WLC). The initiative is restricted to 802.11 standard only. The research can be further expanded to support mobility across WLCs managed by different administrative domains. Future research topics can concentrate on unified access of enterprises wired/wireless LAN and mobility across enterprise buildings that are WAN separated with an effective offload mechanism to make efficient usage of interconnecting bandwidth.

"OpenRadio" designs a programmable wireless dataplane implemented on multi-core hardware platform to re-factor the wireless stack into processing and decision planes. The processing plane includes directed graphs of algorithmic actions while decision plane carries the optimum directed graph logic election for particular packets. Wireless dataplane is compatible with wireless protocols including WiFi and LTE on off-shelf DSP chips with programmatic capability to modify PHY and MAC layers for further optimization (Bansal et al., 2012). This research is not SDN-based but argues that wireless infrastructure needs a programmable dataplane to support such evolvability. "OpenRadio" resembles SDN in decoupling wireless protocol definition from the underlying hardware and designing a software abstraction layer that exposes a modular and declarative interface to program wireless protocols. However, the flexibility in SDN programable infrastructure is far beyond what is proposed as of multiple vendor support and wide range of off-shelf integrated applications.

"OpenRF" project proposes a cross layer architecture for managing multiple-input and multiple-output (MIMO) signal processing with commodity WiFi cards and real applications. "OpenRF" adopts SDN abstraction to create a self-configuring architecture that facilitates for APs on the same channel to cancel their interference at each other's clients transparently without any need for administrators' awareness of MIMO or physical layer techniques (Kumar et al., 2013). Despite being a good initiative, still real deployments of WiFi providers are not expected to rely on low commodity WiFi cards. This can jeopardize the power of SDN controllers in managing software applications for canceling interference instead of the PHY layer. In turn, this can dramatically decrease the overall performance and position the controller as a single point of failure.

"SoftRAN" is the first SDN research to argue that existing LTE distributed control plane is suboptimal when allocating radio resources, balancing load between cells, managing interference and implementing handovers. The research tries to abstract all base stations in a local geographical area as a virtual big-base station comprised of a central controller and radio elements using SDN (Gudipati et al., 2013). "SoftRAN" is based on a valid argument. However, no effective evaluation measure is proposed. In addition, the goals are directed to adopting SDN to enhance existing features performance without solving any existing challenges or proposing new features.

"OpenRAN" argued about closeness and ossification of existing RAN to propose a new architecture that benefits from virtualization and SDN programmatic capability for efficient convergence across heterogeneous networks (Yang et al., 2013). The research proposes four levels of virtualization; The first is application virtualization that divides the flow space to several virtual spaces representing several network carriers or services that are operated and managed independently. The second is cloud virtualization, that enables SDN controller to create virtual Base Band Units (vBBU) and virtual Base Station Controllers (vBSC) through virtualization of physical processors and allocation of appropriate computing and storage resources. The third is radio frequency spectrum virtualization which enables several virtual Remote Radio Units (vRRUs) with different wireless protocols to coexist in one shared physical Remote Radio Units (pRRU). The fourth is cooperation virtualization, which constructs several virtual networks, including virtual nodes and virtual links to benefit from inter-cell interference elimination and facilitate communications across different vBBUs and vBSCs. "OpenRAN" can be considered a creative architecture. However, the research has not stated any evaluation or validation methods, or carriers' cross billing mechanism, or effective measures for resources management.

"AnyFi" is SDN mobility framework for WiFi Carriers to support indoor mobility on available hotspots. In this framework, the SDN controllers of WiFi carriers have full power on home APs after software update. The updated home AP is called SDWN service termination. When MN joins the home AP, details of its set including MAC address are sent to service provider controller to make a virtual WiFi network available on the go. Also, MN stores home SSID and authentication credential of home AP in the network preferred list. When the same MAC address of MN's set comes near a foreign AP, SDWN radio software probes the SDN controller to create a home virtual SSID on visited AP thus MN's device connects automatically to it with the stored home credentials. Then, SDWN tunnels MN's packets to home AP (Anyfi, 2014). "AnyFi" framework has real deployment in the initial launching phase. This is considered a major improvement over previous works. However, three main challenges can hinder wide expansion of "AnyFi" service. The first is lack of integration mode to cellular or LTE networks as it is restricted to WiFi technology only. Thus, both deployment scope and geographical coverage are limited. The second challenge is the enforcement of remote controlled AP without considering customers' security policies. The third challenge is lack of enterprise scale support as it will be hard to create a virtual hotspot in the street per probed MAC. In large scale deployment, it will take huge amount of time till identifying the home network of a nearby MAC. Also, there will be limitations on the numbers of virtual APs to be created on the fly.

"CellSDN" introduces a local agent to make real time decision using centralization of logic in SDN while reducing complexity in the control plan and the cost of purchased equipment during expansion. The paper argues that SDN deployment in cellular networks can facilitate isolation of subscribers' flows using MPLS or VLAN TAG instead of bearer/tunneling signaling overheard. The expectations are tremendous simplification in newer services launching process, in existing capacity expansion for acquiring more subscribers, in real-time services application and fine-grained policies enforcement, as well as in deep packet inspection and in packets header compression (Erran, Mao, & Rexford, 2012). "CellSDN" can be considered a real transformational shift in network by converting traditional mobility bearer and tunneling signaling overheard to lower level overhead TAGs. However, the vision that the SDN controller is managing a group of OpenFlow switches not standard routed network has not been efficiently utilized. Standard routed network drops MN's hardware address after first hop router for this MPLS tags packets to avoid IP conflict cross overlapping subnets. In SDN multi-tenancy, MNs' hardware addresses are not dropped. Thus, packets can be isolated based on (IP, hardware) addresses pair without VLAN TAGs or MPLS Labels. With this fact, further performance improvement over "CellSDN" is expected.

"SoftCell" is the successive research of "CellSDN". This research proposes a scalable architecture to support fine grained policies' enforcement on mobile devices in cellular core networks through sequences of middle boxes using commodity switches and servers. Packets classification occurs at access switches next to base stations to guarantee that sessions belonging to the same connection traverse the same sequence of middle boxes in both directions even in the presence of mobility. The architecture prototype proves scalability and flexibility in real LTE workloads and in large-scale simulations (Jin, Erran, Vanbevery, & Rexford, 2013). In "SoftCell", the centralization of logic and the enforcement of fixed paths for the same connection can have dramatical side effects on performance as well as fault tolerance and load balance strategies. Further expansions of "CellSDN" and "SoftCell" are required to include solutions for existing LTE challenges as offload mechanism, handover problems, core congestion, and roaming across carriers.

Later, software-defined Heterogeneous Cloud Radio Access Network (H-CRAN) presents a centralized large-scale processing for suppressing co-channel interferences. The research introduces a new communication entity, called Node C, to converge existing ancestral base stations and act as base band unit (BBU) pool for managing all accessed remote radio heads (RRHs) (Contreras, Cominardi, Qian, & Bernardos, 2016). "H-CRAN" is considered a good initiative for heterogeneous radio access network aggregation. This represents a future research direction for optimum resource allocation and fast handover evaluation across heterogenous interconnected clouds.

"DISCO" proposes an extensible DIstributed SDN COntrol plane to cope with the distributed and heterogeneous nature of modern overlay networks to support WAN communication with resilient, scalable and extendible SDN control plane. DISCO controller manages its own network domain and communicates with other controllers to provide end-to-end network services. The framework is implemented on top of Floodlight OpenFlow controller and advanced message queuing protocol. Communication across DISCO controllers is based on a lightweight and highly manageable control channel protocol. The feasibility of this approach is evaluated through an inter-domain topology disruption use case (Phemius, Bouet, & Leguay, 2014). This research represents a starting point for negotiation across controllers to ensure resilience and fault free network. DISCO controller represents a future mobility direction for advance customization and orchestration of the services offered to roaming MNs' at visited network.

Another research proposes a new approach based on SDN concept for providing IP mobility in localized network without standard/proxy IP mobility protocol implementation. The objective is reducing loss in packets compared to PMIP and simplifying real implementation through solving overhead problem and decreasing handover latency. In this approach, routers signal the SDN controller of MN's join and handover. In turn, the controller deletes MN's flows with corresponding nodes and creates new flow rules matching the current attached location (Tantayakul, Dhaou, & Paillassa, 2016). This research has direct contribution in prototyping mobility without tunneling overheads. This is considered strong improvement in performance for solving mobility inside a single PDN. However, across geographical separated PDNs, MN's packets need an exit from the new point of attachment. In turn, MN's packets will be mapped to a new public IP address regardless keeping the registered private IP address. This will disconnect all active sessions and destroy the proposed mobility concept. Mobility performance inside single PDN can be further improved provided canceling out-band signaling between routers and the controller. By default, new MN's flow triggers the controller for path determination. Out-band signaling is just redundant latency.

SMARC FRAMEWORK

Overview

This research adopts 5G-PPP vision, introducing IP mobility as service using OpenFlow SDN-based technology. SMARC is a new mobility framework replacing existing MIP solutions for session continuity as PMIP/GTP in LTE. SMARC unique advantages, over existing mobility protocols and SDN related trials, are adopting OpenFlow virtual paths without extra Tunnel/MPLS/VLAN headers as well as solving both core congestion and session continuity problems associating handover in wide area motion and residential/enterprise offerings through effective offload mechanism. SMARC assures higher performance as of signaling overhead elimination, extra headers reduction, and OpenFlow hardware abstraction. PMIP/GTP-C out-band signaling and control messages for bearer setup/release during MN's join/handover are replaced with recursive DHCP relay process occurring in-line with IP allocation from home DHCP server or that at initial PDN. Furthermore, SMARC seamlessly extends indoor residential/enterprise offerings across carriers under Service Level Agreement (SLA) regardless if overlapping configurations exist. This feature is almost infeasible in other proposed solution as of tunneling extensive delay. Across single carrier's PDN, PMIP flat domain uses at least a single tunnel compared to zero tunnels in SMARC while PMIP domain chaining model uses at least three tunnels compared to a single tunnel across carriers with direct SMARC SLA. Overlapping carriers' configurations, as subnet, VLAN, can exist as SMARC routing is based on a new unique identifier not subnet. Thus, no public IP address/DNS name is preserved for mobility as in PMIP. SMARC proposes a hybrid mode deployment for backward integration with 4G LTE and smooth migration to 5G SDN-based. The design is fully aligned with ONF SDN arch 1.0, ONF OpenFlow spec 1.4, and RFC 7426 (Betts et al., 2014; Denazis et al. 2015; Nygren et al. 2013).

This section is organized as follows; the second subsection describes the new identifier format that is proposed to facilitate overlapping configuration presence and instant retrieval of MNs' profiles. The third subsection states MN's IPs allocation process. The fourth subsection illustrates the overlay network structure that transparently handles mobility through three-tiers of distributed OpenFlow switches. The fifth subsection describes how SMARC preserves existing investments with hybrid mode deployment, MN's Join and Handover Phases. The sixth subsection illustrates how MN's packets are categorized into virtual paths for effective offload mechanism, and the last subsection shows how SMARC solves the challenges presented in the second section.

User Equipment Mobility Subscription Identifier

UEMS_ID is a unique identifier assigned to MN during subscription. Either it is send by MN or set by HSS in DHCP_CLIENT_ID field of DHCP_REQUEST/DHCP_DISCOVERY messages after authentication for IP allocation. DHCP_CLIENT_ID, option field number #61, was designed for clients to specify their unique identifier in an administrative domain. DHCP servers used it to index their binding databases (Alexander et al., 1997). Latter, this option was replaced with client's L2/hardware address specified in CHADDR field, MAC/IMSI. DHCP_CLIENT_ID field can be setup manually in UE's DHCP client configuration or HSS sets it in "Mobile-Node-Identifier" #AVP Code 506 of "Diameter Update Location" message as in RFC 5779 (Korhonen, 2010). It is possible to replace UEMS_ID with the International Mobile Subscriber Identity (IMSI). SMARC prefers using a new identifier for uniformity in 3GPP/5G-PPP trusted and untrusted access without EAP-SIM authentication enforcement. This confides with 5G-PPP project's vision 2015 of a unified programmable telecom and IT infrastructure.

UEMS-ID Format:

<Domain ID>-<Carrier ID>-<Optional Sub-Carrier ID>-<Subscription ID>-<User ID>

UEMS_ID design purposes:

1. Instant identification and location of MNs' subscription profile during Join Phase.
2. Recursive establishment of virtual path to the home of MN with residential/enterprise mobility subscription.
3. Handover in wide area motion across multiple PDNs through inverse lookup from MN's home IP to UEMS_ID.

IP Allocation Process

Every MN is offered three IP addresses for instant breakout to avoid core network congestion while ensuring seamless mobility, extension of residential/enterprise service, and handover in wide area motion.

- **Home Address (HA):** Obtained from home DHCP server of MN with residential/enterprise service registration or from DHCP server at initial point of attachment. It is the only address forwarded to MN during Join Phase and is kept by till MN disconnects or DHCP lease expires.
- **SDN Address (SA):** Assigned by SDN controller based on UEMS_ID for carrier internal services accessibility and orchestration based on MN's subscription profile.
- **Care of Address (CoA)**: Assigned to MN from the current point of attachment to offer instant breakout to UDP and new TCP sessions for offloading the core network.

Three-Tiers Mobility Overlay

Overview

SMARC uses a three-tiers overlay network with tree structure to dynamic establish OpenFlow virtual paths between current and previous attachment points as well as to home location of MN with residential/enterprise subscription. OpenFlow SDN-based technology has a unique advantage over standard L2/L3 network which is packets isolation per MN's hardware address without extra tunnel/VLAN header. This facilitates creation of shared OpenFlow virtual path carrying packets of multiple MNs having the same IP address but different L2 address without conflict. SMARC virtual path concept has a dramatically improvement in overall performance over traditional LTE bearer PMIP/GTP setups. Overlay network has dynamic procedure for automatic identification of mobility entities' insertion/removal as well as topology layout. Each mobility entity has a unique identifier called MOBILITY_ID. Links connecting mobility entities can be direct physical links or virtual links established using GRE/IPSEC tunnels.

Mobility Entities

1. **Mobility Access Switch Tier (AS):** Is composed of OpenFlow switches functioning as LTE S-GWs. AS tier forwards packets between base stations and upper tier as well as inter-eNB/inter-3GPP handover.
2. **Mobility Detector Switch Tier (DS):** Is composed of OpenFlow switches for MN's services orchestration at current point of attachment; internet, intranet, and home services. DS resembles LTE P-GW in being MN's exist to external world and intranet services. Unlike P-GW, DS is not responsible for MN's IP allocation. DS relays DHCP messages to MN's home DHCP server or to that at initial point of attachment for HA allocation and to the attached PDN DHCP server for CoA allocation. SDN controller assigns SA based on UEMS_ID regardless the point of attachment. DS is responsible for services orchestration and re-writing of MN's packets' headers to map the three assigned IPs; HA, SA, and CoA.
3. **Mobility Gateway (MG) and Relay Switch (RS) Tier:** Are OpenFlow switches responsible for inter-domain and inter-overlay mobility respectively. Inter-domain mobility refers to interactions between overlays in different PDNs managed by different SDN controllers but under roaming Service Level Agreements (SLA). Intra-domain mobility refers to interactions inside and across overlays managed by the same SDN controller. These are intra-overlay and inter-overlay mobility respectively. RS is responsible for inter-overlay mobility; connecting overlays at different PDNs for ensuring session continuity in wide area motion while DS is responsible for intra-overlay.

SMARC Hybrid Mode Deployment

Layout

Hybrid mode deployment is designed to preserve existing investments through backward compatibility with legacy cellular network for providing smooth migration to next generation networks. Both P-GW/ GGSN and S-GW/SGSN are consolidated to a single box as in small 3G/4G LTE deployment. AS tier are added on top as shown in figure 6. AS tier carries the same S-GW functionality; as inter-eNB and inter-3GPP handover as well as forwarding packets between base stations and next DS tier.

MN Join Phase

Authentication

HSS authenticates MN then instructs AS-WAG/AS-PGW-SGW to set DHCP_CLIENT_ID value to UEMS_ID in DHCP_REQUEST/DHCP_DISCOVERY if MN is not configured for this.

Recursive Relay Procedure

Initially, AS tier has no OpenFlow rules installed for a joining MN, thus it consults the SDN controller which matches UEMS_ID field against AS mobility routing table. The match process output directs DHCP message to MOBILITY ID of parent DS. A temp profile is created for MN in AS database. DS mobility routing table relays the message to either the attached PDN DHCP server for HA allocation or to parent RS/MG if MN is registered to residential/enterprise service. Again, a temp profile is created for MN in DS database. The process is repeated till correctly relaying the message to home DHCP server.

Service Activation

Once a valid DHCP_OFFER for joining MN enters any mobility entity storing a temp profile, the status is updated to active and both input and output ports are stored for forwarding future MN's packets without consulting the mobility routing table. DS, assigned to the attached PDN, requests SA IP from SDN controller and CoA IP from PDN DHCP.

Virtual Path Establishment

With the first MN non-DHCP packet entering any mobility entity, the activated profile is used to install OpenFlow rules based on MN's subscription profile for wire speed forwarding of future packets without further interruption to the controller. This process is recursively repeated till the virtual path is established.

MN Handover Phase

Intra-Overlay Handover

This refers to MN's handover inside a PDN when crossing multiple ASs connected to the same parent DS. Intra-overlay handover introduces minor latency for AS profiles update. The next AS has no flows installed thus it performs reverse lookup of HA IP to UEMS_ID using the SDN global database. If tie

Figure 6. Hybrid Mode Deployment and Virtual Paths

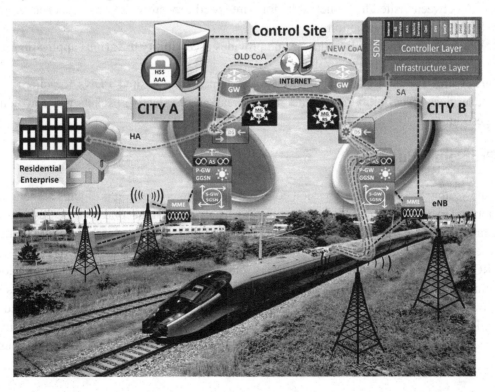

exists, HA IP is allocated to multiple MNs, L2 addresses are used to resolve conflict. Once UEMS_ID is retrieved, MN global profile is updated to reflect current attachment, as well as parent DS profile then OpenFlow rules are re-installed.

Inter-Overlay Handover

This refers to MN's handover across PDNs ruled by a single SDN controller in wide area motion when crossing multiple DSs connected to the same parent RS. Process resembles intra-overlay handover in consulting SDN global database followed by profiles update. Extra steps are performed to ensure session continuity and instant breakout. A new CoA is allocated for MN in the new PDN to avoid core congestion. All MN's active sessions in the switch flow table of DS at previous PDN are filtered by L4 parser. Both MN's subscription profile and parser output create new sets of OpenFlow rules that are installed on both DSs. Thus, active TCP sessions and indoor packets are redirected to previous DS while the new CoA provides instant breakout in current PDN for UDP and new TCP sessions.

Virtual and Relay Paths

After activation of MN's mobility profile, MN's packets are divided into three main OpenFlow virtual paths as shown in figure 7 and another path for relaying DHCP messages.

- **DHCP Path:** Is established for relaying DHCP messages between roaming MN and home DHCP server or that serving initial attached PDN. No OpenFlow rule is installed to keep SDN controller aware of MN's lease status.
- **Home Path:** Is identified by HA subnet. OpenFlow rules are installed on all mobility entities in the path between visited/home networks for wire speed forwarding of MN's packets to avoid future SDN controller interruption.
- **SDN Path:** Is identified by SA address and provides accessibility to various applications and services provided by the SDN controller.
- **Internet and Intranet Paths:** Use the CoA assigned from visited PDNs for internet and intranet services accessibility to provide instant breakout of selective packets while avoiding core congestion.

Initial Problem Solution

Figure 7 shows how SMARC solved session continuity and core congestion problems associating wide area motion and residential/enterprise service offering discussed in the second section. Session continuity is guaranteed as MN's keeps HA IP regardless several PDNs handover and the virtual paths spoof MN's presence in previous location.

SMARC L4 parser converts GTP/PMIP bearers setup/release to profile update and OpenFlow rules reinstallation. No MPTCP proxy is deployed in inter-overlay handover. After handover, MN's packets become divided into four OpenFlow virtual paths to guarantee that only home packets, identified by HA, and active TCP sessions, identified by Old CoA, across the core. The efficient offload mechanism creates instant breakout for new TCP and UDP packets at nearest PDN using New CoA to avoid core congestion.

Figure 7. Efficient Data Forwarding Solution

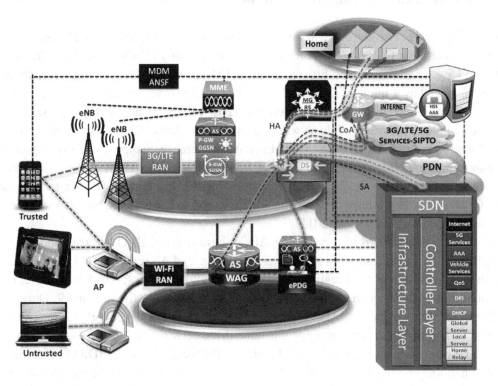

EXPERIMENTS

Overview

A prototype is established for assessing SMARC's unique handover capability in wide area motion with performance equivalent to SDWN handover in WLAN. SMARC ensures wire speed forwarding inside OpenFlow virtual paths with performance improvements over PMIP bearers' setup delays and handover latencies. Handover inside the same carrier with a single SDN controller includes both intra-overlay and inter-overlay handovers. The former refers to mobility handover across two ASs in the same city; that is equivalent to S-GWs handover in LTE while the latter refers to that between two ASs located in two separate cities' PDN with different DSs. Inter-overlay represents handover in high mobility situation across cities' boundaries. This handover type is currently not supported in LTE as of separate PDN per city with different P-GWs. Either MN's active sessions are disconnected as of new bearer's setup to the new city P-GW or session continuity is provided through bearer re-establishment to initial P-GW leading to EPC congestion problem. This problem is discussed in challenges of Mobile IP Background section. Furthermore, the prototype evaluates virtual paths' performance of inter-domain mobility against that of standard network. Inter-domain mobility refers to the capability of extending indoor services across different carriers deploying separate SDN controllers under direct SLA.

Evaluation Methods and Validation Plans

The prototype consists of two carriers with separate SDN controllers; CarrierX and CarrierY. The latter has a Network Operation Centers (NOC) for City C with mobility overlays CY_OC. The former has two NOCs for City A and B with mobility overlays CX_OA and CX_OB respectively. In CarrierX, both overlays are managed by a single SDN controller located at City A while CarrierY has its own controller located at City C. Each overlay is connected to CTRL_VLAN for intranet/internet services and to L3 backbone switch for providing cable network services to residential/enterprise mobility subscribers.

CX_OA overlay has two ASs, ID:OA4/OA9. Each AS has three overlapping APs to compare SMARC intra-overlay handover performance against that of SDWN inside City A. CX_OB overlay has a single AS, ID:OB4, to compare inter-overlay handover performance against that of SDWN when crossing City A AS, ID:OA4 to City B. Moreover, all results are compared to Cisco delay budget for default bearer establishment listed in table 2 (Savic, 2011).

Mobility handover scenarios of overlapping APs in intra-overlay, ID:OA4/OA9, and inter-overlay, ID:OA4/OB4 are shown in figure 9. To avoid interference, each of AS sets with ID:OA4/OA9/OB4 uses different channel and Service Set Identifier (SSID). Two mobile stations; STA1 and STA2, registered for residential mobility service, move across the topology as shown in figure 8 and 9. During motion, both stations establish sessions to SERVER_OA located at their home network VLAN12 of City A. Experiments ensure that both stations cover the same distance with the same velocity but in different directions. STA1 moves in a straight path that handovers the three APs of AS ID:OA4 to evaluate standard SDWN handover performance while STA2 starts at AS ID:OA4 and undergoes SDWN handover followed by either intra-overlay or inter-overlay handover to ASs ID:OA9 or OB4 respectively. Both STA1 and STA2 results are compared to highlight the difference between the three handover types in TCP/UDP packets loss, throughput, latency, UDP jitter and TCP sessions throughput restoration. After inter-overlay handover, STA2 remains stationary at City B. Its performance when communicating with SERVER_OA is compared against HOST_OB at City B VLAN 12 to highlight the wire speed performance of OpenFlow virtual paths.

Table 2. Cisco Delay Budget for Default Bearer

Nodes	Interface Name	Nodes Involved	Delay Budget
eNB	S1-MME/NAS	eNodeB-MME	~50ms
MME	S6a	MME-HSS	~100ms
MME	DNS	MME-DNS (APN)	~50ms
MME	S11	MME-SGW	~50ms
SGW	S5/S8	SGW-PGW	~50ms
PGW	Gx	PGW-PCRF	~100ms
PGW	Gy	PGW-OCS	~100ms
Total Bearer Setup Time			~500ms
eNodeB	X2	eNB-eNB	~20ms

Figure 8. Prototype Layout

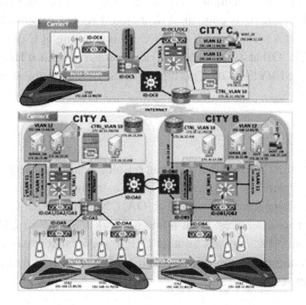

Figure 9. Handover with APs at 80m Apart

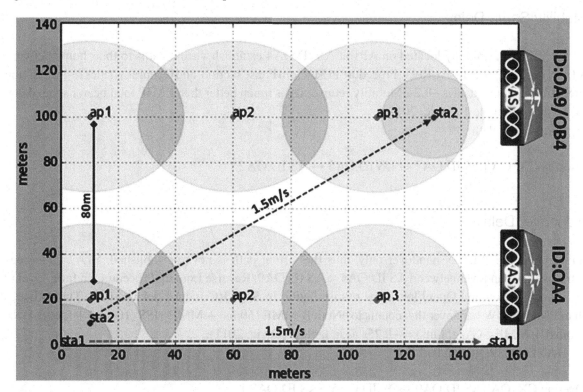

Inter-domain performance of residential/enterprise mobility subscribers, inside the OpenFlow virtual paths; inter CarrierX and CarrierY, is compared to standard routed network for evaluating the effect of single tunnel presence. STA3 from VLAN12 of City B is roaming at City C. It's performance when communicating to SERVER_OB at its home VLAN12 is compared against HOST_OC at City C VLAN 12 when communicating to SERVER_OB. This highlights SMARC feasibility to extend mobility across carriers with direct SLA.

Implementation

The experiments are performed by two Ubuntu Linux 64-bit machine running two SDN controller with mininet to create mobility OpenFlow switches and SDWN to virtualize mobility in radio network. Three 3640 Cisco L3 switches are connected to overlay CX_OA, CX_OB and CY_OC to act as cable networks providing residential/enterprise services. Each home network has its own DHCP server that lease IP addresses to mobile station. NOC contains another DHCP server that leases IP addresses to standard users without residential/enterprise subscription. Each NOC has a Cisco router 3725 with firewall OS to provide internet/intranet services.

Intra-Overlay Results

Mobility Setup Delay

STA1 and STA2 initially located on AP1 at AS ID:OA4 establish virtual paths to their home network VLAN12. DHCP Recursive Relay Procedure in MN Join Phase section, and OpenFlow rules re-installation in **SMARC** induce delays **~0.2s**. Mobility setup delay is much better than **PMIP** total bearer setup delay **~0.5s** stated in table 1 (Savic, 2011).

Join Virtual Path:

Overlay CX_OA [AS ID:OA4 ↔ DS ID:OA5 ↔ AS ID:OA2]

Handover Delay

STA1 undergoes SDWN handover only, thus the virtual path is not re-established. STA2 undergoes intra-overlay handover between AS ID:OA4 ⇔ AS ID:OA9. Reverse Lookup Procedure, in Intra-Ovelay Handover section and OpenFlow rules re-installation in **SMARC** induce delays **~0.1s**. This is better than **PMIP** S-GW handover that counts to eNodeB-MME (50ms) + MME-HSS (100ms) + MME-DNS (50ms) + MME-SGW (50ms) = **~0.25s** as in table 2 (Savic, 2011).

HANDOVER Virtual Path:

Overlay CX_OA [AS ID:OA9 ↔DS ID:OA5↔AS ID:OA2]

ICMP Latency and Packets Loss Comparison

Figure 10 shows results of 250 pings from (SERVER_OA ⇔ STA1) and (SERVER_OA ⇔ STA2). STA1 represents SDWN handover performance at AS ID:OA4. It experiences two handovers; the first between (AP1 ⇔ AP2) at 13s~15s while the second between (AP2 ⇔ AP3) at 47s~49s. On the other hand, STA2 experiences two handovers; the first is SDWN handover at AS ID:OA4 between (AP1 ⇔ AP2) at 9s~13s while the second is intra-overlay handover between (AP2 at AS ID:OA4 ⇔ AP2 at AS ID:OA9) at 46s~48s. Figure 9 and 10 highlight that ~4s are the worst latency and packets loss experienced by STA2 during SDWN handover. The large value is correlated to the large distance crossed in the overlapping region between AP1 and AP2 at AS ID:OA4. On the other hands, STA1's SDWN handover latency and packets loss are ~2s. This is equivalent to that experienced by STA2 in intra-overlay handover.

TCP Performance Comparison

Figure 11 highlights that intra-overlay handover latency is almost equivalent to that of SDWN handover and that several consecutive handovers negotiate lower TCP windows size that decrease the average throughput without TCP sessions disconnect. Between 48s~54s, STA2 undergoes intra-overlay handover while STA1 undergoes SDWN handover between 51s~56s. STA2 throughput ~330KB/s drops to ~25.7KB/s for ~6s while STA1 throughput ~151KB/s drops to ~17.82KB/s for ~5s. At 75s~90s, both stations become stationary. Figure 9 and 11 shows that TCP throughput of STA2, near AP center, be-

Figure 10. ICMP Performance of Intra-Overlay Handover

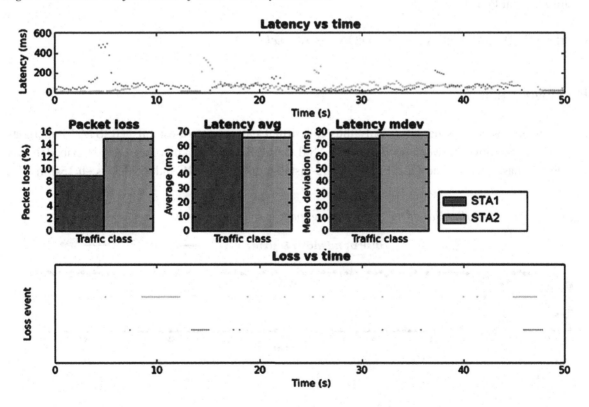

Figure 11. Intra-Overlay Handover TCP Throughput

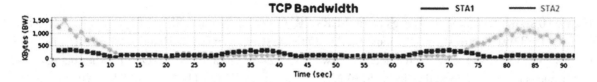

comes much higher than STA1 at edge. This emphasizes that average TCP throughput depends mainly on how far stations are from AP center not on handover.

UDP Performance Comparison

Between 41s~44s, STA1 undergoes SDWN handover while STA2 undergoes intra-overlay handover between 43s~46s, as in figure 12. Either type of handover induces an instant decrease in average UDP throughputs due to some packets drop. During handover, UDP jitters depend mainly on how far station is from AP center rather than handover type.

Inter-Overlay Results

Mobility Setup Delay

The same delay value in range **0.2s** as in intra-overlay.
Join Virtual Path:

Overlay CT_OA [AS ID:OA4 ↔ DS ID:OA5 ↔ AS ID:OA2]

Handover Delay

STA1 undergoes SDWN handover only thus the virtual path is not re-established. STA2 undergoes inter-overlay handover between AS ID:OA4 ⇔ AS ID:OB4. Reverse lookup procedure, described in MN Handover Phase section, and OpenFlow rules re-installation in **SMARC** induce delays **~0.15s**. In LTE

Figure 12. Intra-Overlay Handover UDP Throughput

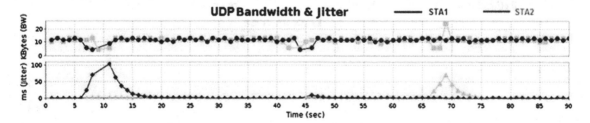

no handover occurs across P-GWs in wide area motion. Two solutions are possible. The first is sessions disconnect followed bearer re-establishment to new PDN P-GW while the second is bearer re-establishment between new PDN S-GW and old PDN P-GW to preserve session continuity. The second solution induces EPC congestion problem. In both scenarios, **PMIP** delay is equal to total bearer setup **~0.5s**.

HANDOVER Virtual Path:

Overlay CX_OB [AS ID:OB4 ↔DS ID:OB5↔RS ID:OB0] ↔

Overlay CX_OA [RS ID:OA0 ↔DS ID:OA5↔AS ID:OA2]

ICMP Latency and Packets Loss Comparison

Figure 13 compares results of 250 pings from (SERVER_OA ⇔ STA1) and (SERVER_OA ⇔ STA2) when STA1 performs SDWN handover AS ID:OA4 (AP2 ⇔ AP3) at 27s~29s and STA2 performs inter-overlay handover between (AP4: AS ID:OA4 ⇔ AP2: AS ID:OB4) at 26s~29.5s. Figure 13 highlights that inter-overlay handover induces ~1.5s latency higher than SDWN.

Figure 13. ICMP Performance for Intra-Overlay Handover

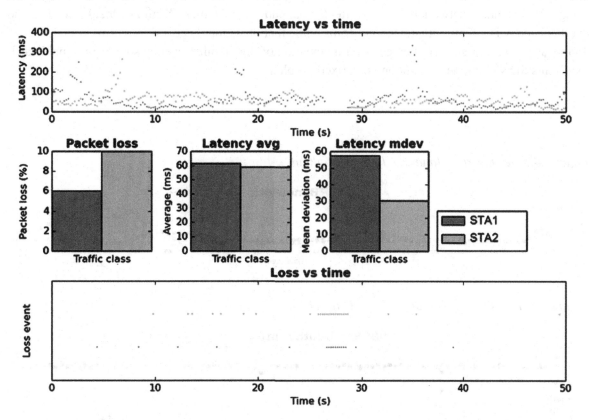

TCP Performance Comparison

Figure 14 highlights that inter-overlay handover introduces ~2s latency more than SDWN handover till resuming moving stations' average throughput. STA2 undergoes inter-overlay handover between 42s~47s while STA1 undergoes SDWN handover between 47s~50s. During handover, STA2's average throughput 276KB/s drops to ~0.848KB/s for 5s while STA1's average throughput 144KB/s drops to ~27.8KB/s for 3s. Again, Figure 14 emphasizes that TCP average throughput depends mainly on how far the stations are from AP center not on handover.

UDP Performance Comparison

Results in figure 15 are similar to intra-overlay handover. STA1 undergoes SDWN handover between 47s~50s while STA2 undergoes inter-overlay handover between 51s~55s. Both handover types induce instant decrease in average UDP throughput due to packets drop. UDP jitters depend mainly on how far stations are from AP center rather than on handover.

Virtual Path Wire Speed Forwarding

This experiment compares the latency in 100 pings sent from STA2 roaming at City B to SERVER2_OA at City A VLAN12 using SMARC overlay network against those sent by HOST_OB at City B VLAN 12 using L2/L3 standard network before and after virtual path establishment. Without OpenFlow rules, the overlay latency is ~47ms with deviation ~15ms while that of standard network is ~17ms with deviation 3~4ms as shown in figure 16. Large standard deviation of the mobility overlay sounds reasonable as it represents SDN controller utilization not a fixed problem.

Figure 14. Inter-Overlay Handover TCP Throughput

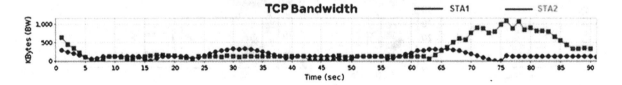

Figure 15. Inter-Overlay Handover UDP Throughput

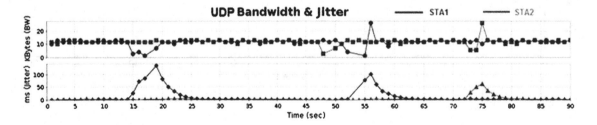

Figure 16. Overlay Performance versus Standard Network without OpenFlow

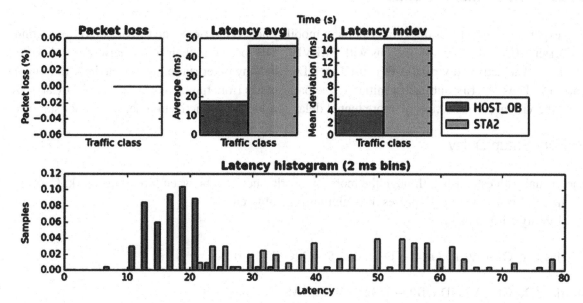

Figure 17 shows that after virtual path establishment, OpenFlow rules make advancement to the latency of mobility overlay. Performance becomes wire speed with *zero latency* and deviation of **~0.1ms** while standard network latency is **~17ms** with **3~4ms** deviation.

Figure 17. OpenFlow Virtual Path Performance versus Standard Network

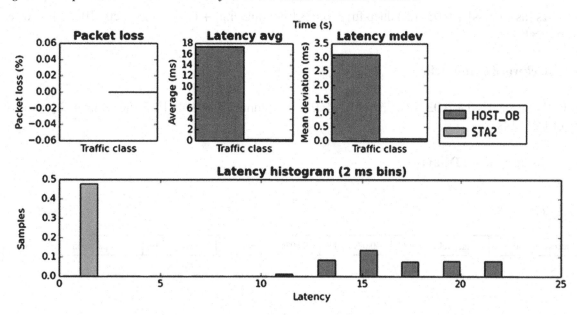

Direct Inter-Domain Results

This experiment prototypes direct inter-domain mobility across CarrierX and CarrierY. STA3, roaming at CarrierY CY_OC, initiates sessions with SERVER_OB located at their home network VLAN12 of CarrierX. The activated virtual path is used by STA3 and any other station from CarrierX roaming at CarrierY. Thus, the first subscriber initiating the path suffers from tunnel delay; CarrierY → CarrierX. Once the virtual path is active, no more tunnel delay is added to the activation delay of other stations.

Mobility Setup Delay

The virtual path established through the mobility overlay network between foreign network, CarrierY, and home network, CarrierX, passes through these mobility entities:
   Mobility Virtual Path:

Overlay CY_OC [AS ID:OC4 ↔ DS ID:OC5 ↔ MG ID:OC0] ↔

Overlay CX_OB [MG ID:OB0 ↔ DS ID:OB5 ↔ AS ID:OB2]

   In direct inter-domain mobility, activation of STA3's profile suffers from the delays listed below. These are average measures of delays calculated from several runs (Table 3).

- **Forward Setup Delay:**
  - **Inactive Tunnel:**

$(0.01s$ [as: access] $+ 0.03s$ [ds: dhcp fd] $+ 1s$ [mg: tunnel.s]) x 2 [cy_oc & cx_ob] $\cong 2.08s$

  - **Active Tunnel:**

$\cong (0.01s$ [as: access] $+ 0.03s$ [ds: dhcp fd] $+ 0.02s$ [mg: tunnel.p] $+ 0.03s$ [mg: dhcp fd]) x 2 [cy_oc & cx_ob] $\cong 0.18s$

- **Backward Setup Delay:**

$\cong (0.01s$ [as: access] $+ 0.02s$ [ds: dhcp fd] $+ 0.02s$ [mg: tunnel.p] $+ 0.02s$ [mg: dhcp bk]) x 2 [cy_oc & cx_ob] $\cong 0.14s$

- **Mobility Setup Delay:**

Table 3.

access	~0.01s	dhcp relay	fd	~0.03s	bk	~0.02s	tunnel	setup	~1s	propagation	~0.02s

- ○ **Inactive Tunnel:** $\cong 2.08s + 0.14s \cong$ **2.22s**
- ○ **Active Tunnel:** $\cong 0.18s + 0.14s \cong$ **0.32s**

Mobility Re-Activation Delay

In direct inter-domain mobility, re-activation of STA3's profile, after expiration of OpenFlow rules as of inactivity, suffers from the delays listed below. These are average delays calculated from several runs (Table 4).

- **Mobility Re-Activation Time:**

$\cong$ (0.01s [as: access] + 0.01s [ds: re-activate] + 0.02s [mg: tunnel.p] + 0.01s [mg: re-activate]) x 2 [cy_oc & cx_ob] $\cong$ **0.10s**

ICMP Latency and Packets Loss

Figure 18 and 19 analyze the performance of direct inter-domain mobility versus that of standard network when connecting foreign carrier, CarrierY, and home network, CarrierX of STA3. This is achieved through comparing performance of STA3 and HOST_OC, located on the same switch OC_SWL3 of CarrierY, when communicating to SERVER_OB. STA3 is an active mobility station using the SMARC overlay while HOST_OC is a standard host using standard L3 network. This experiment compares the latency experienced in 100 ICMP pings sent from HOST_OC to SERVER_OB through standard network and those sent from STA3 to SERVER_OB through the overlay with/without OpenFlow rules installation on mobility entities.

Without OpenFlow Rules

The comparison in figure 18 reveals that standard network induces 14~17ms average latency with deviation of ~4ms while that of mobility overlay is in range of ~52 average latency with deviation of ~13ms without OpenFlow rules installed on the mobility switches. The average latency has increased from ~47ms in inter-overlay to ~52ms as of the induced WAN propagation and tunnel delays. The large value of standard deviation in the mobility overlay sounds reasonable as it represents the utilization of the SDN controller not a fixed problem facing the mobility overlay.

With OpenFlow Rules

The recursive relay feature presented in the mobility framework separates between STA3's standard packets and DHCP messages that require continuous monitoring by the SDN Controller. With this fea-

Table 4.

re-activation	~0.01s	access	~0.01s	tunnel propagation	~0.02s

Figure 18. Direct Inter-Domain ICMP Performance – No OpenFlow

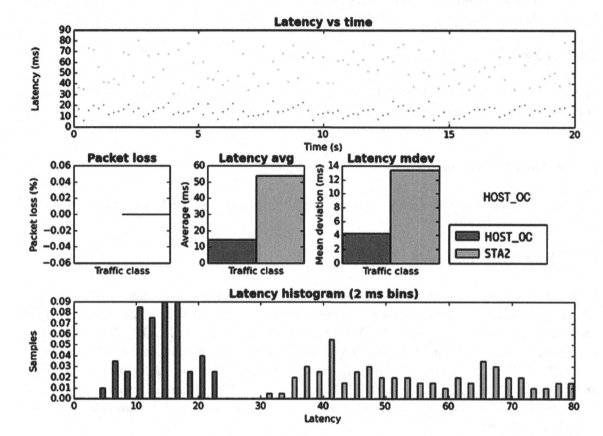

ture, OpenFlow rules make advancement to latency in the mobility overlay as shown in figure 19. After installation of OpenFlow rules, the mobility overlay network's average latency decreases from ~52ms with deviation ~13ms to average latency of ~16ms with deviation ~4.5ms which is almost equivalent or slightly better than that standard network.

TCP and UDP Performance

Without OpenFlow Rules

The comparison in Figure 20 reveals that standard network has higher TCP throughput than direct inter-domain mobility, without installation of OpenFlow rules. Moreover, UDP performance has slightly lower throughput with higher jitter level in direct inter-domain mobility without installation of OpenFlow rules when compared to standard network performance.

With OpenFlow Rules

The comparison in figure 21 reveals that installation of OpenFlow rules dramatically improves the performance of direct inter-domain mobility network over standard networks. Concerning TCP performance, inter-domain mobility's throughput becomes slightly better than standard network. On the hand, UDP throughput and jitter level becomes equivalent to that of standard network.

Figure 19. Direct Inter-Domain ICMP Performance – OpenFlow

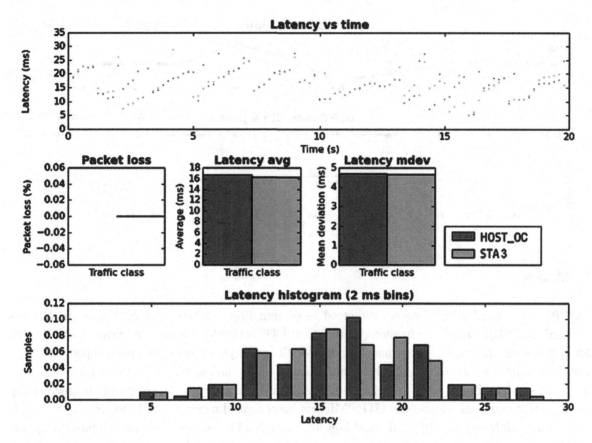

Figure 20. Direct Inter-Domain vs. Standard Network in TCP and UDP - No OpenFlow

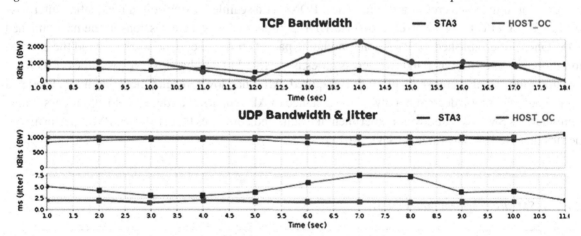

Figure 21. Direct Inter-Domain vs. Standard Network in TCP and UDP - OpenFlow

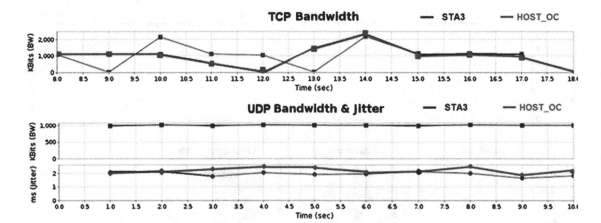

SUMMARY

SMARC wire speed forwarding is guaranteed as of signaling overhead elimination, tunnel headers removal, and SDN-OpenFlow hardware abstraction. LTE GTP/PMIP tunnels is proved to be 2~10% lower than standard network performance while SMARC prototype shows higher throughput with lower latency and more agility over standard network. Inside a single carrier, mobility setup delay is improved from ~0.5s in LTE bearer setup to ~0.2s with SMARC as of recursive virtual path establishment using inline DHCP messages without any GTP/PMIP out-band control messages. SMARC adopts new routing mechanism based on UEMS_ID next hop lookup instead of subnet to simplify intra-overlay and inter-overlay handovers to profile update processes. Intra-overlay handover inside city is ~0.1s which is equivalent to SDWN but better than LTE S-GWs handover of ~0.25s. Inter-overlay handover ensures session continuity when crossing several cities' PDNs with seamless extension of residential/enterprise services using effective L4 parser to offload UDP packets and new TCP sessions at the new attached PDN. Only pre-established TCP sessions and home packets traverse core network to avoid the congestion problem. Inter-overlay handover across cities is ~0.15s. In LTE, handover across cities' P-GWs is not supported. Total bearer setup of ~0.5s is required. After handover, results show that jitters level and throughput restoration depend mainly on the distance to AP center rather than handover process. Table 5 summarizes SMARC handover's experimental results and compares them to Cisco PMIP performance metrics that are listed in Table 2 (Savic, 2011).

Table 5. SMARC Performance vs. PMIP

	SMARC	PMIP
Mobility	Mobility Setup Delay ~**0.2s**	Total Bearer Setup (TBS) ~**0.5s**
Inside City	Intra-Overlay Handover ~**0.1s**	S-GW Handover ~**0.25s**
Across Cities	Inter-Overlay Handover ~**0.15s**	No P-GWs Handover - TBS ~**0.5s**
Forwarding	OpenFlow Wire Speed	Tunnels Overheads **2~10%**

SMARC successfully extends enterprise/residential services across two carriers with direct mobility agreements. This is currently impossible in existing mobility standards as of the tremendous latency imposed by several tunneling requirements and the presence of overlapping configuration in carriers; as IP, VLANs …etc. In direct inter-domain mobility across carriers, TCP throughput becomes slightly better than standard network after installation of OpenFlow rules. On the other hand, ICMP and UDP throughputs and jitter levels become equivalent to that of standard network. Direct inter-domain mobility's performance requires only single shared tunnel across two carriers. This is still much better than existing mobility protocols, as PMIP domain chaining model, that requires at least three separate tunnels per mobility user. SMARC allows overlapping configuration in carriers as of adopting new routing mechanism based on UEMS_ID not subnet. Table 4 summarizes SMARC experimental delays for (re) activating direct inter-domain mobility across carriers. Results show that after activation of the tunnel by the first mobility subscriber for connecting the two carriers, the rest of subscribers suffer from delays that are in range of PMIP total bearer setup delay in a single carrier. This emphasizes the leadership in SMARC performance and the approach feasibility in real deployments.

CONTRIBUTIONS

This chapter proposes a novel framework, called SMARC, for seamless mobility inside and across carriers. Within a single carrier's PDN, SMARC performs the same handover's functions as standard mobile IP protocols but with better performance. Across carrier's PDNs, the framework ensures session continuity during handover across cities' boundaries in wide area motion while ensuring optimum performance, efficient offload mechanism and avoiding the congestion of core networks. Handover across carrier's PDNs is absence in existing standards. Furthermore, SMARC efficiently extends residential/enterprise services across carriers under SLAs even if overlapping configurations exist. The following highlights the keys behind these major contributions:

- **Structure Flexibility:** SMARC is presented by three tiers of OpenFlow switches in fully distributed tree structure. The architecture is scalable to accommodate carrier grade deployments where each tier is represented by a set of multiple load-balanced OpenFlow switches or the three tiers functions are implemented in a single OpenFlow switch for small deployments. Such flexibility facilitates seamless integration with deployed 4G LTE infrastructure to preserve existing infrastructure investment and smooth migration to 5G NGN.

Table 6. SMARC Delays in Direct Inter-Domain Mobility

Direct Inter-Domain			
Forward Path	Tunnel	Inactive	~2.08s
		Active	~0.18s
Backward Path			~0.14s
Setup Delay	Tunnel	Inactive	~2.22s
		Active	~0.32s
Re-Activation Delay			~0.1s

- **Efficient Data Forwarding Plane:** Established OpenFlow virtual paths are well designed to support effective traffic offload mechanisms that ensure efficient data forwarding plane while avoiding the core network congestion problem existing in LTE and the inefficient bandwidth utilization in inter-domain mobility. The following are summary of core principles behind such achievement that dramatically affect NGN scalability and quality of services offered.

  ◦ **New Forwarding Mechanism:** The adopted forwarding mechanism inside the mobility overlay uses the indices of the proposed UEMS_ID to search for next hop MOBILITY_ID. This completely differs from existing routing mechanisms that search for next hop IP address toward destination subnet. The processing delay occurs during virtual path establishment only while latter packets follow the same path after installation of OpenFlow rules to ensure wire speed forwarding. Furthermore, this facilitates the expansion of SMARC mobility across carriers regardless the presence of overlapping configurations.

  ◦ **Seamless Handover Mechanism:** Seamless handover is ensured even in wide area motion with minimum latency as of transforming handover from bearer setup/release to a profile update process and simplifying L4 TCP sessions continuity problem to L2 forwarding of MN's packets.

  ◦ **Traffic Offloading Mechanisms:** SMARC adopts an effective offload mechanism as of its scalable three tiers overlay distributed architecture. The flexibility in the overlay structure is emphasizes in the capability of having multiple breaks out and the usage of at least three IP addresses for fast services accessibility. Figure 6 shows that core network is traversed only for LIPA mobility and for continuity of active TCP sessions while the rest of packets are offloaded through the nearest PDN.

- **Eliminating Signaling Overheads and Restricting Tunnel Headers:** Unlike previous protocols, mobility activation occurs in-line during IP address allocation from home DHCP server without mobility signaling overhead. Moreover, no tunnel/MPLS/VLAN header is required for isolating user's packets as OpenFlow virtual paths isolate traffic based on IP and L2 addresses. Experiments results confirm the agility of OpenFlow SDN-based technology when adopted in SMARC effective architecture and reveal performance improvement over standard network as of hardware abstraction. This emphasizes that SMARC performance is better than existing mobility standard as GTP/PMIP. GTP and GRE tunnel of PMIP is theoretically ~2% performance lower than standard network and experimentally ~10% performance lower as in shown in table 1. In the proposed framework, tunnels are used for connecting MGs of different carriers or connecting mobility entities that are WAN separated to hide L3 routing complexities. These tunnels are established without signaling overhead based on SLAs, pre-defined configurations, and security policies.

- **Across Carriers' Mobility:** SMARC extends residential/enterprise services across carriers under SLA even if overlapping configurations exist. Existing inter-domain solutions, as PMIP domain chaining model, suffers from tremendous delay as of several tunnels establishment. This hinders the adoption of PMIP domain chaining model in real deployment. For direct inter-domain mobility across carriers, SMARC uses single tunnel while PMIP domain chaining model requires at least three tunnels. Experimental results confirm that SMARC performance across carriers is almost equivalent to PMIP performance inside single carriers. This ensures the feasibility of SMARC in real deployments.

CONCLUSION

Seamless mobility is no longer a dream with SMARC. Not only session continuity is guaranteed during handover when crossing carriers' PDNs in wide area motion, but also residential/enterprise indoors services are expanded across carriers under mobility SLA. The framework provides a unique unified architecture that can be adopted over any IP infrastructure; WiFi, LAN, WiMAX ...etc. The research proves that SMARC performance is at least 10% higher than existing mobility standards. Wire speed performance is guaranteed inside carriers as of complete elimination of out-band signaling and tunnel headers.

FUTURE RESEARCH DIRECTIONS

The proposed SMARC framework has several contributions in RAN mobility as stated in the previous sections. Still research work is endless and there are a lot of fields that can expand this topic. The following highlights the main directions in future works.

- Integrating the proposed framework to IEEE 802 OmniRAN to ensure compatibility with heterogeneous access networks.
- Evaluating the feasibility of the proposed framework in load balancing, session continuity, and offload mechanism when multiple MN's cards are active.
- Evaluating across controllers' negotiation for advance value-added service offering to MNs roaming at visited networks.
- Integrating the framework to LTE Proximity Services and evaluating performance against IEEE 802.11p V2X communication.
- Adopting QoS to enhance direct and indirect inter-domain mobility performance.

REFERENCES

5GAmericas. (2016, November). *Network Slicing for 5G and Beyond*. 5G Americas White Paper. Retrieved from http://www.5gamericas.org/files/3214/7975/0104/5G_Americas_Network_Slicing_11.21_Final.pdf

Abdulhussein, M., Abbas, T., Servel, A., Hofmann, F., Thein, C., Bedo, J. S., ...Trossen, D. (2015, October). *5G Automotive Vision*. ERTICO, European Commission, & 5G-PPP.

Al-Surmi, I., Othman, M., & Ali, B. M. (2010, February). Review on Mobility Management for Future IP-Based Next Generation Wireless Networks. *International conference on Advanced Communication Technology (ICACT)*, 989–994.

Alcatel-Lucent. (2009). *The LTE Network Architecture - A Comprehensive Tutorial*. Strategic White Paper.

Alexander, S., & Droms, R. (1997, March). *Dynamic Host Configuration Protocol*. IETF RFC.

Anyfi. (2014). *Software-Defined Wireless Networking: Concepts, Principles and Motivations*. White-paper, Anyfi Networks.

Bailey, S., Bansal, D., Dunbar, L., Hood, D., Kis, Z. L., Mack-Crane, … Varma, E. (2013, December). *SDN Architecture Overview*. Open Networking Foundation, Version 1.0 –draft v08.

Bansal, M., Mehlman, J., Katti, S., & Levis, P. (2012). OpenRadio: A Programmable Wireless Dataplane. *1st Workshop on Hot topics in Software Defined Networks, HotSDN'2012*, 109-114.

Betts, M., Davis, N., Dolin, R., Doolan, P., Fratini, S., Hood, D., … Dacheng, Z. (2014, June). *SDN Architecture*. Issue 1, ONF TR-502.

Cisco. (2016, April). *PMIP: Multipath Support on MAG and LMA, IP Mobility*. Mobile IP Configuration Guide, Cisco IOS XE.

Contreras, L. M., Cominardi, L., Qian, H., & Bernardos, C. J. (2016, April). Software-Defined Mobility Management: Architecture Proposal and Future Directions. *Springer Mobile Netw Appl, 21*(2), 226–236. doi:10.100711036-015-0663-7

Denazis, S., Koufopavlou, O., & Haleplidis, E. (Eds.). (2015, January). Software-Defined Networking (SDN): Layers and Architecture Terminology. RFC 7426.

Elsadek, W. F. (2016). Toward Hyper Interconnected IoT World using SDN Overlay Network for NGN Seamless Mobility. *8th IEEE International Conference on Cloud Computing Technology and Science (CloudCom)*. 10.1109/CloudCom.2016.0078

Elsadek, W. F. & Mikhail, M. N. (2016, October). SRMIP: A Software-Defined RAN Mobile IP Framework for Real Time Applications in Wide Area Motion. *Int. J. of Mobile Computing and Multimedia Communications, 7*(4), 28-49. DOI: 10.4018/IJMCMC.2016100103

Erran, L. L., Mao, Z. Z., & Rexford, J. (2012). Toward Software-Defined Cellular Networks. *European Workshop on Software Defined Networking (EWSDN)*, 7-12.

Esmat, B., Mikhail, M. N., & El Kadi, A. (2000, May). *Enhanced Mobile IP Protocol*. IFIP-TC6/ European Commission NETWORKING 2000 International Workshop, MWCN 2000, Paris, France. 10.1007/3-540-45494-2_13

ETSI. (2016). *Small Cell LTE Plugfest*. Retrieved from http://www.etsi.org/about/10-news-events/ events/1061-small-cell-lte-plugfest-2016

Gudipati, A., Perry, D., Erran, L. L., & Katti, S. (2013). SoftRAN: Software Defined Radio Access Network. *Second ACM SIGCOMM Workshop on Hot topics in Software Defined Networking, (HotSDN)*, 25-30. 10.1145/2491185.2491207

Gundavelli, S. (Ed.). (2008, August). Proxy Mobile IPv6. IETF RFC 5213.

Gupta, R., & Rastogi, N. (2012). *LTE Advanced – LIPA and SIPTO*. White Papers.

Hampel, G., Rana, A., & Klein, T. (2013). Seamless TCP Mobility Using Lightweight MPTCP Proxy. *11th ACM Symposium on Mobility Management and Wireless Access (MobiWac)*, 139-146. 10.1145/2508222.2508226

Jin, X., Erran, L. L., Vanbevery, L., & Rexford, J. (2013). SoftCell: Scalable and Flexible Cellular Core Network Architecture. *Ninth ACM Conference on Emerging Networking Experiments and Technologies (CoNEXT)*, 163-174. 10.1145/2535372.2535377

Johnson, D., Perkins, C., & Arkko, J. (2004, June). *Mobility Support in IPv6*. RFC 3775.

Kolias, C. (Ed.). (2013, September). *ONF Solution Brief OpenFlow™-Enabled Mobile and Wireless Networks*. ONF White Paper.

Koodli, R. (Ed.). (2009, July). Mobile IPv6 Fast Handovers. RFC 5568.

Korhonen, J. (Ed.). (2010, February). Diameter Proxy Mobile IPv6: Mobile Access Gateway and Local Mobility Anchor Interaction with Diameter Server. Proposed Standard RFC 5779.

Korhonen. (2015, February). *5G Vision –The 5G Infrastructure Public Private Partnership: Next Generation of Communication Networks and Services*. European Commission.

Kumar, S., Cifuentes, D., Gollakota, S., & Katabi, D. (2013, October). Bringing Cross-Layer MIMO to Today's Wireless LANs. *ACM SIGCOMM Computer Communication Review*, *43*(4), 387–398.

Mikhail, N. M., Esmat, B., & El Kadi, A. (2001, July). *A New Architecture for Mobile Computing, Mobile and Wireless Computing*. Academic Press.

NMC. (2015a, February). *LTE IP Address Allocation Schemes II: A Case for Two Cities*. Netmanias Technical Document.

NMC. (2015b, February). *LTE IP Address Allocation Schemes I: Basic*. Netmanias Technical Document.

Nygren, A., Pfa, B., Lantz, B., Heller, B., Barker, C., Beckmann, C., … Kis, Z. L. (2013, October). *The OpenFlow Switch Specification*, Version 1.4.0. ONF, Wire Protocol 0x05, ONF TS-012.

Odini, M., Sahai, A., Veitch, A., Gamela, A., Khan, A., Perlman, B., … Lei, Z. (2015, September). *Network Functions Virtualization (NFV); Ecosystem.* Report on SDN Usage in NFV Architectural Framework, ETSI, Draft ETSI GS NFV-EVE 005 V0.2.0.

ONF. (2012, April). *Software-Defined Networking: The New Norm for Networks.* ONF White Paper.

Perkins, C. (Ed.). (2002, August). IP Mobility Support for IPv4. RFC 3344.

Phemius, K., Bouet, M., & Leguay, J. (2014, May). Disco: Distributed Multi-Domain SDN Controllers. *IEEE Network Operations and Management Symposium (NOMS)*, Krakow, Poland.

Ruckus Wireless, Inc. (2013). *The Choice of Mobility Solutions Enabling IP-Session Continuity Between Heterogeneous Radio Access Networks.* Interworking Wi-Fi and Mobile Networks White Paper.

Savic, Z. (2011). *LTE Design and Deployment Strategies.* Cisco Systems Inc.

Schmidt, T. (Ed.). (2014, November). Multicast Listener Extensions for Mobile IPv6 and Proxy Mobile IPv6 Fast Handovers. IETF RFC 7411.

Seite, P., Yegin, A., & Gundavelli, S. (2016, March). *MAG Multipath Binding Option.* Internet-Draft.

Soliman, H. C., Malki, E. K., & Bellier, L. (2005, August). *Hierarchical Mobile IPv6 Mobility Management (HMIPv6).* RFC 4140.

Suresh, L., Schulz-Zander, J., Merz, R., Feldmann, A., & Vazao, T. (2012). Towards Programmable Enterprise WLANS with Odin. *First Workshop on Hot Topics in Software Defined Networks*, 115-120. 10.1145/2342441.2342465

Tantayakul, K., Dhaou, R., & Paillassa, B. (2016, March). Impact of SDN on Mobility Management. *30th IEEE Advanced Information Networking and Applications (AINA)*, 260-265.

Wakikawa, R., Gundavelli, S. (2010, May). *Support for Proxy Mobile IPv4.* IETF RFC 5844.

Wang, H., Chen, S., Xu, H., Ai, M., & Shi, Y. (2015, April). SoftNet: A Software Defined Decentralized Mobile Network Architecture Toward 5G. *IEEE Network*, *29*(2), 16–22. doi:10.1109/MNET.2015.7064898

Weyland, A. (2002, December). *Evaluation of Mobile IP Implementations under Linux.* Academic Press.

Yang, M., Li, Y., Jin, D., Su, L., Ma, S., & Zeng, L. (2013). OpenRAN: A Software-Defined RAN Architecture via Virtualization. *Computer Communication Review*, *43*(4), 549–550. doi:10.1145/2534169.2491732

Yap, K. K., Sherwood, R., Kobayashi, M., Huang, T. Y., Chan, M., Handigol, N., ... Parulkar, G. (2010). Blueprint for introducing innovation into Wireless Mobile Networks. *2nd ACM SIGCOMM Workshop on Virtualized Infrastructure Systems and Architectures*, 25-32. 10.1145/1851399.1851404

Zhang, C., Addepalli, S., Murthy, N., Fourie, L., Zarny, M., & Dunbar, L. (2015, June). *L4-L7 Service Function Chaining Solution Architecture*. ONF White Paper.

Section 2
Mobile Data Access and Management

Chapter 6
Massive Access Control in Machine-to-Machine Communications

Pawan Kumar Verma
Ambedkar National Institute of Technology, India

Rajesh Verma
Raj Kumar Goel Institute of Technology and Management, India

Arun Prakash
Motilal Nehru National Institute of Technology, India

Rajeev Tripathi
Motilal Nehru National Institute of Technology, India

ABSTRACT

This chapter proposes a new hybrid MAC protocol for direct communication among M2M devices with gateway coordination. The proposed protocol combines the benefits of both contention-based and reservation-based MAC schemes. The authors assume that the contention and reservation portion of M2M devices is a frame structure, which is comprised of two sections: contention interval (CI) and transmission interval (TI). The CI duration follows p-persistent CSMA mechanism, which allows M2M devices to contend for the transmission slots with equal priorities. After contention, only those devices which have won time-slots are allowed to transmit data packets during TI. In the proposed MAC scheme, the TI duration follows TDMA mechanism. Each M2M transmitter device and its corresponding one-hop distant receiver communicate using IEEE 802.11 DCF protocol within each TDMA slot to overcome various limitations of TDMA mechanism. The authors evaluate the performance of the proposed hybrid MAC protocol in terms of aggregate throughput, average transmission delay, channel utility, and energy consumption.

DOI: 10.4018/978-1-5225-5693-0.ch006

INTRODUCTION

Machine-to-machine (M2M) communication is a new communication model, where a large number of "intelligent-devices" communicate with each other with the help of wired/wireless networks, and can make collaborative decisions without direct human intervention (Verma et al, 2016, Chen et al, 2012; Igarashi et al, 2012). Machine-to-Machine (M2M) communication has its origin in the supervisory control and data acquisition(SCADA) systems, where sensors and other devices being connected through wired or radio frequency networks are used with computers to monitor and control industrial processes. A key factor behind the growth of M2M communications today is the pervasive accessibility of low cost, ubiquitous connectivity. We have already become used to low cost, high-speed home and commercial internet access. Now-a-days, in many regions around the globe, 3G and LTE mobile networks provide almost similar access speeds at highly competitive prices. The logic behind M2M communications is based on two observations. First, a networked device becomes more valuable than an isolated one. Second, when a large number of machines are networked/interconnected together, they produce entirely new autonomous and intelligent applications. For these reasons, M2M communication has quickly become a new area for research in academics and a market changing force for a wide variety of real-time monitoring applications in industry such as smart grids, home area networks, e-healthcare, intelligent transportation systems, environmental monitoring, and manufacturing applications. The system model of M2M communications is shown in Figure 1, which consists of three interlinked domains. These are M2M device domain, network domain, and application domain (Rongxing et al, 2011). In M2M domain, an M2M area network is formed by the collaboration of a large number of M2M devices (e.g. sensors, actuators, smart meters etc.) and M2M gateways. These M2M devices collect the sensory data by sensing techniques from different part in the M2M domain and collaboratively make intelligent decisions to transmit the sensory and monitored data to an M2M gateway. The M2M gateway itself is an "intelligent device", which receives the sensory data and efficiently manages the received data packets. It forwards the data packets through efficient paths by single-hop or multi-hop channels via a network domain to the back-end server of the application domain. If the M2M domain contains multiple gateways, they can further communicate with each other to make collaborative decisions (P2P model). Furthermore, the network domain acts as an interface between M2M domain and application domain. In network domain, long-range wired/wireless network protocols (e.g. Wi-MAX, 3G/4G cellular networks) are used to provide cost-efficient and reliable channels with wide coverage to transmit the sensory data from M2M domain to the application domain. Lastly, the application domain consists of a back-end server and M2M application clients. The back-end server is the main component of M2M communications model and acts as an integration point to store all the sensory data transmitted from the device domain. It also provides the real-time monitoring data to various client applications for real-time Remote Monitoring Management (RMM), i.e. smart metering, e-health care, and traffic monitoring. The back-end server can also vary for different applications; e.g., in smart grids, the control center acts as the back-end server, whereas in e-healthcare systems, the back-end server is the M2M health-monitoring server.

Limiting our focus on M2M domain only, due to the presence of full mechanical automation among devices, and their ability to support a large number of ubiquitous characteristics (Lien et al, 2011) with better cost efficiency, this communication paradigm has rapidly become a market changing force for a broad range of real-time monitoring and pervasive applications (Cao et al, 2008; Chen et al, 2010; Gau & Cheng, 2013; Lien et al, 2013; Yu et al, 2013; Zhang et al, 2012; Zhang et al, 2011). Some of the important characteristics of M2M communications are,

Figure 1. M2M communications system model

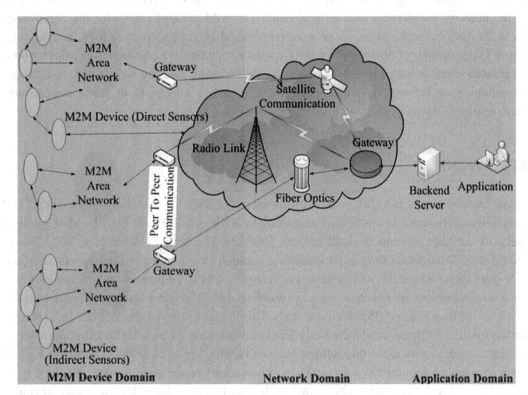

- Low or no human intervention, which means that M2M devices must be self-capable, including self- organization, self-configuration, self-management, and self-healing.
- Within M2M domain, potential communications take place among M2M devices, between M2M devices and the gateway, and between gateway to gateway (peer-to-peer model) to achieve full mechanical automation.
- A large number of devices communicate with each other without any human intervention, resulting in simultaneous random channel access attempts from these devices.

As mentioned above, because of the presence of a large number of M2M devices in service coverage area, one of the most important characteristics of M2M communications is that there are a large number of simultaneous random channel access attempts from M2M devices. Multiple devices which are in each other's radio interference range, cannot access a radio channel simultaneously. This is because neighboring devices may cause signal interference with some devices, which are trying to transmit on the same channel simultaneously. The massive access management of M2M communications over wireless channels generally happens at the Medium Access Control (MAC) layer. In case of wireless communication networks, the channel access control plays a key role in estimating aggregate throughput, and average transmission delay. In this regard, the protocols to be used on M2M device domain side as shown in Figure 1 play a key role in absolute realization of M2M communication paradigm to access the channel simultaneously.

Rest of the chapter is organized as follows. Section II presents the related work. In section III, we propose the network architecture of a densely deployed M2M network. Also, for the communication between M2M devices to take place, we propose a hybrid-MAC frame structure for concurrent channel access from a large number of devices. For more clear understanding, we consider an example of potential multiple transmissions within a single TDMA time-slot, and the gateway architecture to facilitate the direct communication between devices in section IV. Section V comprises of performance study and simulation results followed by the conclusion in section VI.

RELATED WORK

To facilitate massive access in M2M, contention-based (e.g. CSMA) or reservation-based (e.g., TDMA) MAC protocols can be used. In this regard, according to 3GPP and IEEE 802.16, the MAC protocol for M2M networks utilizes contention-based random access (RA) schemes (Hasan et al, 2013; Kim et al, 2012; Wang et al, 2010), facilitating all the devices to contend for the transmission opportunities in entire frame. Devices can dynamically join or leave the network without additional operations. However, the transmission collisions are inevitable when large number of M2M devices attempt to communicate with the gateway or the base station (BS) simultaneously. This increases the number of collisions, resulting in lower throughput, larger delays, and increased energy consumption. To handle this situation, reservation-based schemes such as time-division multiple access (TDMA) are used in M2M network. TDMA is a medium access mechanism for radio networks to reduce interference. Also, it can considerably reduce the hidden/exposed terminal issues. This is due to the fact that the adjacent devices can schedule their transmissions at non-interfering time slots. As a result, TDMA mechanism facilitates a contention-free environment for M2M devices so as to avoid the contention for the radio channels (Hsu & Yen, 2011). But, on the other hand, TDMA has a number of disadvantages also (Ye et al, 2004). First, TDMA is characteristically non-scalable, as it frequently requires a centralized device (access point) to present a collision-free schedule. Further, creating an efficient schedule having a high level of channel reuse or concurrency is very hard (Ramanathan, 1997). Second, TDMA firmly needs clock synchronization. When TDMA clock synchronization failure occurs, there might also be a communication failure between devices, which in turn, will reduce the aggregate network throughput. A strict synchronization results in high energy overhead, as it requires repeated signal exchanges. Third, there might be regular topology changes in M2M networks due to a number of reasons, such as time varying channel conditions, physical environmental changes, device failures, and battery outage. As a result, managing dynamic topology changes is costly, which possibly requires a global transformation. Fourth, it is difficult to find out the interference relation among devices in the vicinity, because signal interference ranges are not identical as data transmission ranges, and several interfering devices may not be in a direct data transmission range. This is called interference irregularity (Zhou et al, 2004).

Therefore, only a contention-based or a reservation-based scheme may not be appropriate to develop a scalable, flexible, and automatic communication structure for a densely deployed M2M network. Ephremides and Mowafi (Ephremides & Mowafi, 1982) first explored impeccable switching between CSMA and TDMA depending upon the contention level for a wireless LAN environment utilizing Probabilistic TDMA (PTDMA). This scheme adapts the MAC behavior between CSMA and TDMA based on the contention level. In case of TDMA, real time is divided into slots, and the relation between access probabilities of owners (a) and non-owners (b) of a specific time slot is expressed as ($a + (M-1)b = 1$), where M is the number of transmitter devices. While the aim of PTDMA and Z-MAC (Zebra-MAC) (Rhee et al., 2008) is very much similar, as compared to TDMA mechanism, PTDMA doesn't address many potential issues in wireless ad hoc networks. These issues are topology changes, interference irregularity, and time synchronization errors, which in turn, can severely reduce the PTDMA performance. According to (Rhee et al, 2008; Zhang et al, 2010), a hybrid-MAC scheme has been proposed, which tries to merge the best characteristics of both reservation-based and contention-based schemes while offsetting their weaknesses. (Rhee et al, 2008) have designed a hybrid-MAC scheme, namely Z-MAC (Zebra-MAC) for sensor networks to adjust to the level of contention in the network. Under low contention, this scheme behaves as carrier sense multiple accesses (CSMA), while under high contention, it behaves as TDMA. But, in case of densely deployed M2M networks with a large number of devices having data to transmit, Z-MAC will essentially behave like TDMA and will no longer maintain its hybrid nature. In the same way, in (Liu et al, 2013), the authors have proposed a hybrid-MAC protocol, where the M2M devices directly transmit data packets to the Base Station (BS). For this to be possible, the BS broadcasts "frames" to all the devices in a regular manner. A frame is a combination of *p*-persistent CSMA and TDMA mechanisms. However, as soon as the contention only period (COP) is over, the devices communicate with the BS directly (and not with each other). This is obviously IEEE 802.11 point coordination function (PCF) mode, as it requires a central unit (BS) to control all the activities in its cell. Additionally, the authors have mentioned only uplink, and some amendments to the protocol are required to downlink, where the M2M devices may opt to receive information from the BS (Liu et al, 2014). (Zhang et al, 2010) present the idea about how to utilize hybrid-MAC schemes for video streaming over wireless networks. These schemes attempt to adjust to different bandwidth conditions based on demand. Moreover, some already existing standards (IEEE 802.15.3, IEEE 802.15.4, and IEEE 802.11ad) assume the hybrid-MAC schemes to assure higher throughput in advanced wireless networking. Nevertheless, some of the existing standards of the hybrid-MAC protocol take into account a range of applications for different users, which may exist in M2M networks. For M2M systems to work efficiently, there are several other hybrid-MAC protocols which have been proposed earlier for various domains merging TDMA (reservation-based), CSMA (contention-based) and FDMA (reservation-based but requires additional circuitry in nodes) schemes (Demirkol et al, 2006). But up to now, none of them have been accepted as a standard, and with more and more devices taking part in communication, this scenario is going to get more disordered, which requires further research in this area. Further, an individual machine node can go to sleep mode for a number of reasons. Therefore, the communication network should be self-adapting to the network topology changes. Therefore, there is a critical need of a massive access control mechanism for developing a scalable, adaptive, flexible, and more robust MAC protocol to enable concurrent transmission among M2M devices (Wang et al, 2010).

802.11ah (C. W. Park et al, 2014) is capable of offloading cellular traffic and supporting M2M communications. This protocol utilizes beacons to split time into frames. These frames are further divided into restricted access window (RAW), and offload traffic (also known as Common Window). Every RAW is again divided into slots and a slot may be randomly selected by the device or it may be allocated to an M2M device by an access point. Binary exponential back-off algorithm is used to poll for the channel. RAW length needs to be optimized for efficient channel access. RAW is divided in to RAW U/L and RAW D/L and the interesting part is U/L. It is necessary to determine RAW U/L size which can be initiated by estimating the number of devices wanting to transmit. This estimate is fulfilled by AP by utilizing the probability of successful transmission in the last frame. The estimated slot in RAW U/L is a linear function of the number of active nodes. This protocol increases the probability of successful transmission.

F.K. Ghaznivi et al, (2017) propose a scalable, hybrid MAC protocol which satisfies quality-of-service (QoS) requirements of the users. The hybrid-MAC protocol presented in this paper categorizes transmissions into frames. Every device is assumed to generate probabilistically one or zero packet per frame. The model facilitates devices with different packet generation probabilities, which in turn, enables modeling of both periodic and non-periodic traffic. The authors have been classified the generated traffic into different classes according to their packet loss tolerances. Each frame has been divided into a number of sub-frames each serving a class of traffic. Further, each sub-frame is divided into two sub-periods one serving contention and the other reserved traffic of that class. The authors derived packet loss probability for each class of traffic. Then, they have formulated a nonlinear optimization problem (NLP) that minimizes frame length subject to packet loss requirements of different classes. The authors compare the performance of the proposed protocol with other protocols proposed for M2M communications, which shows that it achieves better performance for small packet sizes.

OrMAC (E. Shitiri et al, 2017) is a hybrid medium access protocol named orthogonal coded medium access control, which extends the principle of distributed queuing collision avoidance protocol (DQCA) of wireless local area network (WLAN) to delay-sensitive machine-to-machine (M2M) networks. This MAC protocol pre-assigns orthogonal codes, serving as the channel contention signals, to the nodes joining the network. This eliminates contention collisions. Furthermore, This MAC protocol makes use of a prioritized channel access by allowing nodes to control the transmission power of the contention signal depending on the delay sensitivity of the data. The power at which a contention signal arrives at the access point reflects the urgency of the packets waiting for transmission in the buffer. A contention signal with a high received power is assigned a high priority and vice versa for a contention signal with a low received power. The authors have carried out numerical experiments to compare the performance of OrMAC to that of DQCA in terms of the packet delivery ratio, latency, discarded packet ratio, and throughput. The results show that OrMAC can outperform DQCA in terms of all the performance metrics mentioned above.

In comparison to all the existing work in this direction as mentioned above, in this chapter, we develop a hybrid-MAC scheme for a densely deployed ad hoc M2M network, which combines the advantages of both contention-based and reservation-based schemes (A. Rajandekar and B. Sikdar, 2015; Verma et al, 2014). In this scheme, the contention and reservation portion of M2M devices is a frame structure, which is comprised of two sections: contention interval (*CI*) and transmission interval (*TI*). The *CI* duration follows *p*-persistent CSMA mechanism, which allows different devices to contend for the transmission slots. After contention, only those devices, which have won time slots, are allowed to transmit data within their own time slot during *TI*, where *TI* follows TDMA mechanism. Unlike the other hybrid-MAC protocols presented above, in this case, the gateway coordinates the communication between the devices by broadcasting the hybrid-frame structure to all the devices in a regular manner. The potential M2M transmitter devices use IEEE 802.11 DCF protocol within each TDMA time slot to communicate with its one-hop receiver device to alleviate the various disadvantages of TDMA mechanism as mentioned above.

In summary, we make the following contributions in this chapter,

- We propose a high density ad hoc M2M network, where one hop distant M2M devices communicate with each other, and the gateway architecture, which is in the vicinity of all the M2M devices, and coordinates the communication among them.

- We propose a novel scalable hybrid-MAC protocol for inter-M2M communications, which combines the advantages, and at the same time, offsets the disadvantages of p-persistent CSMA, TDMA, and IEEE 802.11 DCF mechanisms.

- We demonstrate the performance of the proposed hybrid-MAC scheme in terms of the aggregate throughput, average transmission delay, channel utility, and energy consumption. The simulation results show the effectiveness of the proposed hybrid-MAC scheme as compared to s-ALOHA and TDMA mechanisms.

To the best of our knowledge, the proposed hybrid-MAC protocol in this chapter is the first protocol which facilitates the M2M devices to share data with each other directly without any human intervention. This is in contrast to the earlier suggested other hybrid-MAC schemes, where, in most of the cases, the communication is between the devices and the base station. In other words, the proposed hybrid-MAC provides full mechanical automation to the M2M network shown in Figure 2 in next section.

SYSTEM MODEL

Network Architecture

Let us consider an M2M network, which consists of a large number of M2M devices as well as the gateway in M2M device domain as shown in Figure 2. It consists of a backbone network and multiple sub-networks. The backbone network itself consists of gateway, which manages the entire network and connecting the sub-networks to rest of the network. An important aspect of M2M environment is their heterogeneity, i.e. different devices with different characteristics (e.g. processing, battery lifetime) and operating in different ways. This also includes different devices using different communication technologies individually or in a cluster/group. For example, any one of the sub-networks shown in Figure 2 may consist of ZigBee enabled devices. Similarly, another sub-network may consist of Bluetooth enabled devices and so on. However, in our proposed system model, we assume that each sub-network consists of IEEE 802.11 DCF enabled M2M devices and works in a self-organized manner and may be designed for a specific application. Since, most of the existing products follow the IEEE 802.11 specifications; our assumption is valid and widely supported. In this domain, there are three types of communications taking place, viz. direct communication among devices, communication between device to gateway, and communication between gateways. Here, we are interested in first two types of communications. In our proposed system model, we assume that all the M2M devices are one-hop away from each other as well as from the gateway and they have same amount of data to transmit with equal priority. The devices are static (mobility is zero), and the density of devices per unit area is very high. We assume that the total number of devices in M2M domain is P, and the active number of devices that have data to transmit during a single frame is Q (the values of P and Q may differ in each frame). Therefore, within each time frame, there will be $(P-Q)$ silent devices, which must go to sleep mode to save the energy consumption. Each active device contends for accessing the channel to transmit data. Out of Q devices, R devices get success during CI, i.e. they secure a transmission slot during TI.

Figure 2. Proposed system model

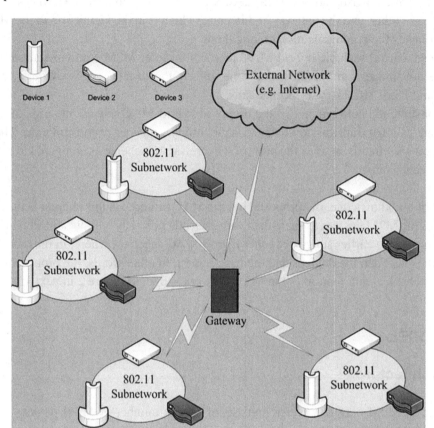

Proposed Hybrid-MAC Protocol

We assume that the gateway is controlling the communication among the devices in M2M device domain, i.e. the gateway broadcasts "frames" repetitively to all the M2M devices. A frame is the basic time unit of the channel access operation, and it consists of four different time intervals. These are Notification Interval (*NI*), Contention Interval (*CI*), Time-slot Assignment Interval (*TAI*), and Transmission Interval (*TI*) as shown in Figure 3.

Each frame starts with *NI*, which is an indication of the commencement of *CI*. All the Q devices that have data to transmit contend for the channel access during *CI* following p-persistent CSMA scheme. During this interval, the devices transmit small transmission request messages to the gateway in an arbitrary manner. While some of these data transmission request messages collide with each other, some of them are successfully received at the gateway. The gateway then broadcasts *TAI* to all the successful devices, declaring the end of the *CI* and the beginning of *TI*. During this interval, the gateway assigns unique time-slots to all the successful M2M devices following TDMA mechanism, so that each sender device transmits data packets to its intended one hop receiver device during its own slot using IEEE 802.11 DCF protocol, as shown in Figure 4. We define each part of the proposed hybrid-MAC frame as follows.

Figure 3. Hybrid-MAC frame structure

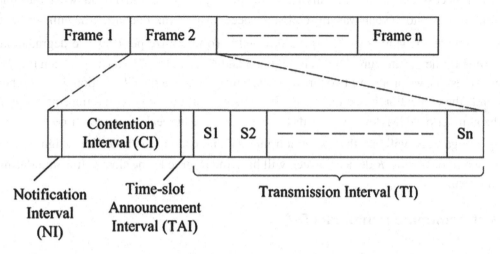

Figure 4. Transmission interval structure

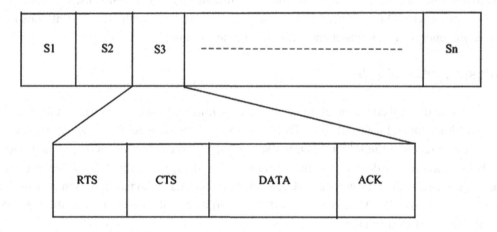

Notification Interval (*NI*)

As the communication between the gateway and the M2M devices commences, the gateway broadcasts a notification message to all the *P* devices at the start of every time frame. The gateway then approximates the number of devices which have data to transmit (value of *Q*) (Liu et al, 2014). Next, the gateway broadcasts the *CI* duration (T_{CI}) and the value of contending probability *p* to all the M2M devices, and the contention among devices to access the channel begins.

Contention Interval (*CI*)

In this interval, the active devices (*Q*) contend for the channel access based on *p*-persistent CSMA protocol, with contending probability *p*. All the *Q* devices start sending the transmission request message (T_{req}) to the gateway. When only a single device sends the T_{req} message to the gateway, the contention is termed as a success, and the device wins a time slot during *TI*. On the other hand, when multiple M2M devices send the T_{req} to the gateway, the collisions occur. Moreover, to control the number of successful devices, we use the value of T_{CI} as the threshold. To minimize performance degradation of the proposed time frame architectures shown in Figure 3 and 4 due to the difference between the analytical and practical results, we propose a two thresholds scheme to control the *CI* duration. In this scheme, the gateway ends the contention based on the number of successful devices *R*, and the contention interval T_{CI}. If the number of M2M devices exceeds the value of *R*, or the real contention time becomes longer than T_{CI}, the gateway will end the contention interval and declare the time slot assignment interval. Hence there will be at most *R* devices, which will have the reserved time slots to transmit during transmission interval.

Time-Slot Assignment Interval (*TAI*)

As soon as the contention period is over, the gateway broadcasts the time-slot assignment message to all the *Q* devices. *TAI* duration has two parts, viz. successful devices' ID, and the transmission schedule. If the device authenticates its ID in the time-slot assignment message, a transmission slot is assigned to it for transmitting data. On the other hand, if the device doesn't authenticate its ID, it goes into sleep mode to save the energy consumption and waits for the next frame.

Transmission Interval (*TI*)

As the transmission interval commences, a successful transmitter M2M device turns-on its radio module within its own time slot and transmits an RTS packet to its intended receiver M2M device. As soon as the receiver device receives the RTS packet, it also turns-on its radio module and responds with a CTS packet to its intended sender device. The transmitter device, then, sends the data packet to the receiver device, and the receiver device sends an ACK packet to its sender device. As soon as this four way handshake (RTS-CTS-DATA-ACK) mechanism gets completed, all the communicating devices turn their radio modules off to save energy.

Further, an important point about the proposed hybrid-MAC frame structure is that if we fix the frame duration T_{frame}, given that ($T_{CI} + T_{TI} \leq T_{frame}$), it is clear that if we increase the *CI* duration, the number of successful devices will also increase. But at the same time, the increase in *CI* duration will reduce the *TI* duration, resulting in reduced number of transmission slots, and hence, less number of transmitter devices. On the other hand, if we increase the *TI* duration, it will result in reduced *CI* duration. This will lead to less number of devices to contend for the channel access. Hence, there will be even lesser number of transmitter devices, winning time slots during *TI*. As a result, a number of TDMA slots may remain unused. Therefore, we need to calculate and fix the *CI* duration based on the number of active devices (*Q*) and their contending probabilities (*p*) to achieve optimum performance of the proposed scheme. From Bruno et al., (2003), we can derive the expression for contention interval T_{CI} as,

$$T_{CI}(R,p) = \sum_{i=1}^{R} \frac{\left(1-p\right)^{(Q-i)} \delta_{idle}}{\left(Q-i\right)p\left(1-p\right)^{(Q-i-1)}} + [\frac{1-\left(1-p\right)^{(Q-i)}}{\left(Q-i\right)p\left(1-p\right)^{(Q-i-1)}} - 1].\, \delta_{coll} + \delta_{succ} \qquad (1)$$

where $\delta_{idle}, \delta_{coll} = E\left[Coll_j\right]$, and $\delta_{succ} = E\left[L_i\right]$ are constants. Here, $E\left[Coll_j\right]$ is the average duration of collisions, given that a collision occurs, and $E\left[L_i\right]$ is the average duration of request messages within successful transmission duration. Thus, it is clear from the above equation, that the value of T_{CI} is not fixed, and depends upon the values of total number of devices, which are successful during contention, i.e. R, and the value of contending probability p, with which, the devices contend for the channel access. In other words, the value of T_{TI} is also not fixed, and can vary based on the number of devices, winning time-slots to transmit their data during *TI*. But the length of the frame (T_{frame}) remains constant according to the constraint ($T_{CI} + T_{TI} \leq T_{frame}$).

Based on the above equation, we propose a practical hybrid-MAC, which operates on frame by frame basis. Every frame consists of four different time intervals. These are notification interval (*NI*), contention interval (*CI*), time slot assignment interval (*TAI*), and transmission interval (*TI*). By combining CSMA, TDMA, and IEEE 802.11 DCF protocol in the way described above, the hybrid-MAC becomes more adaptive (Wang et al, 2010) and robust in terms of time varying channel conditions, slot assignment failures, time varying channel conditions, timing failures, and topology changes as compared to *p*-persistent CSMA or TDMA alone. Additionally, in comparison to the frame architecture discussed in (Liu et al, 2014), where the authors proposed a hybrid-MAC protocol for communication between the base station and other devices, we have proposed our hybrid-MAC protocol for direct communication among M2M devices with the gateway coordination. In our protocol, the sender and its corresponding one hop receiver M2M device communicate with each other using the IEEE 802.11 DCF within each TDMA slot for several reasons. First, within each time slot, the sender M2M device uses RTS packet to make its one-hop idle receiver M2M device active in the sense that it will turn-on its radio module. After hearing RTS, the receiver M2M device responds with CTS. Further, as there can be multiple gateways within M2M domain, the RTS/CTS exchange among one-hop M2M devices will again reduce hidden/exposed terminal problems with M2M devices in neighboring regions belonging to other gateways. Lastly, the exchange of RTS/CTS also makes sure that clock synchronization failures in TDMA do not lead to communication failure between M2M devices.

GATEWAY ARCHITECTURE

To coordinate the communication among M2M devices, the proposed gateway architecture is as shown in Figure 5. All the network related functionalities are realized at the gateway. These functionalities include application management, machine management, network management, and network interconnection. The descriptions of all these blocks are as follows.

Machine Management

The entire basic information about a device (i.e. its type, status, and capabilities) is incorporated in a description file. Controlling of signaling and messaging among a large number of devices is called machine management, indicating how to send, receive and resolve the device description file. The network observes the registration, cancellation, and status updates of devices through machine management. Device clustering (dividing networked machines into several groups) is an important function which

Figure 5. Proposed gateway architecture

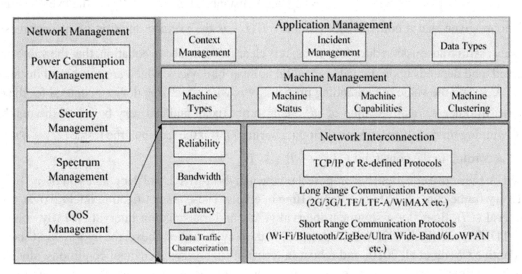

also provides scalability so as to manage a large number of devices. In this case, each group of devices is self-organized and co-operates to achieve a particular task.

Application Management

Application management disintegrates an application into a series of M2M communications, discovering the network settings, and commencing the M2M transmissions. Specifically, context management aims to adapt to diverse conditions such as application type, machine status, and the network status. Incident management activates the suitable M2M communications (e.g. routine or emergency events). The data type management converts the data format in the transmitting device into a proper format for communication, when the data is transmitted via heterogeneous devices.

Network Management

The aim of network management is to improve M2M communications performance in terms of power consumption, security, spectrum efficiency, and QoS. All these network management functional units operate across the entire protocol stack. Power consumption management facilitates to extend the machine working lifetime for battery operated devices. The task of spectrum management is to optimize the overall utilization of spectrum resources in time, frequency and spatial dimensions to enhance the spectrum efficiency.

Spectrum Cognition

Spectrum cognition keeps the devices aware of the surrounding radio environment. It is comprised of three important functional units, namely sensing, sharing, and configuration. The spectrum sensing makes the devices being able to spot the available frequency bands. Spectrum sharing controls and coordinates the spectrum resource allocations among the large number of devices which coexist in a local area.

These two functionalities (spectrum sensing, and sharing) can be implemented in devices as well as in gateways in a centralized M2M network. Lastly, the spectrum configuration helps devices to schedule the spectrum handling in local areas and enabling the spectrum resource reuse in wide areas. Spectrum configuration can be implemented in each device in a decentralized M2M network.

Network Interconnection

This block aims to provide network connectivity among devices, and between devices and the gateways through the use of transport layer protocols (TCP/IP or re-defined protocols). Depending upon the type of M2M application, long range (2G/3G/LTE/LTE-A/WiMAX) and short range (Wi-Fi/Bluetooth/ZigBee/Ultra Wide-Band/6LoWPAN etc.) communication protocols are used.

In the next section, we evaluate the performance of the proposed hybrid-MAC protocol with respect to a contention-based and a reservation-based MAC scheme in terms of aggregate throughput, average transmission delay, channel utility, and energy consumption.

PERFORMANCE EVALUATION

In this section, we compare the proposed hybrid-MAC protocol with a contention-based protocol (slotted-ALOHA) and a reservation-based protocol (TDMA) in terms of aggregate throughput, average transmission delay, channel utility, and energy consumption. We simulate our proposed scenario in network simulator (version ns-2.35) (http://www.isi.edu/nsnam/ns/) on UBUNTU-12.04 LTS platform. Our motivation of using ns-2 simulator stems from the fact that many researchers around the world are developing modified versions of ns-2 in order to introduce new features such as protocols, algorithms, etc. The standard practice adopted in doing this is to get an official version of the ns-2 source distribution, make the needed modifications on the source code, add new files somewhere in the existing code tree, and finally build everything into the ns-2 executable. The simulation parameters are as shown in the following table 1. In this table, T_{ACK} is the time, gateway takes to acknowledge the devices that their transmission request message has been received successfully.

The proposed protocol also has the overload control ability, keeping in mind that as the number of devices increases, the value of p decreases. Also, it is quite obvious that if we increase the size of T_{frame}, the number of successful devices will also increase. This in turn also shows that our proposed scheme is efficient enough, as T_{frame} can be set to be small, and still it works well for any number of contending devices. Additionally, the value of R increases as the frame duration (T_{frame}) is increased, which in turn, shows the efficiency of the proposed scheme. Additionally, although our approach includes a phase of p-persistent CSMA, the M2M devices transmit small transmission request messages (T_{req}) to the gateway to request for the channel access only, and the devices do not transmit their data packets during this phase. Therefore, p-persistent CSMA phase do not contribute in calculating the throughput or average transmission delay. Once, a device wins a time-slot during TDMA, it transmits its data packets during its own time-slot only. Therefore, we compare the performance of our proposed hybrid-MAC protocol in terms of throughput and average transmission delay with that of slotted-ALOHA, and TDMA, as the devices transmit their data packets during TDMA phase only.

Table 1. Simulation parameters

T_{NI}	10.5 µs	Notification time interval
T_{TAI}	10.5 µs	Transmission announcement time interval
T_{frame}	1500ms	Frame length
T_{tran}	2.5 ms	Transmission time for each device
T_{req}	22.5 µs	Length of transmission request message
T_{ACK}	8 µs	Acknowledgement message duration
T_{SIFS}	2.5 µs	Short inter-frame space duration
T_{BIFS}	7.5 µs	Back-off inter-frame space duration
RTS	40 bytes	RTS packet size
CTS	39 bytes	CTS packet size
DATA	500 bytes	DATA packet size
ACK	39 bytes	ACK packet size
P_{tr}	1.5 W	Transmission power of a device
P_{rv}	1 W	Receiving power of a device
P_i	0.5 W	Idle power of a device

Throughput

Here, we compare the aggregate throughput in terms of the total number of active devices Q for $= 1500$ms. We set the transmission probability p in case of slotted-ALOHA equal to 0.08. Under this condition, the relation between Q and R is $R = 10\%Q$, $R = 30\%Q$, and $R = 50\%Q$. As shown in Figure 6 below, the aggregate throughput of the proposed protocol remains always higher than that of TDMA, and as the values of Q and R are increased, it is also higher than slotted-ALOHA. This means that while s-ALOHA works well at low-contention level (LCL), and the TDMA does well in high-contention level (HCL), our proposed hybrid-MAC protocol is able to optimize the contention probability p and the number of devices winning the time slots during *TI* so as to maximize the aggregate throughput.

Figure 6. Throughput comparison in terms of total number of active devices, when R = 10%Q

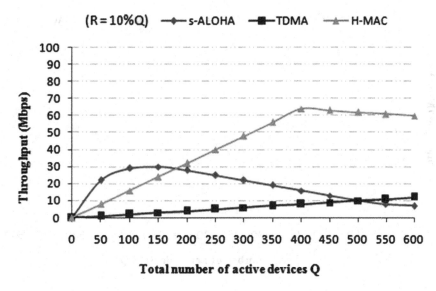

Figure 7. Throughput comparison in terms of total number of active devices, when R = 30%Q

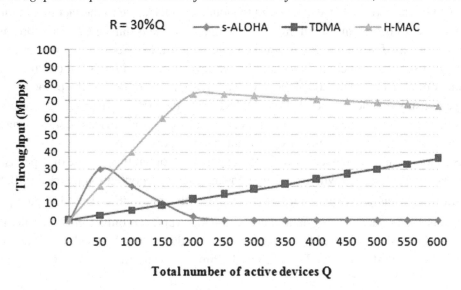

Average Transmission Delay

We compare the average transmission delay of our proposed hybrid-MAC protocol with that of slotted-ALOHA and TDMA. Average transmission delay is the amount of time it takes for a frame to travel from the device to the gateway. It can be calculated as the ratio between the connection length and the transmission speed over the medium. We set the device transmission probability in s-ALOHA, $p = 0.08$, and $T_{frame} = 1500$ms, so that the relation between Q and R is $R = 10\%Q$, $R = 30\%Q$, and $R = 50\%Q$. Figure 7 shows the comparison between the average transmission delays of slotted-ALOHA, TDMA,

Figure 8. Throughput comparison in terms of total number of active devices, when R = 50%Q

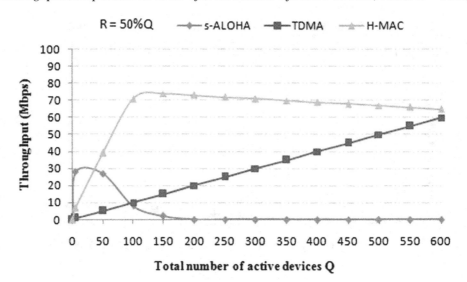

and our proposed hybrid-MAC protocol. This comparison clearly shows that the proposed protocol achieves considerable lower delay as compared to slotted-ALOHA due to the fact that in our proposed scheme, the devices only transmit small transmission request packets during *CI*. Therefore, as collisions occur, the waiting time for the devices is considerably reduced. Further, as the number of devices increases, our proposed protocol controls the number of devices during *TI*. Also, it controls the transmission probability of each device to reduce the device congestion. Additionally, the proposed protocol has close average transmission delay with that of TDMA, due to the fact that in our proposed protocol, during contention, the devices have to wait a bit more, and during *TI*, the devices first exchange RTS/CTS before transmitting data, which results in significant signaling overhead.

Note: As a large number of devices communicate with each other, the *p*-value of *p*-persistent CSMA mechanism cannot be large, and will always remain small. During contention interval (*CI*), the devices only send small transmission request messages (T_{req}) to the gateway, and not the data packets. The devices exchange data packets during data transmission interval (*DTI*) only. Hence, smaller the *p*-value, more and more devices will contend for the channel access during *CI*. This will result in more number of devices winning time slots during *TI*, which will ultimately result in greater channel utility.

Channel Utility

The channel utility can be defined as the ratio of transmission interval (T_{DTI}) to the frame interval (T_{frame}). Let us suppose that C_U represents the channel utility, then,

$$C_U = \frac{T_{DTI}}{T_{frame}} \quad (2)$$

Figure 9. Average transmission delay in terms of total number of active devices, when R = 10%Q

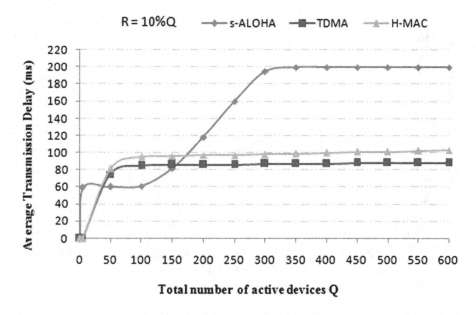

Figure 10. Average transmission delay in terms of total number of active devices, when R = 30%Q

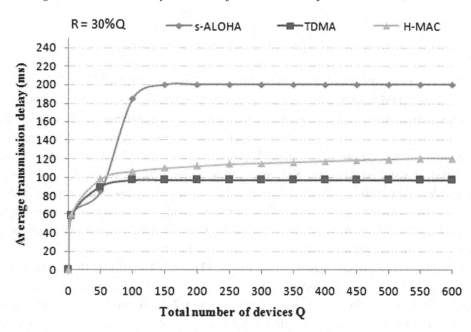

Figure 11. Average transmission delay in terms of total number of active devices, when R = 50%Q

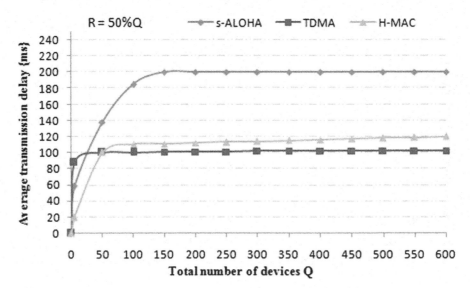

Replacing the values of T_{DTI} and T_{frame}, after solving this equation, we get,

$$C_U = \frac{T.R}{\left[T_{CNI} + T_{CI} + T_{TAI} + T_{DTI}\right]} = \frac{R.\left(T_{RTS} + T_{CTS} + T_{ACK} + 3T_{SIFS}\right)}{\left[T_{CNI} + T_{CI} + T_{TAI} + T_{DTI}\right]} \qquad (3)$$

We compare the performance of channel utility of our proposed hybrid-MAC scheme with that of slotted-ALOHA and TDMA. Figure 8 shows the utility performances of all the three schemes mentioned above, in terms of total number of contending M2M devices. Under these circumstances, the relation between Q and R is, $R = 10\% \ Q$, $30\% \ Q$, and $50\% \ Q$. From Figure 8, it is clear that initially, when the value of Q is less (up to 150-200), the channel utility of the proposed hybrid-MAC scheme remains lower than that of the slotted-ALOHA. But, as the total number of active M2M devices increases (Q becomes larger than 200), two important events take place. Due to increase in the number of collisions in slotted-ALOHA, its channel utility degrades abruptly. On the other hand, as our proposed hybrid-MAC scheme uses IEEE 802.11 DCF mechanism within each TDMA slot for data transmission, the successful M2M devices transmit data packets without any collisions. Further, the channel utility of our proposed hybrid-MAC protocol is significantly higher than that of TDMA. This is due to the fact that our proposed hybrid-MAC scheme facilitates only those M2M devices to contend for the channel access, which have data to transmit. In this way, the transmission slots during *DTI* can be fully utilized. This is in contrast with that of TDMA, where the slot assignment is static and fixed for each device, without considering the topology changes in the network, which significantly reduces its channel utility.

Energy Consumption

In this Section, we analyze the energy consumption of the proposed M2M network during a single frame. Our proposed M2M network consists of a gateway and other M2M devices, which may be battery operated. It is to be noted that in each frame,

Figure 12. Channel utility in terms of total number of active devices (Q)

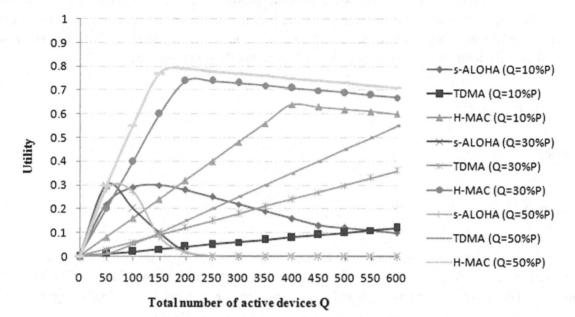

- During contention notification interval *(CNI)*, the gateway and all the other devices *(P)* may be active;
- During contention interval (*CI*) and time-slot assignment interval (*TAI*), the gateway and only the contending M2M devices (*Q*) will be active, and
- During data transmission interval (*DTI*), both the transmitter as well as their corresponding one-hop away receiver M2M devices will be active. This means that if total *R* transmitter devices successfully contend the channel and win time slots, then there will be also equal number of receiver devices (i.e. *R*), and these devices will communicate with each other within their respective time slots during *DTI*.

Let us assume that for all the M2M devices (including gateway), the power consumption during transmission mode is P_{tr} W; the power consumption during receiving mode is P_{rv} W; and the power consumption during idle mode is P_{id} W. We define the energy consumption of the M2M network during each frame as follows.

During *CNI*, the gateway broadcasts the notification message to all the M2M devices with power P_{tr} W for the notification message length of T_{CNI}. Also, during this interval, each M2M device receives a notification message with power P_{rv} W. Let E_{CNI} represents the total energy used for receiving the contention notification messages during *CNI*. Hence,

$$E_{CNI} = P_{tr} \cdot T_{CNI} + P \cdot P_{rv} \cdot T_{CNI} \tag{4}$$

Here, P is the total number of M2M devices in the network and T_{CNI} is the length of contention notification message. During *CI*, there are Q devices which send the transmission request messages to gateway, out of which, the gateway receives the transmission request messages of R devices successfully. Hence, the total energy used for sending and receiving these messages is,

$$E_{CI} = \sum_{m=1}^{Q} E_{(Q,i)} + P_{rv} \cdot T_{CI} \tag{5}$$

where,

$$E_{(Q,i)} = \sum_{j=1}^{N_{Q,i}^c} \left[Idle_{(Q,j)} \cdot P_{id} + Coll_j \cdot P_{tr} \right] + Idle_{(N_{Q,i}^c+1)} \cdot P_{tr} + S_{(Q,i)} \cdot P_{tr} \tag{6}$$

During *TAI*, the gateway broadcasts an announcement message to all the active M2M devices to declare the end of the contention interval and commencement of transmission interval. Therefore, Q numbers of devices receive the message, and let us suppose that E_{TAI} represents the total energy required to receive the announcement message from the gateway during *TAI*. Hence,

$$E_{TAI} = P_{tr} \cdot T_{TAI} + Q \cdot P_{rv} \cdot T_{TAI} \tag{7}$$

During *DTI*,

- The gateway remains idle, and each sender M2M device transmits its data packets to its one-hop away receiver M2M device on its scheduled time slot T_r. Therefore, if R is the number of total sender M2M devices winning time slots during *DTI*, the number of receiver M2M devices will also be same, i.e. R. Hence, the total energy consumption by all the devices during *DTI*,

$$E_T = R \cdot (P_{tr} + P_{rv}) \cdot T_r \tag{8}$$

Moreover, the devices which have been unsuccessful during contention interval go to idle mode and keep their radio module OFF during *DTI*. Considering the entire M2M network, there will be *(P-Q)* devices which will remain idle for the entire frame duration. Then, there will be *(Q-R)* devices which failed during contention interval and will remain idle for entire *DTI* duration. In addition, out of R devices which have won time slots during *DTI*, there will be *(R-1)* devices which will remain idle in each time slot (as there will be only one transmission during each time slot). Therefore, these devices consume the following energy over *DTI* duration,

$$E_I = (P-Q) \cdot P_{id} \cdot T_{frame} + R \cdot (Q-R) \cdot P_{id} \cdot T_r + R \cdot (R-1) \cdot P_{id} \cdot T_r \tag{9}$$

Note that *DTI* duration = R. T_r. Hence, the total energy consumed by all the devices during *DTI* is,

$$E_{DTI} = E_T + E_I \tag{10}$$

Therefore, the total energy consumption during a single frame is,

$$E_f = [\, E_{CNI} + E_{CI} + E_{TAI} + E_{DTI} \,] \tag{11}$$

In simulation, we compare the energy consumption in a single frame among *p*-persistent CSMA, TDMA, and our proposed hybrid-MAC protocol. Figure 9 shows the simulation results and comparison of energy consumptions during a single frame. It is clear that our proposed scheme consumes less energy than *p*-persistent CSMA due to the fact that our hybrid-MAC scheme allows the M2M devices to transmit only small transmission request messages during *CI*. This helps in reducing energy consumption during collisions. Additionally, as soon as *CI* is over, our proposed hybrid-MAC scheme controls the number of M2M devices which failed to contend for the channel access, and turns them to idle mode to further save the energy. On the other hand, our proposed protocol consumes a little higher more energy than that of TDMA due to the fact that in our protocol, the M2M devices consume more energy during each time slot (RTS/CTS/ACK exchange between one-hop devices). But, at the same time, this additional energy consumption results in higher channel utility.

Figure 13. Energy consumption comparison of TDMA, H-MAC, and p-persistent CSMA in terms of total number of active devices (Q)

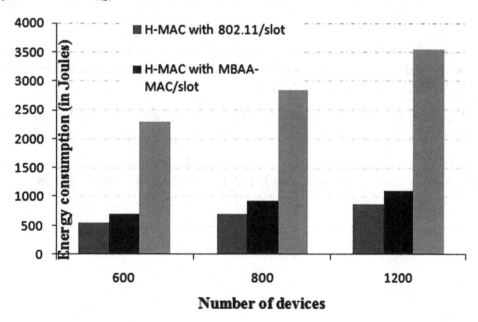

CONCLUSION

In this chapter, we have focused on massive access control scheme for densely deployed M2M network. In this protocol, each frame operates in two parts: contention interval (*CI*) and transmission interval (*TI*). The devices transmit the transmission request messages during *CI* to contend for the channel access, and during *TI*, the transmitter devices transmit data packets to their one-hop receiver devices using IEEE 802.11 DCF protocol within their own TDMA time-slots. In this scenario, the gateway controls the communication among M2M devices in such a way that it maximizes the aggregate throughput by controlling the duration of *CI* and *TI*, given that $(T_{CI} + T_{TI} \leq T_{frame})$. To implement the scheme, we have presented a hybrid-MAC protocol for the M2M networks. We have analyzed the aggregate throughput, average transmission delay, channel utility, and energy consumption to show the efficiency of the proposed hybrid-MAC protocol. In future, we will consider the concurrent transmission mechanism for our proposed hybrid-MAC protocol, where the M2M devices will be able to transmit DATA and ACK packets concurrently. This will further scale down the size of the frame structure, and in turn will help us to obtain higher performance of the hybrid-MAC protocol in terms of the performance parameters mentioned above.

REFERENCES

Bruno, R., Conti, M., & Gregori, E. (2003). Optimal Capacity of *p*-persistent CSMA Protocols. *IEEE Communications Letters*, *7*(3), 139–141. doi:10.1109/LCOMM.2002.808371

Cao, X., Chen, J., Zhang, Y., & Sun, Y. (2008). Development of an integrated wireless sensor network micro-environmental monitoring system. *Elsevier ISA Trans.*, *47*(3), 247–255. doi:10.1016/j.isatra.2008.02.001 PMID:18355827

Chen, J., Cao, X., Cheng, P., Xiao, Y., & Sun, Y. (2010). Distributed collaborative control for industrial automation with wireless sensor and actuator networks. *IEEE Transactions on Industrial Electronics*, *57*(12), 4219–4230. doi:10.1109/TIE.2010.2043038

Chen, M., Wan, J., & Li, F. (2012). Machine-to-Machine Communications: Architectures, Standards, and Applications. *KSII Transaction on Internet and Information Systems*, *6*(2), 480-497.

Demirkol, I., Ersoy, C., & Alagoz, F. (2006). MAC protocols for wireless sensor networks: A survey. *IEEE Communications Magazine*, *44*(4), 115–121. doi:10.1109/MCOM.2006.1632658

Ephremides, A., & Mowafi, O. A. (1982). Analysis of a hybrid access scheme for buffered users-probabilistic time division. *IEEE Transactions on Software Engineering*, *SE-8*(1), 5261. doi:10.1109/TSE.1982.234774

Gau, R. H., & Cheng, C. P. (2013). Optimal tree pruning for location update in machine-to-machine communications. *IEEE Transactions on Wireless Communications*, *12*(6), 2620–2632. doi:10.1109/TWC.2013.040413.112086

Ghazvini, F. K., Ali, M. M., & Doughan, M. (2017). Scalable hybrid MAC protocol for M2M communications. *Computer Networks. Elsevier Publications*, *127*, 151–160.

Hasan, M., Hossain, E., & Niyato, D. (2013). Random access for machine-to-machine communication in LTE-advanced networks: Issues and approaches. *IEEE Communications Magazine*, *51*(6), 86–93. doi:10.1109/MCOM.2013.6525600

Hsu, T. H., & Yen, P. Y. (2011). Adaptive time division multiple access-based medium access control protocol for energy conserving and data transmission in wireless sensor networks. *IET Communications*, *5*(18), 2662–2672. doi:10.1049/iet-com.2011.0088

Igarashi, Y., Ueno, M., & Fujisaki, T. (2012). Proposed Node and Network Models for an M2M Internet. *World Telecommunications Congress (WTC)*, 1-6.

Kim, S., Cha, J., Jung, S., Yoon, C., & Lim, K. (2012). Performance evaluation of random access for M2M communication on IEEE 802.16 network. *Proc. ICACT*, 278–283.

Lien, S. Y., Chen, K. C., & Lin, Y. (2011). Toward Ubiquitous Massive accesses in 3GPP Machine- to-Machine Communications. *IEEE Communications Magazine*, *49*(4), 66–74. doi:10.1109/MCOM.2011.5741148

Lien, S. Y., Liau, T. H., Kao, C. Y., & Chen, K. C. (2012). Cooperative access class barring for machine-to-machine communications. *IEEE Transactions on Wireless Communications*, *11*(1), 27–32. doi:10.1109/TWC.2011.111611.110350

Liu, Y., Yuen, C., Cao, X., Hassan, N. U., & Chen, J. (2014). Design of a Scalable Hybrid MAC Protocol for Heterogeneous M2M Networks. *IEEE Internet of Things Journal*, *1*(1), 99–111. doi:10.1109/JIOT.2014.2310425

Liu, Yi., Yuen, C., Chen, J., & Cao, X. (2013). A Scalable Hybrid MAC Protocol for Massive M2M Networks. *IEEE conference on Wireless Communications and Networking,* 250 – 255.

Lu, R., Li, X., Liang, X., Shen, X., & Lin, X. (2011). GRS: The green, reliability, and security of emerging machine to machine communications. *IEEE Communications Magazine, 49*(4), 28–35. doi:10.1109/MCOM.2011.5741143

Park, C. W., Hwang, D., & Lee, T. J. (2014). Enhancement of IEEE 802.11ah MAC for M2M communications. *IEEE Communications Letters, 18*(7), 1151–1154. doi:10.1109/LCOMM.2014.2323311

Rajandekar, A., & Sikdar, B. (2015). A survey of MAC layer issues and protocols for machine-to-machine communications. *IEEE Internet of Things Journal, 2*(2), 175–186. doi:10.1109/JIOT.2015.2394438

Ramanathan, S. (1997). A unified framework and algorithms for (T/F/C)DMA channel assignment in wireless networks. *Proceedings - IEEE INFOCOM,* 900–907.

Rhee, I., Warrier, A., Aia, M., Min, J., & Sichitiu, M. L. (2008). Z-MAC: A hybrid MAC for wireless sensor networks. *IEEE Trans. Netw., 16*(3), 511–524. doi:10.1109/TNET.2007.900704

Shitiri, E., Park, I., & Cho, H. (2017). OrMAC: A Hybrid MAC Protocol Using Orthogonal Codes for Channel Access in M2M Networks. *Sensors. MDPI Publications, 17*(9), 1–10.

The Network Simulator home page. (n.d.). Retrieved from http://www.isi.edu/nsnam/ns/

Verma, P. K., Tripathi, R., & Naik, K. (2014). A Robust Hybrid-MAC Protocol for M2M Communications. *5th IEEE International Conference on Computer and Communication Technology (ICCCT),* 267–271. 10.1109/ICCCT.2014.7001503

Verma, P. K., Verma, R., Prakash, A., Agrawal, A., Naik, K., Tripathi, R., ... Abogharaf, A. (2016). Machine-to-Machine (M2M) Communications: A Survey. *Journal of Network and Computer Applications. Elsevier Publications, 66,* 83–105.

Wang, G., Zhong, X., Mei, S., & Wang, J. (2010). An adaptive medium access control mechanism for cellular based machine to machine (M2M) communication. *Proc. IEEE Int. Conf. Wireless Inf. Technol. Syst. (ICWITS),* 1-4. 10.1109/ICWITS.2010.5611820

Ye, W., Heidemann, J., & Estrin, D. (2004). Medium access control with coordinated adaptive sleeping for wireless sensor networks. *IEEE/ACM Transactions on Networking, 12*(3), 493–506. doi:10.1109/TNET.2004.828953

Yu, R., Zhang, Y., Gjessing, S., Xia, W., & Yang, K. (2013). Toward cloud-based vehicular networks with efficient resource management. *IEEE Network, 27*(5), 48–55. doi:10.1109/MNET.2013.6616115

Zhang, R., Ruby, R., Pan, J., Cai, L., & Shen, X. (2010). A hybrid reservation/contention-based MAC for video streaming over wireless networks. *IEEE Journal on Selected Areas in Communications, 28*(3), 389–398. doi:10.1109/JSAC.2010.100410

Zhang, Y., Yu, R., Nekovee, M., Liu, Y., Xie, S., & Gjessing, S. (2012). Cognitive Machine to Machine Communications: Visions and Potentials for the Smart Grid. *IEEE Network, 26*(3), 6–13. doi:10.1109/MNET.2012.6201210

Zhang, Y., Yu, R., Xie, S., Yao, W., Xiao, Y., & Guizani, M. (2011). Home M2M networks: Architecture, standards and QoS improvements. *IEEE Communications Magazine, 49*(4), 44–52. doi:10.1109/MCOM.2011.5741145

Zhou, G., He, T., Krishnamurthy, S., & Stankovic, J. A. (2004). Impact of radio irregularity on wireless sensor networks. *Proceedings of ACM MobiSys*, 125-138. 10.1145/990064.990081

Chapter 7

Adaptive Power–Saving Mechanism for VoIP Over WiMAX Based on Artificial Neural Network

Tamer Emara
Shenzhen University, China

ABSTRACT

The IEEE 802.16 system offers power-saving class type II as a power-saving algorithm for real-time services such as voice over internet protocol (VoIP) service. However, it doesn't take into account the silent periods of VoIP conversation. This chapter proposes a power conservation algorithm based on artificial neural network (ANN-VPSM) that can be applied to VoIP service over WiMAX systems. Artificial intelligent model using feed forward neural network with a single hidden layer has been developed to predict the mutual silent period that used to determine the sleep period for power saving class mode in IEEE 802.16. From the implication of the findings, ANN-VPSM reduces the power consumption during VoIP calls with respect to the quality of services (QoS). Experimental results depict the significant advantages of ANN-VPSM in terms of power saving and quality-of-service (QoS). It shows the power consumed in the mobile station can be reduced up to 3.7% with respect to VoIP quality.

INTRODUCTION

Recently, Voice over Internet Protocol (VoIP) has increasingly become a common service for wireless networks such as worldwide interoperability for microwave access (WiMAX). As the VoIP technology offers Mobile WiMAX clients the ability to use voice services with lower expenses compared with Public Switched Telephone Network (PSTN). Skype has earned more and more popularity since it is seen as the best VoIP software (Adami, Callegari, Giordano, Pagano, & Pepe, 2012). For VoIP calls, packets are required to be sent constantly. Considering the real-time nature of voice, the radio cannot easily save power. As it is difficult to transition to sleep state while sending inter-packets intervals (Zubair, Fisal, Abazeed, Salihu, & Khan, 2015).

DOI: 10.4018/978-1-5225-5693-0.ch007

A mobile station (MS) usually depends on a portable power source like batteries. These batteries have limited lifetime. Thus, all wireless systems offer power-saving classes (PSCs), such as in IEEE 802.16e (IEEE, 2006), to save the power consumed in MSs through sleep and active mode (Ghosh, Wolter, Andrews, & Chen, 2005).

Regarding IEEE 802.16e, an MS operates alternately between awake and sleep mode. It wakes up to exchange data with a base station (BS) in a listening period. The sleep mode includes sleep and listening periods. In sleep periods, BS buffers incoming packets to the MS, and BS sends it to the MS when MS switches to listen periods. To fit services and applications properties, three PSCs are offered by the IEEE 802.16e for different types of traffic for the purposes of power saving. Thus, MS can associate the sleep parameters for the suitable PSC when connected to a BS. The sleep parameters comprise the starting of sleep time, initial sleep period, final sleep period and the listen period. Clearly, the parameters of each class should be carefully determined to minimize the power consumed in the MS with restriction to the quality-of-service (QoS) requirements of that connection.

This chapter focuses on power saving in short latency periods included in active time which is sufficient for VoIP services. This chapter considers two states of VoIP conversation: mutual silent state and talk-spurt state. In mutual silent periods, there are not any packets transmitted between MS and BS. Therefore, the sleep period in mutual silence can be increased more than in the talk-spurt period to reduce the power consumption. Hence to increase energy-efficient, it is necessary to design a new algorithm that can place its sleep intervals flexibly in mutual silence periods.

In this chapter, a power conservation algorithm is proposed based on artificial neural network (ANN-VPSM). The proposed algorithm can be applied to VoIP service over WiMAX systems. Artificial Intelligent model using feed forward neural network with a single hidden layer has been developed to predict the mutual silent period that used to determine the sleep period for power saving class mode in IEEE 802.16. Artificial Neural Networks (ANNs) is a simulation of the biological nervous system in the human brain. That network can regulate neurons and learn through experience. The development of ANN using computer programs to identify patterns of data sets using training data through supervised learning. ANNs has a flexible learning system and adaptive ability allow them to learn from linear and non-linear function (Haykin, 2007).

An overview of previous attempts in power saving for VoIP services is addressed in the next section. Then the proposed prediction mechanism is presented. The simulation results of the proposed mechanism with discussion are introduced in experiment and results. Finally, the conclusion is discussed.

BACKGROUND AND RELATED WORK

WiMAX (IEEE 802.16e) introduces three types of PSCs suitable to several applications that produce different traffic characteristics (IEEE, 2006). There are various QoS requirements, for various types of connections between MS and BS. Subsequently, various connections are classified into varied PSCs to achieve their requirements of QoS.

1. **The Power Saving Class Type I (PSC-I):** Starts with initial sleep window (S_{min}), then this sleep window is doubled, if MS doesn't receive any message about the existence of data packets during its listening window. This process is repeated consistently till reaching the maximum sleep window period (S_{max}), after that, the next keeps the same.

2. **The Power Saving Class Type II (PSC-II):** Has a sleep window (S_T) with the same size that is repeated consistently.
3. **The Power Saving Class Type III (PSC-III):** Has a sleep cycle with one predefined sleep interval (S_M) and without listening interval.

In Power Saving Mode (PSM), a trade-off exists between the performance of QoS for transmitting packets and conserving power of an MS (Nga, Kim, & Kang, 2007). In VoIP service, that trade-off occurs in accordance with the length of the sleep periods. To satisfy VoIP QoS, the sleep period can be equal to the time that is needed to generate packets of the VoIP codec, which is in a short period of 10 to 30 ms (Cox & Kroon, 1996), but it is not easy to get enough power conservation because of the comparatively short sleep period. On the contrary, if the sleep period's length is increased, certainly, more power will be saved, however, it leads to an additional delay to VoIP traffic which breaks up VoIP QoS. Thus, to estimate the exact length of the sleep periods, it should be considered the performance of PSM within the requirements of QoS.

There have been recent works that support VoIP service in the environment of WiMAX. In (Choi & Cho, 2007; Choi, Lee, & Cho, 2007; J. R. Lee, 2007; J. R. Lee & Cho, 2009; Lin, Liu, Wang, & Kwok, 2011), the authors introduce many power saving mechanisms, which mainly depend on applying PSC-II in a talk-spurt periods, however for saving more power, they applied PSC-I during silence periods. Emara (2016) applies PSC-III at mutual silence state, while authors in (Emara, Saleh, & Arafat, 2014) combine between PSC-III and PSC-I in mutual silence to increase power saving.

In (J. Lee & Cho, 2008), the authors discuss the methodology to choose the optimal sleep periods of the PSC-II, that have the most efficient energy. In (Choi, Lee, & Cho, 2009), the VoIP hybrid power saving mechanism (HPSM) performance had been estimated, according to the model of network delay. In (Namboodiri & Gao, 2010), an algorithm is presented for emerging dual-mode mobiles in order to save power by computing sleep/wake-up schedules, making the radio remain in the lower power sleep mode for curtail periods throughout a phone call.

Many scheduling algorithms are proposed in (Chen, Deng, Hsu, & Wang, 2012; Cheng, Wu, Yang, & Leu, 2014; Kao & Chuang, 2012; Ke, Chen, & Fang, 2014; Teixeira & Guardieiro, 2013), the packets are categorized according to the characteristics of their traffic. The objective of these algorithms is to transmit the packets in appropriate transmission time with the appropriate sleep-mode operation. These algorithms try to schedule packets for minimizing listening intervals and maximizing sleep intervals within the QoS requirement. In (Liao & Yen, 2009) the authors proposed a scheduling algorithm which considered the QoS jitter and delay types should be synchronized.

In case that the delay limit is less stringent and the traffic load is light, the authors in (Jin, Chen, Qiao, & Choi, 2011) examine sleep mode management suitable for IEEE802.16m that gains good performance. The authors (Kalle et al., 2010) introduce procedures of sub-frame sleep, variable listening interval, and discrete listening window extension to enhance VoIP traffic power saving in IEEE 802.16m. The analysis of power consumed in Transmission Control Protocol (TCP) data transfer was in (Hashimoto, Hasegawa, & Murata, 2015).

Problem Definition and Plane of Solution

The most famous standard speech codecs utilized in VoIP applications are G.723.1A (ITU-T, 2006), G.729B (ITU-T, 2012) and AMR (Abreu-Sernandez & Garcia-Mateo, 2000). These standards use the function of silence suppression. Silence suppression function does not permit the speech codec to generate VoIP packets in silent periods, so as to save system bandwidth. Anyway, considering the speech model suggested by ITU (ITU-T, 1993), human speech comprises interchanging silent periods and talk-spurt, Brady (Brady, 1969) concluded that about 60 percent of the total VoIP conversation duration is occupied by the silent periods. Therefore, it is not efficient to use PSC-II only, as PSC-II does not deal with the silent periods, its sleep interval has fixed length. To increase energy-efficient, it is necessary to design a new algorithm that can place its sleep intervals flexibly in mutual silence periods.

If each terminal uses the speech codec with silence suppression functionality, the codec generates VoIP packets at specific times continuously during talking periods, while the packets are not generated during silent periods. Instead of not generating packets in silent periods, it produces a Silent Insertion Descriptor (SID) frame (Estepa, Estepa, & Vozmediano, 2004). The SID packet is a small packet, in comparison with the voice packet. It contains only background noise information, which can help the receiving side's decoder to generate artificial noise during the silent period. SID is generated at the beginning of the silent period and sent only once. And so on, the SID frames receipt of both parts A and B could confirm the starting time of alternating silent. When SID frame is transmitted only at the starting of the mutual silence, bandwidth utilization increases (Hussain, Marimuthu, & Habib, 2014) and thus the network traffic is reduced.

Brady (Brady, 1969) shows that about 60% of the total VoIP conversation duration is occupied by the silent periods and about 19% for mutual silent periods and the mutual silence duration average is approximately 306 ms. Therefore, for saving more energy, MS can sleep during these mutual silence periods, yet the standard PSC-II has nothing to do with the VoIP traffic silent periods. Thus, MS must be activated at specific times continuously during those periods even if it doesn't receive any voice packets. Consequently, the main focus of the proposed mechanism (ANN-VPSM) is to apply another PSM depending on each side's voice activity. As shown in Figure 1, PSC-II is used during talk-spurt periods. Throughout the mutual silent intervals, PSC-III with S_M period is predicted using Artificial Neural Networks (ANNs) techniques. All the notations used in this chapter are described in Table 1.

Figure 1. An example of the proposed ANN_VPSM mechanism operation

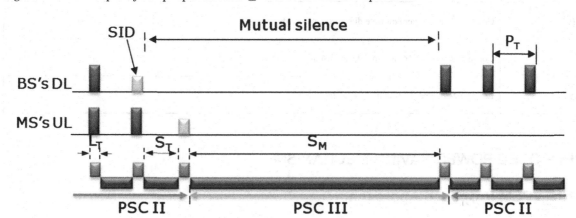

Table 1. Table of notations

Notation	Explanation
P_T	Packet interval
1 frame	One frame duration
L_T	Listen interval
S_T	Sleep interval for PSC-II
S_M	Sleep interval for PSC-III
S_{min}	Minimum sleep interval for PSC-I
S_{max}	Maximum sleep interval for PSC-I
X	Distribution of the mutual silent length
λ	Arrival rate
p_i	The probability of arriving the first voice packet after the mutual silent state started in the i-th sleep cycle
B_f	Buffering delay
E_1	Consumed energy per unit time in PSC-I
E_3	Consumed energy per unit time in PSC-III
E_m	Consumed energy per unit time during mutual silent periods
E_S	Consumed energy per unit time during sleep periods
E_L	Consumed energy per unit time during listen periods
S_i	Length of i-th sleep period in the i-th sleep cycle

PROPOSED POWER SAVING MECHANISM

ANN_VPSM Proposed Approach

In recent times, ANN technology has developed to make it possible to solve most complicated non-linear systems with variable parameters. ANN simulates a human nervous system that is working as the human brain. It is built with artificial neurons meanwhile; the interconnections are similar to the nervous system (Haykin, 2007). ANN has been effectively used to predict the mutual silence in a VoIP conversation.

This work utilizes a neural network to predict the mutual silent periods, which determines the value of the sleep cycle for PSC-III to achieve the maximum power saving. At the moment that ANN is built, it becomes ready to train. Since these initial weights are chosen randomly (Simoes & Bose, 1995). The proposed ANN-VPSM algorithm consists of two modes: the first is training mode; while the ANN trained, VPSM (Emara et al., 2014) mechanism is executed. The second mode is prediction mode; after ANN training finished, it is ready to predict the sleep cycle for PSC-III. ANN has the two inputs, last voice duration's time and last mutual silent duration's time and has one output, optimum sleep cycle for PSC-III.

In this study, the used ANN structure is Multilayer Perceptron (MLP); feeding input data to the neural layer for producing the desired output. The learning method is back-propagation method which feeds the output's error data to the input. The ANN constructs three layers with on hidden layer. The nodes numbers or Processing Elements (PEs) are chosen based on the complexity of the data and studies.

Training Mode

After constructing ANN, it needs to be trained. So, the training mode starts firstly. While the conversation is running, the needed data for training are collected. To save more power in training mode our previous work VPSM (Emara et al., 2014) is chosen.

The main aim of adopting the VPSM is using the PSC-II during talk-spurt intervals with a fixed sleep interval's length. While, throughout the mutual silent intervals, PSC-III with S_M less than 300 ms is launched at the outset, then PSC-I is applied till the end of mutual silence. Figure 3 illustrates the procedures of VPSM. Algorithm 1 is the proposed algorithm for training mode. It works on the frames which received or sent.

Sent Data

At the moment of the speech coder encode current data, the Voice Activity Detection (VAD) (ITU-T, 2012) is operated. The VAD result could be 1 or 0 depending on whether the voice activity is presented or absence respectively. The current encoded frame becomes SID if VAD result was equal to 0, while VAD result was equal to 1 if the current encoded frame is a voice.

The PSC-II has requested if the current encoded frame is active voice. However, to start the mutual silent state, the two sides confirm it by generating SID frame at the beginning of the silent period for each side. Therefore, if the current encoded frame is SID, the last received frame from the other side must be checked. If it is SID, the mutual silent state starts, otherwise, PSC-II is requested.

Received Data

When the frame is received, the G.729 decoder is used to decode the received packets. In case VAD result was 1, the active voice frames are reconstructed. Else, the comfort noise generator (CNG) module (ITU-T, 2012) reproduces the non-active voice frames.

The PSC-II is requested when the frame is for active voice. However, if the received frame is SID, current MS state must be in silence to start the mutual silent state. Thus, MS state must be checked, whether it is on silent or not. In case MS is in talking activity, then it is not the mutual silent state. Therefore, the PSC-II is launched. On the contrary, the mutual silent state is initialized if the last encoded frame for the current MS is SID.

Thus, we can summarize the work of this algorithm to the following points:

1. Mutual silence occurs when both sides in silence (i.e. current VAD = 0 and last received frame was SID)
2. Mutual silent sleep mode is initialized by launching PSC-III in case the current PSM is PSC-II.
3. If no frames sent or received during the sleep cycle of PSC-III, The PSC-I has launched after PSC-III termination.
4. The silent mode is over when a voice frame is received or encoded.

After finishing the mutual silent state, the network is being trained. By the end of this test for the current frame, it will be repeated for the coming frame (received or sent) to the call end.

VPSM Numerical Analysis

For analyzing the VPSM during the mutual silent interval, there are two cases. Case 1 discusses the analysis of receiving active voice packets during S_M period, while case 2 discusses the receiving of active voice packets during any sleep cycle of PSC-I.

Case 1: For any active voice packet sent or received, PSC-II is applied. i.e., the MS sleeps with fixed S_T size and wakes up with fixed L_T size. While, the mutual silent state is started by launching PSC-III with S_M period. If there is any voice packet transmitted on uplink or downlink throughout the listen period of PSC-III, it backs again to PSC-II.

Let X be the distribution of a mutual silent length. According to Brady model, the cumulative distribution function of X is defined by:

$$F_X\left(t\right) = 1 - e^{-\lambda t} \tag{1}$$

λ is the arrival rate. Let p_i be the probability of arriving the first voice packet after starting the mutual silent state at the i-th sleep cycle.

Figure 2. An example of VPSM operation (Emara et al., 2014)

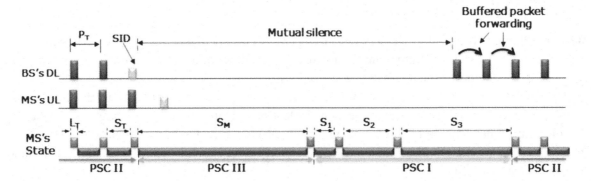

$$p_i = P\{X \in [0, S_M]\} = F_X(S_M)$$

(2)

$$p_i = 1 - e^{-\lambda(S_M + L_T)}$$

(3)

Hence, the average duration of PSC-III, $E[S_M]$ is calculated by

$$E[S_M] = (S_M + L_T)$$

(4)

Assume that all voice packets that are buffered through the sleep period S_M will be transmitted immediately in the following listen period, the buffering delay is obtained by $Bf = S_M - X$, therefore, the average buffering delay becomes

$$E[Bf] = E[S_M] - E[X] = (S_M + L_T) - \frac{1}{\lambda}$$

(5)

For calculating the total consumed energy when the VPSM is used as a mutual silent sleep mode in this case. Let E_S and E_L represent the energies consumed per unit time during sleep and listening periods, respectively. Therefore, the consumed energy per unit time in PSC-III, E_3, is calculated as follows:

$$E_3 = \frac{(S_M E_S + L_T E_L)}{S_M}$$

(6)

Consequently, the energy consumption throughout mutual silent periods per unit time, E_m, is defined as follows:

$$E_m = E_3$$

(7)

Case 2: if the sleep cycle of PSC-III finished without receiving or generating any active voice packet, the MS applies PSC-I directly. Considering S_i be the length of i-th sleep period in PSC-I, the length of the i-th sleep cycle, $S_i + L_T$, increases exponentially as follows:

$$S_i + L_T = \begin{cases} 2^{i-1} \cdot P_T & , \ if \ 1 \le i < N \\ S_{max} & , \ if \ \quad i \ge N \end{cases}$$

(8)

where the maximum sleep period is $[S_{max} = 2^N S_1]$, N is the number of doubled sleep window till the sleep window size reaches S_{max}.

Algorithm 1. The proposed algorithm in training mode

```
1:   Start Call
2:   for each frame do
3:      if Data sent then
4:         Coding data using voice coder
5:         if VAD=0 then
6:            Check last received frame
7:            if last received frame = SID then
8:               Call Mutual_Silent_Case
9:            end if
10:        else
11:           Call Talking_Case
12:        end if
13:     else if Data received then
14:        Decode received frame using voice coder
15:        if received frame = SID then
16:           Check VAD for last sent frame
17:           if VAD=0 then
18:              Call Mutual_Silent_Case
19:           end if
20:        else
21:           Call Talking_Case
22:        end if
23:     end if
24:  end for
25:  End Call
26:
27:  procedure Mutual_Silent_Case
28:     Check current PSC
29:     if current PSC = PSC-II then
30:        Request PSC-III
31:     else if current PSC= PSC-III then
32:        Request PSC-I
33:     end if
34:  end procedure
35:
36:  procedure Talking_Case
37:     Check current PSC
38:     if current PSC = PSC-III or current PSC = PSC-I then
39:        Request PSC-II
40:        Calculate silence period
41:        Train ANN
```

continued on following page

Algorithm 1. Continued

```
42:      end if
43:          end procedure
```

VPSM mechanism is ended according to the arrival of a voice packet at the i-th sleep cycle. Let X be the distribution of the length of a mutual silent period. According to Brady model, the cumulative distribution function of X is defined by

$$F_X\left(t\right) = 1 - e^{-\lambda t} \tag{9}$$

where λ is arrival rate. Let p_i be the probability of arriving at the first voice packet after the beginning of mutual silence period in the i-th sleep cycle.

$$p_i = P\left\{X \in \left[L_{i-1}, L_i\right]\right\} = F_X\left(L_i\right) - F_X\left(L_{i-1}\right) \tag{10}$$

$$= e^{-\lambda L_{i-1}}\left(1 - e^{-\lambda\left(S_i + L_T\right)}\right) \tag{11}$$

where $L_i = \sum_{j=1}^{i}(S_j + L_T)$ then, the average duration of PSC-I, $E\left[L\right]$, will be;

$$E\left[L\right] = \sum_{i=1}^{\infty} p_i.L_i = \sum_{i=1}^{\infty} p_i \sum_{j=1}^{i}(S_j + L_T) \tag{12}$$

Assume that all voice packets that are buffered through the sleep period S_i will be transmitted instantly throughout the following listen period. The buffering delay is $Bf_i = L_i - X$. Consequently, the average buffering delay is calculated by

$$E\left[Bf\right] = E\left[L\right] - E\left[X\right] = \sum_{i=1}^{\infty} p_i \sum_{j=1}^{i}\left(S_j + L_T\right) - \frac{1}{\lambda} \tag{13}$$

The total consumed energy for VPSM during this case is an aggregate of the consumed energy in PSC-III and in PSC-I. Let E_S and E_L indicate the consumed energies per unit time in the sleep and listening periods, respectively. Then, the consumed energy per unit time in PSC-III, E_3, is obtained as follows:

$$E_3 = \frac{\left(S_M E_S + L_T E_L\right)}{S_M} \tag{14}$$

And the consumed energy per unit time in PSC-I, E_I, is obtained as follows:

$$E_1 = \frac{\sum_{i=1}^{\infty} p_i \sum_{j=1}^{i} \left(S_j E_S + L_T E_L \right)}{E\left[L\right]} \tag{15}$$

Therefore, the consumed energy during mutual silent periods per unit time, E_m, will be:

$$E_m = E_1 + E_3 \tag{16}$$

Prediction Mode

After finishing the training, the network is ready to run. A flowchart of the predicted mode is shown in Figure 3. The mechanism works on the frames which are received or sent.

Sent Data

VAD (ITU-T, 2012) is checked after the speech codec encoded the current frame. If its result was 0, it means the current frame is SID. While the current frame is active voice, in case the result was 1. When the current frame is active voice, the PSC-II is demanded. However, the mutual silence starts when the two sides generate SID frames at the beginning of the silent periods on each side to confirm the mutual silent state. Thus, if the current frame is SID, it doesn't mean this case is mutual silence without checking the last received frame. Therefore, if the last received frame from the other side was a voice that indicates this case is not mutual silence, as a result, the PSC-II is demanded. However, the mutual silent state starts when the last received frame was SID, therefore mutual silent sleep mode is initialized by launching PSC-III with sleep interval is predicted using the ANN.

Received Data

Once the frame is received, the VAD is checked as the first step to decode it by G.729 decoder. In case that its value is equal 1, active voice frames will be reconstructed. But, if its value is equal 0, the non-active voice frames will be reproduced by CNG module (ITU-T, 2012).

If the received frame is active voice frame, the PSC-II is requested. However, if it is SID, the current MS state must be silent to start the mutual silent state. Thus, MS state must be checked, whether it is on silent or not. In case that MS state was active talk, it indicates the case is not mutual silence, as a result, the PSC-II is demanded. In contrast, if MS state was silent that shows this case is mutual silence. Consequently, the mutual silent sleep mode is initialized by launching PSC-III with sleep interval which is predicted using ANN.

By the end of this test for the current frame, it will be repeated for the following frame received or sent until the end of the call.

Figure 3. Flowchart of ANN-VPSM in predicting mode

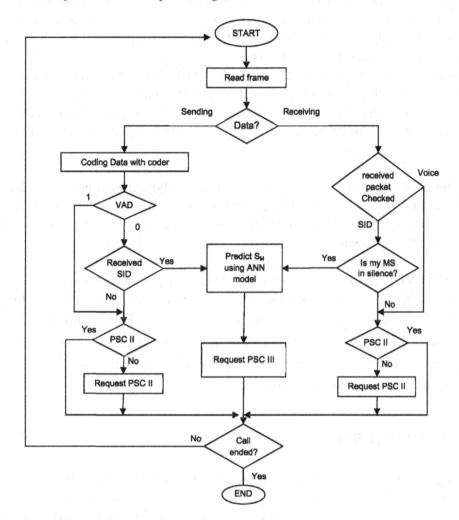

EXPERIMENT AND RESULTS

Simulation Setup

To design a neural network, a sufficiently large amount of data is needed for training, testing, and cross-validation. So, the simulation algorithm consists of three stages:

1. **Getting Training Data:** A program coded with C++ language and G.729B speech codec to get the required data for training the neural network.
2. **Building ANN:** The training is automated with the Neuro Solution (version 5) which trains the network using the back-propagation method. 60% of data is used for training, 15% for cross-validation and the last 25% for testing.

Neuro Solutions software is used to obtain the results and accordingly, simulations are carried out on voice time and silent time as inputs and next mutual silence, which determines the sleep cycle of PSC-III as the desired output. The voice time and silent time were input to MLP with one hidden layer of MLPs were tested with maximum epoch 1000.

Numerous learning rules like Conjugate Gradient (CG), Delta Bar Delta (DBD), Levenberg Marquardt (LM), momentum, and Quick prop (QP) are used for training and better performance parameters are observed. Results are observed while using different Transfer Functions like Tanh Axon, Linear Tanh Axon, Sigmoid Axon, and Linear Sigmoid Axon to find that minimize the power consumed in MS while still guaranteeing VoIP QoS.

3. **Performance Evaluation:** The ANN performance is validated by simulation coded with C++ language and G.729B speech codec to calculate the power consumption for the mutual silent period and mean opinion score (MOS). The MOS score has a scale of 1-5 with 1 representing the worst quality and 5 the best quality.

ITU-T G.729B (ITU-T, 2012) codec is considered as of being frequently used in VoIP applications for its low bandwidth requirements. The packet, P_T, is generated every 40ms. Therefore, the MS should wake up every 40 ms. So, $L_T = 1\,frame$ was decided. Since the VoIP packet size is very small (8kbps in G.729B) in comparison with the 802.16 bandwidth (about 30Mbps), L_T for 1 frame (10ms) is enough to receive the buffered VoIP packets. So, S_T, becomes 3 frames (30ms) because $P_T = S_T + L_T$. Table 2 summarizes the initial values for simulation factors.

RESULTS AND DISCUSSION

Using transfer function Tanh Axon to build the neural network. After constructing the neural network, training begins by changing learning rules. To see more clearly the effects of the neural network with a transfer function Tanh Axon throughout a call time, Figure 4-a plots the power consumed every 30s, Figure 4-b plots MOS every 30s. From these figures, the best performance made when using learning

Table 2. Initial values of simulation's factors

Parameter	Description	Value
P_T	Packet interval	40 ms
1 frame	One frame duration	10 ms
L_T	Listen interval	10 ms
S_T	Sleep interval for PSC-II	30 ms
S_M	Sleep interval for PSC-III	

rule Momentum, less power is consumed with MOS greater than 3.8. Simulation results, when using learning rule Momentum, show that the power consumed in MS can be reduced up to 3.74%.

Using transfer function Sigmoid Axon to build the neural network. After constructing the neural network, training begins by changing learning rules. To see more clearly the effects of the neural network with a transfer function sigmoid axon throughout a call time, Figure 5-a plots the power consumed every 30s, Figure 5-b plots MOS every 30s. From these figures, the best performance made when using learning rule Quickprop. Simulation results, when using learning rule Quickprop, show that the power consumed in MS can be reduced up to 3.73% with MOS more than 3.8.

Using transfer function Linear Tanh Axon to build the neural network. After constructing the neural network, training begins by changing learning rules. To see more clearly the effects of the neural network with a transfer function Linear Tanh axon throughout a call time, Figure 6-a plots the power consumed every 30s, Figure 6-b plots MOS every 30s. From these figures, the best performance made when using learning rule Quickprop or Momentum. Simulation results show that the power consumed in MS can be reduced up to 3.76% with MOS more than 3.8.

Using transfer function Linear Sigmoid Axon to build the neural network. After constructing the neural network, training begins by changing learning rules. To see more clearly the effects of the neural network with a transfer function Linear Sigmoid axon throughout a call time, Figure 7-a plots the power consumed every 30s, Figure 7-b plots MOS every 30s. Simulation results indicate that the power consumed in MS can be reduced up to 3.75% with MOS more than 3.8.

Table 3 summarizes the result achieved when various transfer functions are applied for the proposed ANN-VPSM.

Figure 4. Transfer function Tanh Axon a) Power consumption b) MOS

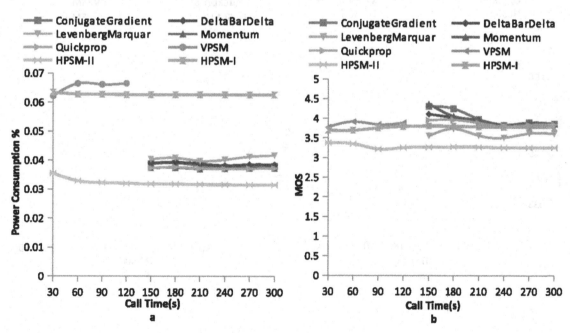

Figure 5. Transfer function Sigmoid Axon a) Power consumption b) MOS

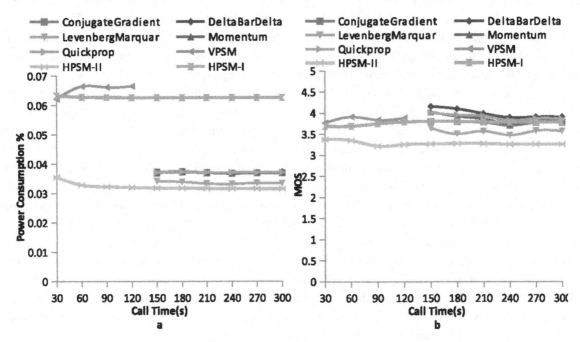

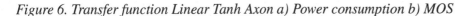

Figure 6. Transfer function Linear Tanh Axon a) Power consumption b) MOS

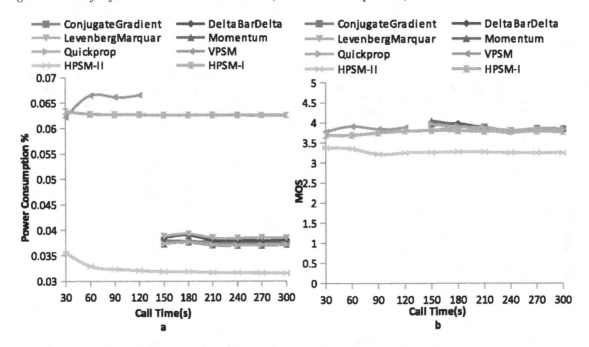

Figure 7. Transfer function Linear Sigmoid Axon a) Power consumption b) MOS

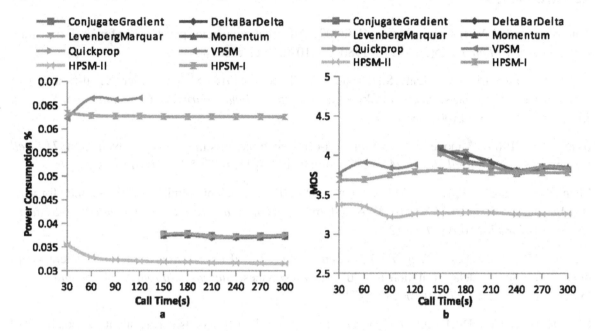

Table 3. The result achieved when various transfer functions are applied for the proposed ANN-VPSM

Transfer Function	Best Learning Rule	Power Consumption	MOS
Tanh axon	Momentum	3.7%	3.8:4.2
Sigmoid axon	Quickprop	3.7%	3.8:4
Linear Tanh axon	Quickprop or Momentum	3.6%:3.7%	3.7:3.9
Linear Sigmoid axon	Quickprop or Momentum	3.7%	3.8:4

CONCLUSION

Recently, VoIP is applied in broadband wireless access systems for their voice services, therefore, the MS requires an effective power saving mechanism for VoIP services to increase its battery lifetime. In this chapter, the power saving mechanism based on artificial neural network (ANN-VPSM) is proposed for VoIP services over WiMAX systems. The proposed ANN-VPSM mechanism can maintain more power in mutual silent periods. The experimental findings indicate that ANN-VPSM can minimize the power consumed in MS efficiently. Furthermore, it can increase the utilization of the bandwidth and decrease the network traffic. Experimental results show that the power consumed in MS can be reduced up to 3.7% with respect to QoS when applying the ANN-VPSM mechanism.

REFERENCES

Abreu-Sernandez, V., & Garcia-Mateo, C. (2000). Adaptive multi-rate speech coder for VoIP transmission. *Electronics Letters*, *36*(23), 1978–1980. doi:10.1049/el:20001344

Adami, D., Callegari, C., Giordano, S., Pagano, M., & Pepe, T. (2012). Skype-Hunter: A real-time system for the detection and classification of Skype traffic. *International Journal of Communication Systems*, *25*(3), 386–403. doi:10.1002/dac.1247

Brady, P. T. (1969). A model for generating on-off speech patterns in two-way conversation. *The Bell System Technical Journal*, *48*(7), 2445–2472. doi:10.1002/j.1538-7305.1969.tb01181.x

Chen, Y. S., Deng, D. J., Hsu, Y. M., & Wang, S. D. (2012). Efficient uplink scheduling policy for variable bit rate traffic in IEEE 802.16 BWA systems. *International Journal of Communication Systems*, *25*(6), 734–748. doi:10.1002/dac.1206

Cheng, T. K., Wu, J. L. C., Yang, F. M., & Leu, J. S. (2014). IEEE 802.16e/m energy-efficient sleep-mode operation with delay limitation in multibroadcast services. *International Journal of Communication Systems*, *27*(1), 45–67. doi:10.1002/dac.2342

Choi, H. H., & Cho, D. H. (2007). Hybrid energy-saving algorithm considering silent periods of VoIP traffic for mobile WiMAX. *2007 IEEE International Conference on Communications,* 1-14, 5951-5956. 10.1109/ICC.2007.986

Choi, H. H., Lee, J. R., & Cho, D. H. (2007). Hybrid power saving mechanism for VoIP services with silence suppression in IEEE 802.16e systems. *IEEE Communications Letters*, *11*(5), 455–457. doi:10.1109/LCOMM.2007.070035

Choi, H. H., Lee, J. R., & Cho, D. H. (2009). On the Use of a Power-Saving Mode for Mobile VoIP Devices and Its Performance Evaluation. *IEEE Transactions on Consumer Electronics*, *55*(3), 1537–1545. doi:10.1109/TCE.2009.5278024

Cox, R. V., & Kroon, P. (1996). Low bit-rate speech coders for multimedia communication. *IEEE Communications Magazine*, *34*(12), 34–41. doi:10.1109/35.556484

Emara, T. Z. (2016). Maximizing Power Saving for VoIP over WiMAX Systems. *International Journal of Mobile Computing and Multimedia Communications*, *7*(1), 32–40. doi:10.4018/IJMCMC.2016010103

Emara, T. Z., Saleh, A. I., & Arafat, H. (2014). Power saving mechanism for VoIP services over WiMAX systems. *Wireless Networks*, *20*(5), 975–985. doi:10.100711276-013-0650-5

Estepa, A., Estepa, R., & Vozmediano, J. (2004). A new approach for VoIP traffic characterization. *IEEE Communications Letters*, *8*(10), 644–646. doi:10.1109/LCOMM.2004.835318

Ghosh, A., Wolter, D. R., Andrews, J. G., & Chen, R. H. (2005). Broadband wireless access with WiMax/802.16: Current performance benchmarks and future potential. *IEEE Communications Magazine*, *43*(2), 129–136. doi:10.1109/MCOM.2005.1391513

Hashimoto, M., Hasegawa, G., & Murata, M. (2015). An analysis of energy consumption for TCP data transfer with burst transmission over a wireless LAN. *International Journal of Communication Systems*, *28*(14), 1965–1986. doi:10.1002/dac.2832

Haykin, S. (2007). *Neural Networks: A Comprehensive Foundation* (3rd ed.). Prentice-Hall, Inc.

Hussain, T. H., Marimuthu, P. N., & Habib, S. J. (2014). Supporting multimedia applications through network redesign. *International Journal of Communication Systems*, *27*(3), 430–448. doi:10.1002/dac.2371

IEEE. (2006). IEEE Standard for Local and Metropolitan Area Networks Part 16: Air Interface for Fixed and Mobile Broadband Wireless Access Systems Amendment 2: Physical and Medium Access Control Layers for Combined Fixed and Mobile Operation in Licensed Bands and Corrigendum 1. In IEEE Std 802.16e-2005 and IEEE Std 802.16-2004/Cor 1-2005 (Amendment and Corrigendum to IEEE Std 802.16-2004) (pp. 0_1-822).

ITU-T. (1993). P.59: Artificial conversational speech. In ITU-T P.59 (03/93).

ITU-T. (2006). G.723.1: Dual rate speech coder for multimedia communications transmitting at 5.3 and 6.3 kbit/s. In ITU-T G.723.1 (05/06).

ITU-T. (2012). Coding of speech at 8 kbit/s using conjugate-structure algebraic-code-excited linear prediction (CS-ACELP). In ITU-T G.729

Jin, S., Chen, X., Qiao, D. J., & Choi, S. (2011). Adaptive sleep mode management in IEEE 802.16m wireless metropolitan area networks. *Computer Networks*, *55*(16), 3774–3783. doi:10.1016/j.comnet.2011.03.002

Kalle, R. K., Gupta, M., Bergman, A., Levy, E., Mohanty, S., Venkatachalam, M., & Das, D. (2010). Advanced Mechanisms for Sleep Mode Optimization of VoIP Traffic over IEEE 802.16m. *2010 IEEE Global Telecommunications Conference Globecom 2010*. 10.1109/GLOCOM.2010.5683895

Kao, S. J., & Chuang, C. C. (2012). Using GI-G-1 queuing model for rtPS performance evaluation in 802.16 networks. *International Journal of Communication Systems*, *25*(3), 314–327. doi:10.1002/dac.1242

Ke, S. C., Chen, Y. W., & Fang, H. A. (2014). An energy-saving-centric downlink scheduling scheme for WiMAX networks. *International Journal of Communication Systems*, *27*(11), 2518–2535. doi:10.1002/dac.2486

Lee, J., & Cho, D. (2008). An optimal power-saving class II for VoIP traffic and its performance evaluations in IEEE 802.16e. *Computer Communications*, *31*(14), 3204–3208. doi:10.1016/j.comcom.2008.04.029

Lee, J. R. (2007). A hybrid energy saving mechanism for VoIP traffic with silence suppression. *Network Control and Optimization. Proceedings*, *4465*, 296–304.

Lee, J. R., & Cho, D. H. (2009). Dual power-saving modes for voice over IP traffic supporting voice activity detection. *IET Communications*, *3*(7), 1239–1249. doi:10.1049/iet-com.2008.0300

Liao, W. H., & Yen, W. M. (2009). Power-saving scheduling with a QoS guarantee in a mobile WiMAX system. *Journal of Network and Computer Applications*, *32*(6), 1144–1152. doi:10.1016/j.jnca.2009.06.002

Lin, X. H., Liu, L., Wang, H., & Kwok, Y. K. (2011). On Exploiting the On-Off Characteristics of Human Speech to Conserve Energy for the Downlink VoIP in WiMAX Systems. *2011 7th International Wireless Communications and Mobile Computing Conference (Iwcmc)*, 337-342.

Namboodiri, V., & Gao, L. X. (2010). Energy-Efficient VoIP over Wireless LANs. *IEEE Transactions on Mobile Computing*, *9*(4), 566–581. doi:10.1109/TMC.2009.150

Nga, D. T. T., Kim, M. G., & Kang, M. (2007). Delay-guaranteed energy saving algorithm for the delay-sensitive applications in IEEE 802.16e systems 1339. *IEEE Transactions on Consumer Electronics*, *53*(4), 1339–1347. doi:10.1109/TCE.2007.4429222

Simoes, M. G., & Bose, B. K. (1995). Neural-Network-Based Estimation of Feedback Signals for a Vector Controlled Induction-Motor Drive. *IEEE Transactions on Industry Applications*, *31*(3), 620–629. doi:10.1109/28.382124

Teixeira, M. A., & Guardieiro, P. R. (2013). Adaptive packet scheduling for the uplink traffic in IEEE 802.16e networks. *International Journal of Communication Systems*, *26*(8), 1038–1053. doi:10.1002/dac.1390

Zubair, S., Fisal, N., Abazeed, M. B., Salihu, B. A., & Khan, A. S. (2015). Lightweight distributed geographical: A lightweight distributed protocol for virtual clustering in geographical forwarding cognitive radio sensor networks. *International Journal of Communication Systems*, *28*(1), 1–18. doi:10.1002/dac.2635

KEY TERMS AND DEFINITIONS

Base Station (BS): A BS is informed of the MS capabilities, security parameters, service flows, and full MAC context information. The MS transmits/receives data to/from the BS.

Mobile Station (MS): A station in the mobile service intended to be used while in motion or during halts at unspecified points. An MS is usually a subscriber station (SS).

Mutual Silence: The both sides of a conversation are in silence at the same moment.

Silent Insertion Descriptor (SID): For a VoIP conversation, if each side uses the speech codec with silence suppression functionality, the codec generates VoIP packets at specific times continuously during talking periods, while the packets are not generated during silent periods. Instead of that, it produces a SID packet. The SID is a small packet, in comparison with the voice packet. It contains only background noise information, which can help the receiving side's decoder to generate artificial noise during the silent period.

Speech Codec: An algorithm is designed to operate with a digital signal obtained by first performing telephone bandwidth filtering of the analog input signal, then sampling it.

Talk-Spurt: For at least one side of a conversation is talking at a moment.

Voice Over Internet Protocol: (VoIP): A service that allows users to communicate with each other through the internet. It has a lower expense compared with public switched telephone network (PSTN).

Chapter 8
Optimizing Channel Utilization for Wireless Broadcast Databases

Agustinus Waluyo
Monash University, Australia

ABSTRACT

A very large number of broadcast items affect the access time of mobile clients to retrieve data item of interest. This is due to high waiting time for mobile clients to find the desired data item over wireless channel. In this chapter, the authors propose a method to optimize query access time and hence minimize power consumption. The proposed method is divided into two stages: (1) The authors present analytical models and utilize the analytical models for both query access time over broadcast channel and on-demand channel; (2) they present a global index, an indexing scheme designed to assist data dissemination over multi broadcast channel. Several factors are taken into account, which include request arrival rate, service rate, number of request, size of data item, size of request, number of data item to retrieve, and bandwidth. Simulation models are developed to find out the performance of the analytical model. Finally, the authors compare the performance of the proposed method against the conventional approach.

INTRODUCTION

The development of wireless technology has led to mobile computing, a new era in data communication and processing (Barbara, 1999; Badrinath and Phatak; Imielinski and Viswanathan, 1994). With this technology, people can now access information anytime and anywhere using portable size wireless computer powered by battery (e.g. PDAs). These portable computers communicate with central stationary server via wireless channel. Mobile computing provides database applications with useful aspects of wireless technology, and mobile database is a subset of mobile computing that focuses on query to central database server.

DOI: 10.4018/978-1-5225-5693-0.ch008

The main properties of mobile computing include mobility, severe power and storage restriction, frequency of disconnection is much greater than in a traditional network, bandwidth capacity and asymmetric communications costs. Radio wireless transmission usually requires approximately 10 times power as compared to the reception operation (Zaslavsky and Tari, 1998). Moreover, the life expectancy of a battery (e.g. nickel-cadmium, lithium ion) was estimated to increase the time of effective use by only another 15% in several years to come (Paulson, 2003). This is amplified with the fact that most applications involve read operations rather than write operations (Huang, Sistla and Wolfson, 1994). Thus, efficient query optimization and processing is definitely one of the main issues.

Broadcast mechanism refers to periodically transmit database items to mobile clients through one or more broadcast channels. Mobile clients filter their desired data on the fly. This strategy is known as an effective way to disseminate database information to a large set of mobile clients. In this way, mobile client enables to retrieve information without wasting power to transmit a request to the server. Other characteristics of data broadcasting includes scalability as it supports a large number of queries, query performance is not affected by the number of users in a cell as well as the request rate, and effective to a high-degree of overlap in user's request.

With the increase number of data items to be broadcast, the access time also increase accordingly as access to data item is sequential (Imielinski, Viswanathan, and Badrinath, 1997; Lee and Lo, 2003). This situation may cause some mobile clients to wait for a substantial amount of time before receiving desired data item. Consequently, the advantages of broadcast strategy will be eliminated. Alternatively, mobile client can send a request via point-to-point channel or on-demand channel to the server. The server processes the query and sends the result back to the client. This situation severely affects the query response time and power consumption of mobile clients, as it involves queuing and cannot scale over the capacity of the server or network. Thus, in this chapter we try to outperform the query access time over on-demand channel in any situation by optimizing the broadcast channel.

Figure 1 illustrates two mechanisms of mobile clients to obtain desired data. Figure 1 (a) is utilizing on-demand channel, while Figure 1 (b) is through broadcast channel. These two mechanisms are also referred as pull-based and push-based approach respectively (Aksoy et al, 1999).

Figure 1. The architecture of on-demand and data broadcast mechanism

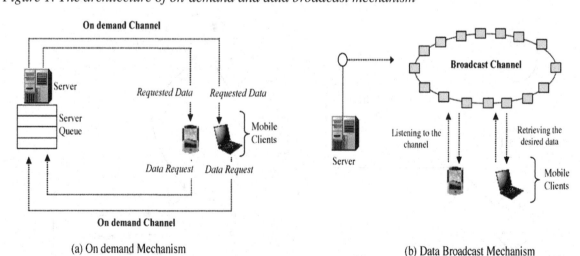

(a) On demand Mechanism (b) Data Broadcast Mechanism

Background

In general, each mobile user communicates with a Mobile Base Station (MBS) or also known as Mobile Support Station (MSS) to carry out any activities such as transaction and information retrieval. Activities that focus on query to central database server refer to mobile databases. MBS has a wireless interface to establish communication with mobile client and it serves a large number of mobile users in a specific region called cell.

Mobile units or mobile clients in each cell can either connect to the network via wireless radio or wireless Local Area Network (LAN). Wireless radio bandwidth has asymmetric communication behavior in which the bandwidth for uplink communication is smaller than downlink communication (Imielinski, Viswanathan and Badrinath, 1997). Uplink bandwidth is applicable when mobile clients send a request or query to the server, while downlink bandwidth is from the server back to the clients. In mobile environment architecture, each MBS is connected to a fixed network as illustrated in Figure 2.

MAIN FOCUS OF THE CHAPTER

In this chapter, we aim to minimize the query response time or access time of mobile client to retrieve database information from broadcast channel.

In order to optimize the data access time from broadcast channel, we utilize the query access time over on-demand channel as a threshold point to determine the optimum number of database items to be broadcast in a channel. When the optimum number is located, the number of broadcast channel should be increased. Subsequently, the length of broadcast cycle is split, and broadcast over multiple channels.

To achieve our objective, we propose analytical models for both on-demand and broadcast channel. Subsequently, we use simulation model to verify the results of our analytical models in finding the optimum number of broadcast items. Furthermore, a Global indexing scheme is to be applied (Waluyo, Srinivasan, and Taniar, 2003). This is used to reduce the tuning time or the time mobile client need to

Figure 2. Mobile environment architecture

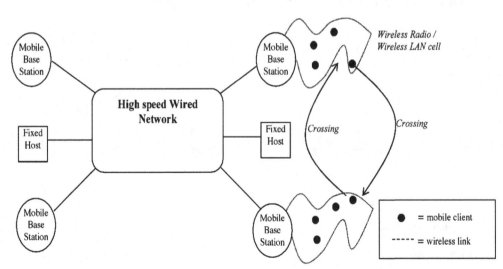

listen to the channel, to download or retrieve the desired data item which is used to indicate energy consumption. The indexing scheme provides accurate information for a client to tune in at the appropriate time for the required data to arrive in the channel. It allows mobile clients to switch into "doze mode" or power saving mode while waiting for the data and safe power consumption.

In general, broadcast indexing technique causes a trade-off between optimizing the client tuning time and the query access time. The consequence of minimizing one of them is the increase of the other. For instance, to minimize the access time is to reduce the length of broadcast cycles. In this case, the index can be broadcast once in each cycle but it will make the tuning time suffer since the client will have to wait for the index to arrive which happens only once in each broadcast cycle. On the other hand, when the index directory is frequently broadcast in each broadcast cycle to reduce the tuning time, the access time will be greatly affected due to the occupancy of the index in the cycle. Thus, it is necessary to find the optimal balance of these two factors. This chapter includes a comprehensive description on the approach as well as thorough performance evaluations, which is an extension of our previous work in (Waluyo, Srinivasan, and Taniar, 2004).

Figure 3 illustrates access time when index segment and data segment are in different channel. Data item corresponds to database record or tuples, and data segment contains a set of data item. The subsequent sections in this chapter are organized as follows. Section 2 contains the related work of the proposed technique. It is then followed by description of the analytical models for both broadcast and on-demand channel, and its application in section 3. Section 4 presents our simulation-based experiments in order to validate the analytical models as well as to evaluate the proposed strategy as compared to conventional one. Finally, section 5 concludes the chapter.

Present your perspective on the issues, controversies, problems, etc., as they relate to theme and arguments supporting your position. Compare and contrast with what has been, or is currently being done as it relates to the chapter's specific topic and the main theme of the book.

Figure 3. Query access time

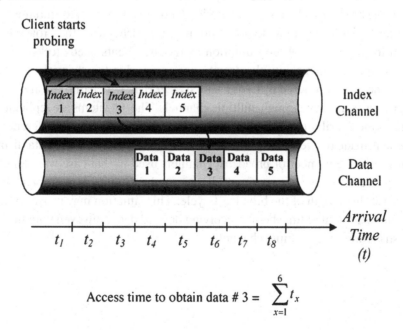

$$\text{Access time to obtain data \# 3} = \sum_{x=1}^{6} t_x$$

RELATED WORK

Broadcast strategy is known to be effective when involving a large number of users in the cell, which corresponds to a high frequency of requests (Acharya, et al, 1995;Leong and Si, 1997; Si and Leong, 1999).

Hybrid and dynamic channel allocation method, which manipulates both the on-demand channels and broadcast channels, has been presented in (Lee, Hu and Lee, 1997; Hu, Lee and Lee, 1999; Hu, Lee and Lee, 1998). The hybrid allocation method provides a solution by determining the number of data to be served from on-demand channel and broadcast channel to enhance the query performance. Dynamic hybrid method is the improvement of the hybrid method, which change the number of channel allocated for on-demand mode and broadcast mode dynamically by considering the load of the server and request access pattern. However, in this scheme, a vigorous cost calculation introduces a high degree of complexity in the algorithm before it can decide the best channel allocation. Furthermore, the utilization of on-demand channel forces the client to consume large amount of energy especially after considering the asymmetric communication cost as indicated earlier. Thus, it is desirable to have a better strategy in utilizing broadcast channel.

Acharya, et al (1995) proposed broadcast disk as an information system architecture, which utilizes multiple disks of different sizes and speeds on the broadcast medium. The broadcast forms of chunks of data from different disks on the same broadcast channel. The chunks of each disk are evenly scattered with each other. However, the chunks of the fast disks are broadcast more frequently than the chunks of the slow disks. With differing broadcast frequency of different items, hot items or items that are retrieved frequently can be broadcast more often than others. The server is assumed to have the indication of the clients' access patterns so that it can determine a broadcast strategy that will give priority to the hot items. The drawback in this scheme is that the broadcast cycle may be longer as several data items are broadcast more than once within a single cycle. Consequently, the access time of some clients will be severely affected.

(Huang and Chen, 2002; Huang and Chen, 2003) proposed some algorithms to identify the most effective organization of broadcast data items. They concern with broadcast data organization scheme in the context of multiple broadcast channels and employ Genetic Algorithm to find the best organization of broadcast data items. However, they do not apply indexing scheme in the broadcast program. This situation may lead to wasteful power consumption as mobile clients needs to keep listening into the channel and filtering the data items until the desired ones arrived in the channel.

Another strategy has been proposed by Leong and Si (1995), in which the data are broadcast based on replication and partition technique through multiple channels. The data items are replicated and organized in a way that reflects the distribution of the data. Thus each channel can have different distribution pattern. This technique is made to overcome problems such as data distortion and data distribution (Leong and Si, 1995). However, the number of broadcast channels used in that case was static. Consequently, when there are very large numbers of users in a cell, the number of data item of interest increase, which subsequently increases the length of the broadcast cycle. This situation may cause some mobile clients to wait for a substantial amount of time before receiving desired data item even with the help of different pattern of data distribution over multiple channels.

OPTIMISING CHANNEL UTILISATION OVER BROADCAST CHANNEL:

Proposed Method

This section describes the proposed method to optimise overall query access time over broadcast channel. We divide this section into two stages: (*i*) presenting analytical models to calculate optimal broadcast channel, and (*ii*) presenting Global Indexing approach.

Analytical Models

In general, the data items are broadcast over a single channel with underlying reason that it provides the same capacity (bit rate/bandwidth) as multiple channels, and it is used to avoid problems such data broadcast organization and allocation while having more than one channel (Imielinski, Viswanathan and Badrinath, 1997). Nevertheless, the use of single channel will show its limitation when there is very large number of a data item to be broadcast.

The proposed strategy splits the length of the broadcast cycle when the number of broadcast items reaches an optimum point. This process continues until the access time is above the optimal point. Figure 4 illustrates the situation when a broadcast cycle is split into two channels.

The allocation of broadcast cycle into an optimal number of broadcast channels will eliminate the chance of long delay before obtaining the desired data items. As mobile clients got the index of the required data item, they can determine which channel contains the relevant data items. Once it has been found, they may wait to retrieve the data items in the corresponding channel. Since the length of the broadcast cycle is optimal, the waiting time will be considerably short. Figure 5 illustrates a situation when the optimum number of items in a channel is known. A simple example, assume the size of the each data items are: 50 bytes. Consider a mobile client who wants to retrieve data item #8. The following two cases analyze the access time of mobile client when the data is broadcast over a single channel as compared to two channels. It is assumed that the client probes into a first data item in a channel.

Figure 4. Multiple channels architecture

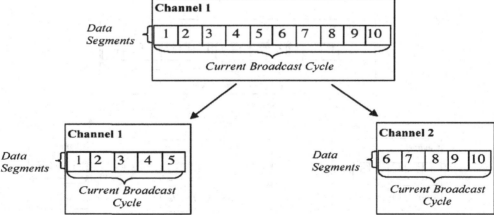

Case 1: Data items (1-10) are broadcast in a single channel

With this case, all data items are available in a single channel. Client who wants to retrieve data item #8, has to wait until the desired item arrives in the channel. The total access time required will be (50 × 8) = 400 bytes.

Case 2: Data items (1-10) are split into two broadcast channels: channel one and channel two with 6 and 4 data items respectively. Since data items are split into two broadcast channel, the access time will also be different. In this case, client can switch to another channel and wait for the data item of interest to arrive. The total access time for client to retrieve data item #8 would be (50 × 2) = 100 bytes. Thus, as compared to the first case, this time client can have a much more efficient data access.

Based on the above reason, we will reduce the broadcast channel's data access overhead by determining the optimum number of data items in a channel. To find the optimum number of broadcast items, we develop analytical models to calculate average access time of broadcast and on-demand channel. In this context, a mobile client can only initiate a single request at a time, the next request has to wait until the first request has been completely responded.

Analytical Model of Broadcast Channel

To calculate the average access time for retrieving data (TB) using broadcast channel, we consider the following scenario (Figure 6).

In Figure 6, the broadcast cycle is partitioned into three areas: area A contains data segments preceding data segments in area B, area B contains the number of desired data segments, and area C includes the rest of data segments. x_t is the total number of broadcast data that makes up of $A+B+C$. There are three different scenarios when mobile client probe into one of these area.

Probe A: When mobile client probes into area A, the total access time is given:

Figure 5. Broadcast channel partitioning: an example

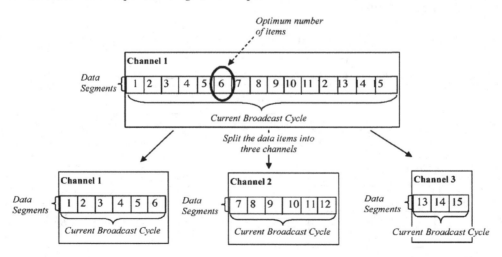

Figure 6. Broadcast cycle partitioned area

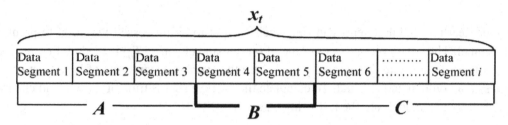

$$\frac{\sum_{i=0}^{A-1}(A-i+B)\times s}{b_s} \text{ or similarly } \frac{\left[A\times(A+2B+1)\right]\times s}{2\times b_s} \tag{1}$$

Probe B: When mobile client probes into area B, then the total access time is equal to the total length of broadcast cycle (A+B+C):

$$\frac{(A+B+C)\times B\times s}{b_s} \tag{2}$$

Probe C: When mobile client probes into area C, the total access time can be calculated from:

$$\text{or equally } \frac{\sum_{i=0}^{C-1}(C-i+A+B)\times s}{b_s} \text{ or equally } \frac{C\times(2A+2B+C+1)\times s}{2\times b_s} \tag{3}$$

We need to calculate the average access time from equation (1), (2), and (3) as follows:

$$\frac{\left[(A\times(A+2B+1))+(2B\times(A+B+C))+(C\times(2A+2B+C+1))\right]\times s}{2\times x_t\times b_s} \tag{4}$$

Equation (4) can be rewritten as:

$$T_B\approx\frac{\left[(x_t-B)^2+(2\times B\times x_t)+(2\times B\times(x_t-B))+(x_t-B)\right]\times s}{2\times x_t\times b_s} \tag{5}$$

s corresponds to the size of data item, we consider the size is uniform. The downlink bandwidth is denoted by b_s.

Analytical Model of On-Demand Channel

As mentioned earlier, mobile clients can send their request to be processed via point-to-point or on-demand channel. Thus, the average access time of on-demand channel (T_D) is made for comparison. A comparison is made to locate the optimal point where the average access time of broadcast channel is equal or less than on-demand channel. To calculate the average access time of on-demand channel, we classify the analytical model into the following categories:

Pre-Determined Number of Request

In this scenario, the total number of request has been known. The average access time can be calculated using the following analytical model:

$$T_D \approx \frac{r}{b_r} + \frac{\sum_{i=1}^{n} i \times \frac{1}{\mu}}{n} + \frac{s}{b_s} \quad (6)$$

The analytical model to calculate the access time of on-demand channel in equation (6) is based on the architecture that is depicted in Figure 1(a). The access time of on-demand channel comprises of the time needed to transmit a request to the server, process the request that involves the time spent in the server queue and retrieve the relevant data, and send the result back to the client. The transmission of request to the server and send the requested data item back to mobile client are affected by uplink bandwidth, b_r and downlink bandwidth, b_s. Size of a request and the size of requested data item are indicated by r and s respectively, n denotes the total number of request, i reads the request number and the server service rate is given by μ. However, in real situation the number of requests is hardly known. Thus, we consider arrival rate of request, which is described in the subsequent strategy.

Request Arrival Rate

We replace the determined number of request with the arrival rate of request (λ). Thus, the average access time of on demand channel (T_D) is now calculated using the following analytical model:

If the server is at light load $\rho = \left(\frac{\lambda}{\mu} \leq 1 \right)$, which means there is no waiting time then:

$$T_D \approx \left(\frac{r}{b_r} + \frac{1}{\mu} + \frac{s}{b_s} \right) \quad (7)$$

Else if the server at heavy load $\rho = \left(\frac{\lambda}{\mu} > 1 \right)$, then:

$$T_D \approx \left\lfloor \frac{r}{b_r} \right\rfloor + \frac{\sum\limits_{i=1}^{q\,\max}\left[\left(i \times \frac{1}{\mu}\right) - \left[(i-1) \times \frac{1}{\lambda}\right]\right]}{q\,\max} + \left(\frac{s}{b_s}\right) \tag{8}$$

The average time in server queue and average processing time by the server are dynamically changed depending on the arrival rate of request (λ) and service rate (μ). The maximum number of request in the server queue is defined by q_{max}.

Uniform and Non Uniform Request

This category involves a request that returns multiple data items. However, the order of the data retrieval is made arbitrary, which means first arrival of any data of interest will be firstly retrieved. We left the data ordering retrieval for future work.

Let $d_1, d_2 \ldots d_{max}$ specifies the number of relevant items in a request. We classify the number of relevant data items into uniform request that is when the occurrence rate of $d_1, d_2 \ldots d_{max}$ is equally distributed and non-uniform request, when the occurrence rate is otherwise. The non-uniform request is indicated by the corresponding frequency of occurrence f, for each d. $\sum f = f_1 + f_2 + \cdots + f_{d\,\max} = 1$. We define the range size of request for each d, then we calculate the average size (($r_{(d)min} + r_{(d)max}$)/2). Similarly, for server service rate we determine the rate for each d, and find the average service rate $\bar{\mu}$. The formulas in Table 1 are employed to find the $\bar{d}, \bar{r}$, and $\bar{\mu}$ for uniform and non-uniform request.

We apply the formula in Table 1 in the analytical model of both broadcast channel and on-demand channel:

$$T_B \approx \frac{\left[\left(x_t - \bar{d}\right)^2 + \left(2 \times \bar{d} \times x_t\right) + \left(2 \times \bar{d} \times (x_t - \bar{d})\right) + (x_t - \bar{d})\right] \times s}{2 \times x_t \times b_s} \tag{9}$$

Table 1. Uniform and non-uniform formulas

Uniform Request	Non Uniform Request
$\bar{d} = \dfrac{\sum\limits_{i=1}^{d\,\max} i}{d\,\max}$	$\bar{d} = \dfrac{\sum\limits_{i=1}^{d\,\max} i \times f_i}{d\,\max}$
$\bar{r} = \dfrac{\sum\limits_{i=1}^{d\,\max} \dfrac{r_{i\min} + r_{i\max}}{2}}{d\,\max}$	$\bar{r} = \dfrac{\sum\limits_{i=}^{d\,\max} \left(\dfrac{r_{i\min} + r_{i\max}}{2}\right) \times f_i}{d\,\max}$
$\bar{\mu} = \dfrac{\sum\limits_{i=1}^{d\,\max} \mu_i}{d\,\max}$	$\bar{\mu} = \dfrac{\sum\limits_{i=1}^{d\,\max} \mu_i \times f_i}{d\,\max}$

Pre-determined number of request:

$$T_D \approx \frac{\overline{r}}{b_r} + \frac{\sum_{i=1}^{n} i \times \frac{1}{\overline{\mu}}}{n} + \frac{\overline{d} \times s}{b_s} \tag{10}$$

If the server is at light load $\rho = \left(\frac{\lambda}{\mu} \leq 1\right)$, then:

$$T_D \approx \left(\frac{\overline{r}}{b_r} + \frac{1}{\overline{\mu}} + \frac{\overline{d} \times s}{b_s}\right) \tag{11}$$

Else if the server at heavy load $\rho = \left(\frac{\lambda}{\mu} > 1\right)$, then:

$$T_D \approx \left(\frac{\overline{r}}{b_r}\right) + \frac{\sum_{i=1}^{q\,max} \left[\left(i \times \frac{1}{\overline{\mu}}\right) - \left[(i-1) \times \frac{1}{\lambda}\right]\right]}{q\,max} + \left(\frac{\overline{d} \times s}{b_s}\right) \tag{12}$$

Global Indexing Scheme

This indexing strategy is designed to accommodate multi data channels as well as minimizing the tuning time. Global index is designed based on $B+$-tree structure (Bayer and McCreight, 1972; Comer, 1979). It consists of non-leaf nodes, and leaf node. Leaf node is the bottom most index that consists of up to k keys, where each key point to actual data items, and each node has one node pointer to a right-side neighbouring leaf node. Unlike leaf node, non-leaf node may consist of up to k keys and $k+1$ pointers to the nodes on the next level on the tree hierarchy (i.e. child nodes). All child nodes, which are on the left-hand side of the parent node, have the key values less than or equal to the key of their parent node. On the other hand, keys of child nodes on the right-hand side of the parent node are greater than the key of their parent node.

Having all data pointers stored on the leaf nodes is considered better than storing data pointers in the non-leaf nodes like the original B trees (Elmasri and Navathe, 2003). Moreover, by having node pointers in the leaf level, it becomes possible to trace all leaf nodes from the left most to the right most nodes producing a sorted list of keys. When being broadcast, each physical pointer to the neighbouring leaf node as well as actual data item are replaced by a time value, which indicates when the leaf node or data item will be broadcast.

Our simple scenario is to broadcast weather condition for all cities in Australia. There are 30 cities altogether to be broadcast, and we assume the optimum number of cities in a data channel is 10. Subsequently, we end up with 3 data channels, each channel contains 10 cities. In this case, we use a table consisting of records of IDs, city, weather condition, and temperature as shown in Figure 7. The index is inserted based on the order of the data item in the table. Similarly, to construct our data channels,

Figure 7. Data channel structure and index using B+ tree

Index (B+ Tree):

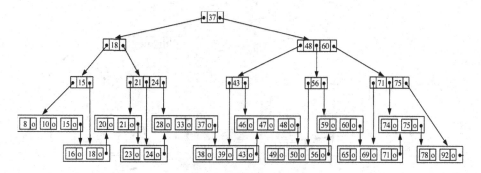

Table (ID, City, Weather Condition, Temperature):

Data Channel 1			
23	Adelaide	Mostly Sunny	24°C
65	Albany	Local Shower	15°C
37	Ballarat	A few showers	12°C
60	Barton	Mostly Fine	19°C
46	Beechworth	Shower or Two	14°C
92	Belmont	Partly Cloudy	10°C
48	Brisbane	Fine	17°C
71	Broome	Shower or two	15°C
56	Bunbury	Mostly sunny	20°C
59	Busselton	Shower	21°C

Data Channel 2			
18	Cairns	Shower	16°C
21	Canberra	Partly Cloudy	23°C
10	Cervantes	Fine	26°C
74	Darwin	A few shower	21°C
78	Gosford	Mostly fine	23°C
15	Grafton	Thunderstorm	21°C
16	Kingscote	Possible Hail	13°C
20	Kingston	Fine	15°C
24	Lismore	Partly cloudy	17°C
28	Maitland	Shower	15°C

Data Channel 3			
39	Melbourne	Mostly fine	25°C
43	Miranda	Thunderstorms and Hail	12°C
47	Newcastle	A few shower or two	15°C
50	Northbridge	Fine	25°C
69	Philip	Cloudy	26°C
75	Queensland	Fine	30°C
8	Queenstown	Mostly sunny	28°C
49	Surfers Paradise	Shower	22°C
33	Sydney	Cloudy	21°C
38	Torguay	Fine	23°C

the data item is placed based on the order of the table, and once it reaches optimum number of cities, a new data channel is created. The data channel is illustrated in Figure 7. Assume that in the index tree, the maximum number of node pointers from any non-leaf node is 4, and the maximum number of data pointers from any leaf node is 3.

Using the same example (shown in Figure 7), Global index is exhibited in Figure 8. In this example, we use the ID attribute as the index-partitioning attribute, which is different from the table partitioning. We assume the range partitioning rules used are that index channel 1 holds data IDs between 1 to 30, index channel 2 holds data IDs between 31 to 60, and the rest go to index channel 3. Notice from Figure 4 that the fifth leaf node (28, 33, 37) is replicated to channel 1 and 2 because key 28 belongs to index channel 1, while keys 33 and 37 belong to index channel 2. Also notice that some non-leaf nodes are replicated whereas others are not. For example, the non-leaf node 15 is not replicated and located only in index channel 1, whereas non-leaf node 18 is replicated to index channel 1 and 2. It is also clear that the root node is fully replicated. We must stress that the location of each leaf node is the not same as where the actual data is broadcast. Our index is broadcast separately with the data, and each index key points to the relevant data channel. Thus, once the right index is found in a specific index channel, mobile client switch to the right data channel and wait for the data of interest to arrive.

The index structure for our Global index employs a single node pointers model, which means each node pointer has only one outgoing node pointer.

Figure 8. Global index model

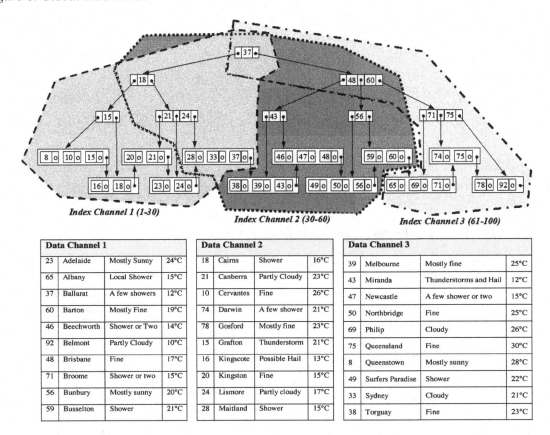

Index Channel 1 (1-30)

Index Channel 2 (30-60)

Index Channel 3 (61-100)

Data Channel 1			
23	Adelaide	Mostly Sunny	24°C
65	Albany	Local Shower	15°C
37	Ballarat	A few showers	12°C
60	Barton	Mostly Fine	19°C
46	Beechworth	Shower or Two	14°C
92	Belmont	Partly Cloudy	10°C
48	Brisbane	Fine	17°C
71	Broome	Shower or two	15°C
56	Bunbury	Mostly sunny	20°C
59	Busselton	Shower	21°C

Data Channel 2			
18	Cairns	Shower	16°C
21	Canberra	Partly Cloudy	23°C
10	Cervantes	Fine	26°C
74	Darwin	A few shower	21°C
78	Gosford	Mostly fine	23°C
15	Grafton	Thunderstorm	21°C
16	Kingscote	Possible Hail	13°C
20	Kingston	Fine	15°C
24	Lismore	Partly cloudy	17°C
28	Maitland	Shower	15°C

Data Channel 3			
39	Melbourne	Mostly fine	25°C
43	Miranda	Thunderstorms and Hail	12°C
47	Newcastle	A few shower or two	15°C
50	Northbridge	Fine	25°C
69	Philip	Cloudy	26°C
75	Queensland	Fine	30°C
8	Queenstown	Mostly sunny	28°C
49	Surfers Paradise	Shower	22°C
33	Sydney	Cloudy	21°C
38	Torguay	Fine	23°C

If a child node exists locally, the node pointer points to this local node only, even when this child node also replicated to other index channels. For example, from node 37 at index channel 1, there is only one node pointer to the local node 18. The child node 18 at index channel 2 will not receive an incoming node pointer from the root node 37 at index channel 1; instead it will receive one node pointer from the local root node 37 only.

If a child node does not exist locally, the node pointer will choose one node pointer pointing to the nearest child node (in case if multiple child nodes exist somewhere else). For example, from the root node 37 at index channel 1, there is only one outgoing right node pointer to child node (48,60) at index channel 2. In this case, we assume that index channel 2 is the nearest neighbour of index channel 3. The child node (48,60), which also exists at index channel 3, will not receive a node pointer from root node 37 at index channel 1.

Using this single node pointer model, it is always possible to trace a node from any parent node. For example, it is possible to trace to node (71,75) from the root node 37 at index channel 1, although there is no direct link from root node 37 at index channel 1 to its direct child node (48,60) at index channel 3. Tracing to node (71,75) can still be done through node (48,60) at index channel 2.

A more formal proof for the single node pointer model is as follows. First, given a parent node is replicated when its child nodes are scattered at multiple locations, there is always a direct link from whichever copy of this parent node to any of its child nodes. Second, using the same methodology as the first statement above, given a replicated grandparent node, there is always a direct link from whichever copy of this grandparent node to any of the parent nodes. Considering the first and the second statements above, we can conclude that there is always a direct link from whichever copy of the grandparent node to any of its child nodes.

Figure 9 shows an example of a single node pointer model. It only shows the top three levels of the index tree exhibited previously in Figure 8.

Data retrieval mechanism in this scheme can be described as follows:

- Mobile client tunes in one of the index channel (i.e. can be any index channel).
- Mobile client follows the index pointer to the right index key. The pointer may lead to another index channel that contains the relevant index. While waiting for the index to arrive, mobile clients can switch to 'doze' mode.
- Mobile client tunes back on at the right index key, which point to the data channel that contains the desired data item. It indicates a time value of the data to arrive in the data channel.
- Mobile client tunes into the relevant data channel, and switch back to 'doze' mode while waiting for the data item to come.
- Mobile client switches back to active mode just before the desired data item arrives, then retrieves the information.

Global Index Maintenance

In this section, we describe the process for maintaining Global Index. The maintenance refers to the Global index restructuring process which includes an insertion and deletion of index node. This maintenance is not regularly required as all data items are broadcasted and made available for mobile clients to obtain whatever data items they are interested in over wireless channel without having to send any queries to the server. This process is only necessary to carry out whenever there are new data inserted in the databases. Whilst the index structure is being reconstructed, the server can keep broadcasting the earlier version of the broadcast structure until the new structure is ready to be applied.

Figure 9. Single node pointers model

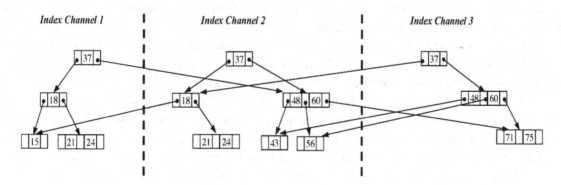

Node Insertion

When a new key (record) is inserted to the global index, the Global index needs restructuring. The following example illustrates the index restructuring process. It is an insertion of entry 21 to the existing index. In this example, we show three stages of index insertion process. The stages are (*i*) the initial index tree and the desired insertion of the new entry to the existing index tree, (*ii*) the splitting node mechanism, and (*iii*) the restructuring of the index tree.

The initial index tree position is shown in Figure 10(a). When a new entry of 21 is inserted, the first leaf node becomes overflow. A split of the overflow leaf node is then carried out. The split action also causes the non-leaf parent node to be overflow, and subsequently, a further split must be performed to the parent node (see Figure 10(b)).

Notice that when splitting the leaf node, the two split leaf nodes are replicated to index channel 1 and 2, although the first leaf node after the split contains entries of the first channel only (18 and 21 – the range of index channel 1 is 1-30). This is because the original leaf node (18, 23, 37) is already replicated to both channels 1 and 2. The two new leaf nodes have a node pointer linking them together.

When splitting the non-leaf node (37, 48, 60) into two non-leaf nodes (21; and 48, 60), index channel 3 is involved because the root node is replicated to channel 3 too. The final step is the restructuring step. This step is necessary because we need to ensure that each node has been allocated to the correct index channels. Figure 10(c)) shows a restructuring process. In this restructuring, index allocation is updated. This is done by performing an in order traversal of the tree, finding the range of the node (min, max), determining the correct index channel(s), and reallocating to the designated channel(s). When reallocating the nodes to channel(s), each channel will also update the node pointers, pointing to its local or neighbouring child nodes. Notice that in the example, as a result of the restructuring, the leaf node (18, 21) is now located in channel 1 only (instead of channel 1 and 2).

Node Deletion

To illustrate how node deletion works, consider the following example. We would like to delete entry 21, expecting to get the original tree structure shown previously before entry 21 is inserted. Figure 11 shows the current tree structure, the merge and collapse processes.

As shown in Figure 11(a), after the deletion of entry 21, the leaf node (18) becomes underflow. A merging with its sibling leaf node is necessary to be carried out. When merging two nodes, the index channel(s) which own the new node are the union of all channels owning the two old nodes. In this case, since node (18) is located in channel 1 and node (23, 37) is in channel 1 and 2, the new merged node (18, 23, 37) should be located in channel 1 and 2. Also as a consequent to the merging, the immediate non-leaf parent node entry has to be modified to identify the maximum value of the leaf node, which is now 37, not 21. As shown in Figure 11(b), the right node pointer of the non-leaf parent node (37) becomes void. Because the non-leaf node (37) has the same entry with its parent node (root node (37)), they have to be collapsed together, and consequently a new non-leaf node (37, 48, 60) is formed (see Figure 11(c)).

The restructuring process is the same as that in the insertion process. In this example, however, index channel allocation has been done appropriately and hence a restructuring is not needed.

Figure 10. Index entry insertion

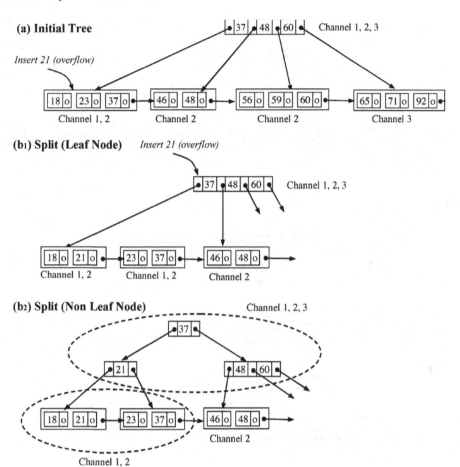

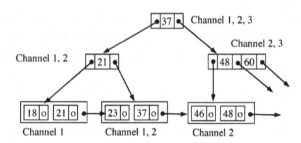

PERFORMANCE EVALUATIONS

This section evaluates the performance of the proposed method. First, we start with analysing the analytical model by comparing the result with the simulation model that we develop. Next, we find out the query access time using our strategy as compared to conventional technique. The simulation is carried out using a simulation package *Planimate*, animated planning platforms (Seeley, 1997).

Figure 11. Index entry deletion

(a) Initial Tree

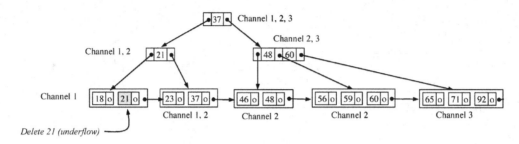

(b) Merge

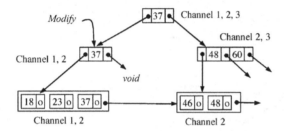

(c) Collapse

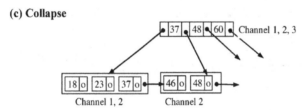

Validation of Analytical Model

In this section, we introduce three cases, and employ the analytical model to determine the optimum number of broadcast data items. We develop some simulations to verify the results of our analytical model.

Simulation Setup

The simulation environment is set to apply random delays for both arrival rate and server service time with given average value. In the simulation model, we incorporate three facilities as multiclient, we run the simulation process for five iteration times, and derive the average result accordingly. In each case, we consider request that return one or more number of items. Subsequently, we classify the request into two scenarios, one is uniform request and the other is non-uniform request. Uniform request is when the frequency of occurrence for all type of request is equal. Whilst, non-uniform request is when the frequency of occurrence varies. Table 2 and 3 contains set of parameter of concerns. The bandwidth values are determined according to common transmission rates for packet-switching in GSM cellular network (Pitoura, and Samaras, 1998).

Table 3. Set of parameters for uniform and non-uniform request

Number of Returned Data Items (d)	1	2	3	4
Non-Uniform Request				
Frequency of Occurrence (f)	0.4	0.3	0.2	0.1
Uniform Request				
Frequency of Occurrence (f)	0.25	0.25	0.25	0.25
Non-Uniform and Uniform Request				
Size of Request (r)	50-100 bytes	101-150 bytes	151-200 bytes	201-250 bytes
Service Rate (μ)	4 request per sec	3 request per sec	2 request per sec	1 request per sec

Table 2. Parameters of concern

Parameter	Description	Initial Value
Broadcast channel		
x_t	*Total number of Broadcast Items*	450
On demand channel		
n	Number of request (pre-determined number)	50
b_r	*Uplink Bandwidth*	9600 bytes
Broadcast and On demand channel		
b_s	*Downlink Bandwidth*	19200 bytes
s	Size of each data item	1000 bytes

Performance Results

Case 1: To find the optimum number of broadcast items with pre-determined number of request (uniform and non-uniform request). This case is categorized into pre-determined number of requests with uniform and non-uniform request. As shown in Figure 12, we try to find the cross line of on-demand channel and broadcast channel. The average access time of on-demand channel is the threshold point, which appears in a straight line since the number of broadcast items does not affect its value. The intersection point indicates the optimum number of broadcast items in a channel.

From Table 4, we can see quite clearly (please refer to section c for accurate results) the optimum number of broadcast item with uniform request, that is between 390-400 data items (analytical result), and 400-410 data items (simulation result). Similarly, from Table 5, we can find the optimum broadcast items for non-uniform request.

Case 2: Introducing a new parameter, arrival rate of request (λ). We specify $\lambda = 1$ per two seconds, to obtain server utilization. As shown in Figure 13, the on-demand channel performs well when the traffic request is low. The broadcast channel seems to have to allocate its data items in every 10-15 data items to a new channel to keep up with the performance of on demand channel. However,

Figure 12. Optimum number of broadcast items with pre-determined number of request (uniform and non-uniform request)

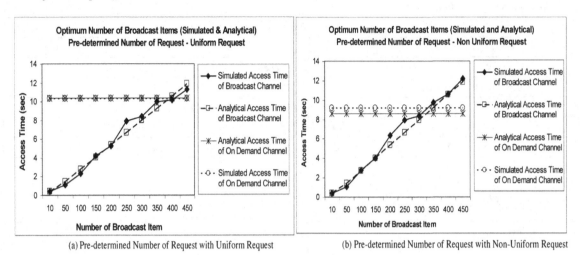

(a) Pre-determined Number of Request with Uniform Request   (b) Pre-determined Number of Request with Non-Uniform Request

Table 4. Average access time of on demand and broadcast channel (pre-determined number of request with uniform request)

Access Time of Broadcast Channel and On Demand Channel (Simulated & Analytical)
Pre-determined Number of Request – Uniform Request

On-Demand Channel		Broadcast Channel		
Analytical Average Access Time (sec)	Simulated Average Access Time (sec)	Number of Broadcast Items	Analytical Average Access Time (sec)	Simulated Average Access Time (sec)
10.3458	10.3013	10	0.394	0.396
		50	1.454	1.0991
		100	2.758	2.326
		150	4.061	4.183
		200	5.363	5.279
		250	6.666	7.8675
		300	7.968	8.373
		350	9.270	9.943
		400	10.572	10.173
		450	11.874	11.295

this is not a good case since the number of channels may be excessive and the bandwidth may be too much wasted.

Case 3: In this case, we specify request arrival rate, $\lambda = 4$ per second to obtain server utilization $\rho > 1$. Figure 14 denotes optimum number of items when the server is at heavy load. Figure 14 (a) shows the optimum number of broadcast item with uniform request, which is approximately 300 data items from both analytical and simulation result. However, there is a gap between the analytical and simulated result in Figure 14 (b) when involving non-uniform request. The explanation for this might be due to the exponential distribution of arrival rate affect the result quite substantially to the on-demand channel especially when the server is overloaded in processing non-uniform request.

Table 5. Average access time of on demand and broadcast channel (pre-determined number of request with non-uniform request)

Access Time of Broadcast Channel and On Demand Channel (Simulated & Analytical)
Pre-determined Number of Request – Non Uniform Request

On-Demand Channel		Broadcast Channel		
Analytical Average Access Time (sec)	Simulated Average Access Time (sec)	Number of Broadcast Items	Analytical Average Access Time (sec)	Simulated Average Access Time (sec)
8.617	9.196	10	0.375	0.396
		50	1.429	1.0466
		100	2.733	2.794
		150	4.035	4.0075
		200	5.338	6.339
		250	6.640	7.925
		300	7.942	8.3725
		350	9.244	9.725
		400	10.546	10.63
		450	11.849	12.189

Figure 13. Optimum number of broadcast items with server utilisation (uniform and non-uniform request)

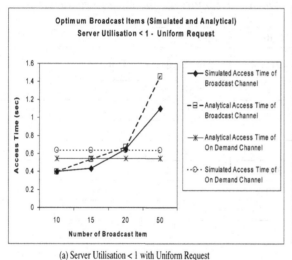 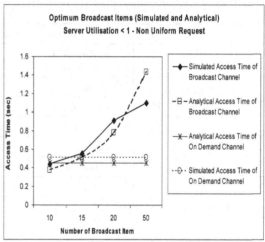

(a) Server Utilisation < 1 with Uniform Request   (b) Server Utilisation < 1 with Non Uniform Request

Performance Analysis

Table 6 shows the optimum number of broadcast items derived from our analytical and simulation results on the three cases. Our analytical calculation differ about 5 percent average from the simulation. It mostly presents a less value than the simulation. Thus, the analytical models are considered very close to accurate. Subsequently, having known the optimum number of items in a broadcast channel, we can decide how many channels are required to broadcast a certain amount of data items. Consequently, the average access time of broadcast channel is minimized.

Figure 14. Optimum number of broadcast items with server utilisation (uniform and non-uniform request)

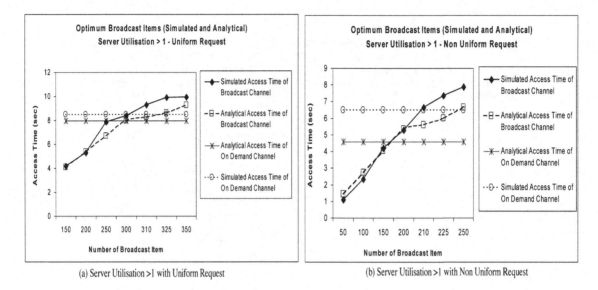

(a) Server Utilisation >1 with Uniform Request   (b) Server Utilisation >1 with Non Uniform Request

Table 6. Simulation and analytical performance in determining optimum number of broadcast items

		Optimum Number of Broadcast Items				
		Uniform/Non Uniform	Analytical	Simulated	Difference	Error Rate
Case 1	Pre-determined Number of Request	Uniform	391	400	9	2.3%
		Non Uniform	325	311	14	4.5%
Case 2	Server Utilization < 1	Uniform	15	16	1	6.2%
		Non Uniform	12	12	0	0%
Case 3	Server Utilization > 1	Uniform	300	305	5	1.6%
		Non Uniform	170	205	35	17.1%
Average Error Rate (per case basis): 5.28%						

Proposed vs. Conventional Method

In this section, we analyse the query access time performance of the proposed method as compared to conventional method. We refer conventional method as widely used broadcasting scheme that integrate index and data into a single channel as applied in (Imielinski, Viswanathan and Badrinath, 1997). For comparison purposes, we use the weather scenario as illustrated in Figure 7, and use the same set of data items. Assuming we found the optimum number of broadcast item is 10 items per channel, and we use the same index partitioning attribute as in Figure 7. The model of the proposed method looks exactly as in Figure 8. The indexing scheme in the conventional method is designed to employ B+-tree structure.

We introduce three cases: (i) to retrieve a single data item, (ii) to retrieve two data items where both items are in the same channel, and (iii) to retrieve two data items where the item can be in the same channel and different channel. We run the simulation model for fifty iteration times, and calculate the average access time for given number of request: 5, 10, 15, 20 and 25 requests. The simulation environment is set to apply random delays for broadcasting rate of data item and index periodically with given an average value. The parameters of concern are given in Table 7.

Case 1: To compare the performance of proposed method and conventional method in the context of single data item retrieval. Figure 15(a) indicates the result of the simulation. It shows that the proposed optimisation strategy outperforms the conventional method with about two and a half time lower access time.

Case 2: To compare the performance of proposed method and conventional method in the context of two data items retrieval. It is designed that both item are located in the same channel. We found from Figure 15 (b) that the proposed strategy still provides approximately two times lower access time as compared to conventional approach.

Case 3: To compare the performance of proposed method and conventional method in the context of two data items retrieval. With regard to the proposed method, we design in a way so that the two items are located in different channel. For each number of requests, half of it find the two data items in the same channel location, while the other half in a different channel. Since the proposed method incorporates three broadcast channel, the location of two data items in two different channels is equally distributed. We can see from Figure 15 (c), our strategy offers a better access time with about one and a half time lower access time as compared to conventional method.

Table 7. Parameters of Concern

Parameters	Value
Index Channel	
Global Index	
Index page in Channel 1	17
Index page in Channel 2	22
Index page in Channel 3	12
Non-Global Index (single channel)	
Index page	41
Global and Non-Global Index	
Node Pointer Size	5 bytes
Data Pointer Size	5 bytes
Indexed Attribute Size	4 bytes
Data Channel	
Total number of Broadcast Items	30
Size of each data item	1000 bytes
Data and Index Channel	
Data item and index arrival rate	4 per sec
Bandwidth	19200 bytes

Figure 15. Optimized vs conventional strategy - one and two data item(s) retrieval

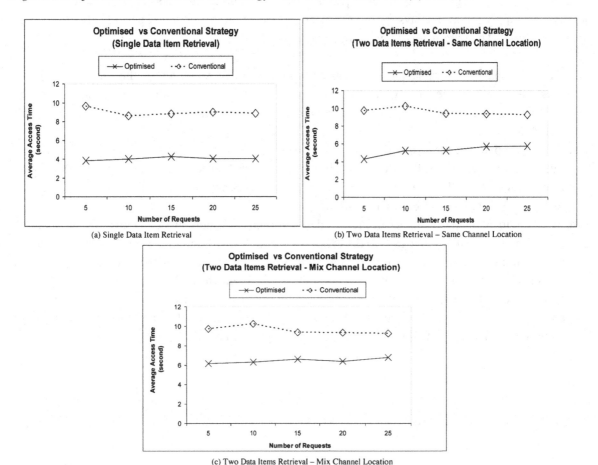

(a) Single Data Item Retrieval   (b) Two Data Items Retrieval – Same Channel Location

(c) Two Data Items Retrieval – Mix Channel Location

Remarks

The proposed approach optimizes the number of broadcast channel required to disseminate a set of database items. This is 'optimized' in a sense that the number of channel is determined upon knowing the query access time over on-demand channel. Therefore, the number of broadcast channel can be regarded as adaptively dependent on the access time of on-demand channel. In this way, the scalability is maintained and mobile client does not have to transmit any request which result in efficient power consumption.

It should be noted that in our cases, when more than one data items are returned as a result from a request, the first data item that appear in the channel is retrieved first. Thus, there is no specific order in obtaining the desired data items. Furthermore, although the advantage of using multi broadcast channel is obvious, exhaustive number of channels may also cause disadvantages such as waste of bandwidth and resources. Therefore, when there is large number of data item to broadcast, the only way to outperform on-demand channel is by allocating a very small percentage of the broadcast data into each channel. However, an on-demand channel may be considered a better scheme due to an enormous number of channel may be required for data broadcast.

CONCLUSION

Data dissemination or known as data broadcasting strategy is an effective way to keep up with number of clients in a cell and their frequency of requests. To maintain the performance of broadcast strategy over a large set of broadcast items, we introduce optimal broadcast channel to determine the optimum number of database items to be broadcast in a channel. The optimum number is used as an indication of when the broadcast cycle needs to be split into different channel, forming multiple channels.

We have presented analytical models for both broadcast and on-demand channel to find the average query access time over the channel. To locate the optimum value of broadcast items, we utilize the access time of on-demand channel as a threshold point. The analytical model of broadcast channel is calculated based on variety of factors like number of data items to be broadcast, number of data item to be retrieved by each request, size of data item, and bandwidth. As for on-demand channel, it takes into account the server service rate, request arrival rate, number of requests, size of request, number of data item to be retrieved by each request, uplink and downlink bandwidth. The optimum number of broadcast items is changed dynamically depending on the value of these factors.

We have also presented Global Index to assist data dissemination in multi broadcast channels. The use of broadcast indexing is to reduce the tuning time by providing accurate information for a client to tune in at the appropriate time for the required data.

We have conducted simulation-based experiments to see the performance of our analytical model in determining the optimum broadcast items. It is found that our analytical model is considerably very close to accurate as compared to the simulation results. Following the verification of the analytical model, we evaluate the performance of the proposed strategy as compared to conventional one. Conventional strategy is widely used broadcasting scheme which incorporates single channel to broadcast the entire index and data segments. We found our proposed strategy provides a substantial lower access time.

REFERENCES

Acharya, S., Alonso, R., Franklin, M., & Zdonik, S. (1995). Broadcast Disks: Data Management for Asymmetric Communication Environments. *Proc. of ACM Sigmod International Conference on Management of Data*. 10.1145/223784.223816

Aksoy, D., Altinel, M., Bose, R., Cetintemel, U., Franklin, M., Wang, J., & Zdonik, S. (1999). Research in Data Broadcast and Dissemination. *LNCS*, *1554*, 194–207.

Badrinath, B. R., & Phatak, S. H. (n.d.). *An Architecture for Mobile Databases*. Technical Report DCS-TR-351. Department of Computer Science, Rutgers University.

Barbara, D. (1999). Mobile Computing and Databases-A Survey. *IEEE TKDE*, *11*(1), 108–117.

Bayer, R., & McCreight, E. M. (1972). Organization and Maintenance of Large Ordered Indices. *Acta Informatica*, *1*(3), 173–189. doi:10.1007/BF00288683

Comer, D. (1979). The Ubiquitous B-Trees. *ACM Computing Surveys*, *11*(2), 121–137. doi:10.1145/356770.356776

Elmasri, R., & Navathe, S. B. (2003). *Fundamentals of Database Systems* (4th ed.). Addison Wesley.

Hu, Q., Lee, D. L., & Lee, W. C. (1998). Optimal Channel Allocation for Data Dissemination in Mobile Computing Environments. *Proc. of 18th International Conference on Distributed Computing Systems*, 480-487.

Hu, Q., Lee, W. C., & Lee, D. L. (1999). Indexing Techniques for Wireless Data Broadcast under Data Clustering and Scheduling. *Proc. of the 8th ACM International Conference on Information and Knowledge Management*, 351-358. 10.1145/319950.320027

Huang, J. L., & Chen, M.-S. (2002). Dependent Data Broadcasting for Unordered Queries in a Multiple Channel Mobile Environment. *Proc. of the IEEE GLOBECOM*, 972-976.

Huang, J. L., & Chen, M.-S. (2003). Broadcast Program Generation for Unordered Queries with Data Replication. *Proc. of the 8th ACM Symposium on Applied Computing*, 866-870. 10.1145/952532.952704

Huang, Y., Sistla, P., & Wolfson, O. (1994). Data Replication for Mobile Computers. *Proc. of the ACM SIGMOD*, 13-24.

Imielinski, T., & Viswanathan, S. (1994). Adaptive Wireless Information Systems. *Proc. of SIGDBS (Special Interest Group in Database Systems)*, 19-41.

Imielinski, T., Viswanathan, S., & Badrinath, B. R. (1997). Data on Air: Organisation and Access. *IEEE TKDE, 9*(3), 353–371.

Lee, G., & Lo, S.-C. (2003). Broadcast Data Allocation for Efficient Access on Multiple Data Items in Mobile Environments. *Mobile Networks and Applications, 8*(4), 365–375. doi:10.1023/A:1024579512792

Lee, W. C., Hu, Q., & Lee, D. L. (1997). Channel Allocation Methods for Data Dissemination in Mobile Computing Environments. *Proc. of the 6th IEEE High Performance Distributed Computing*, 274-281. 10.1109/HPDC.1997.626430

Leong, H. V., & Si, A. (1995). Data Broadcasting Strategies Over Multiple Unreliable Wireless Channels. *Proc. of the 4th International Conference on Information and Knowledge Management*, 96-104. 10.1145/221270.221339

Leong, H. V., & Si, A. (1997). Database Caching Over the Air-Storage. *The Computer Journal, 40*(7), 401–415. doi:10.1093/comjnl/40.7.401

Paulson, L. D. (2003). Will Fuel Cells Replace Batteries in Mobile Devices? *IEEE Computer Magazine, 36*(11), 10–12. doi:10.1109/MC.2003.1244525

Pitoura, E., & Samaras, G. (1998). *Data Management for Mobile Computing*. London: Kluwer Academic Publishers. doi:10.1007/978-1-4615-5527-8

Seeley, D. (1997). *Planimate^tm-Animated Planning Platforms*. InterDynamics Pty Ltd.

Si, A., & Leong, H. V. (1999). Query Optimization for Broadcast Database. *Data & Knowledge Engineering, 29*(3), 351–380. doi:10.1016/S0169-023X(98)00040-8

Waluyo, A. B., Srinivasan, B., & Taniar, D. (2003). Global Index for Multi Channels Data Dissemination in Mobile Databases. *Proc. of the 18th International Symposium on Computer and Information Sciences (ISCIS'03)* (vol. 2869, pp. 210-217). Springer.

Waluyo, A. B., Srinivasan, B., & Taniar, D. (2004). Optimising Query Access Time over Broadcast Channel in Mobile Databases. In Lecture Notes in Computer Science: Vol. 3207. *Proc. of the Embedded and Ubiquitous Computing* (pp. 439–449). Springer-Verlag. doi:10.1007/978-3-540-30121-9_42

Zaslavsky, A., & Tari, S. (1998). Mobile Computing: Overview and Current Status. *Australian Computer Journal*, *30*(2), 42–52.

Section 3
Mobile Data Visualization

Chapter 9
Visualization–Driven Approach to Fraud Detection in the Mobile Money Transfer Services

Evgenia Novikova
SPIIRAS, Russia & St. Petersburg Electrotechnical University "LETI", Russia

Igor Kotenko
SPIIRAS, Russia & St. Petersburg ITMO University, Russia

ABSTRACT

Mobile money transfer services (MMTS) are widely spread in the countries lacking conventional financial institutions. Like traditional financial systems they can be used to implement financial frauds. The chapter presents a novel visualization-driven approach to detection of the fraudulent activity in the MMTS. It consists in usage of a set of interactive visualization models supported by outlier detection techniques allowing to construct comprehensive view on the MMTS subscriber behavior according to his/her transaction activity. The key element of the approach is the RadViz visualization that helps to identify groups with similar behavior and outliers. The scatter plot visualization of the time intervals with transaction activity supported by the heat map visualization of the historical activity of the MMTS subscriber is used to conduct analysis of how the MMTS users' transaction activity changes over time and detect sudden changes in it. The results of the efficiency evaluation of the developed visualization-driven approach are discussed.

INTRODUCTION

Firstly introduced in 2000 in Philippines, mobile money, the electronic cash card associated with a mobile phone account, has gained a solid market segment especially in the developing countries. Currently the mobile money transfer services (MMTS) are available in over 80 countries and work on smartphones and basic feature phones offering a good alternative to bank accounts. For example, M-PESA, mobile

DOI: 10.4018/978-1-5225-5693-0.ch009

money transfer service supported by Safaricom, Kenya's largest mobile network operator is arguably the most developed mobile money transfer system in the world. It has over 18.2 million subscribers in a country of 43.2 million (Kamana, 2014). Other M-PESA markets such as Gabon, Ghana, Kenya, Namibia, Tanzania, India and Romania have more than 40% of active mobile money users. (Corkin, 2016). Another popular mobile money transfer service is Orange Money that is deployed in 10 countries and gathers around 14% of the mobile subscribers of these countries (Orange Money, 2017).

In general case, the MMTS allows users to deposit money into an account stored on their cell phones and called mWallet, to transfer mobile money to other users, to accept payments for goods and services from the customers via their mobile phones, and to withdraw deposits for regular money. People in Kenya can now also use M-Pesa to top up cashless travel cards for public transport (Kamana, 2014). Users are charged a small fee for sending and withdrawing money using the service. In such services, transactions are made with electronic money, called mMoney (Merrit, 2010).

The usage of the MMTS from one side makes business owners less susceptible to the risks of handling cash, such as theft and fake currency. From the other side, the mobile financial services create novel opportunities for fraud (CGAP Report, 2017). For example, the MMTS provide higher level of anonymity, speed and portability compared to traditional banking systems. Thus, additional risks caused by the large number of non-bank participants are introduced (Merrit, 2010). Therefore, it is required to determine new approaches to detect frauds in mobile money transfer services.

In this chapter the authors present *a novel visualization-driven approach to the analysis of the MMTS subscribers' transactions* that can assist in anomaly detection. It allows an analyst to form overall understanding of the MMTS users' activity by providing a metaphoric presentation of their behavior in the system and then focus on users of the particular interest by drilling down into their contacts and transactional activity. Its key elements are (1) *the RadViz visualization* (Ankerst et al., 1998) of the MMTS users that helps to determine groups of similarities and outliers among them, (2) *the scatter plot presentation* of the time periods with the subscriber's transactional activity targeted to explore suspicious changes in it supported by the graph-based presentation of the subscriber's contacts and (3) *the heat map presentation* of the transaction attributes used to form temporal profile of the user's activity.

The main contribution of the authors is the interactive visual representation of the *MMTS subscribers allowing detection of the groups of users with similar behavior and outliers*. To the best of authors' knowledge, this work is the first to exploit the RadViz visualization technique to visualize MMTS subscribers and the scatter plot presentation to analyze transactional activity of the particular user. This chapter is extended version of the paper (Novikova et al, 2014). It contains detailed description of the proposed approach, including its enhancement targeted to detect short-term types of the behavior frauds (such as theft of mobile phones), suggested analysis workflow as well as discussion of the introduced modification and its influence on the efficiency of the proposed approach.

The rest of the chapter is structured as follows. Section "SUBJECT AREA AND RELATED WORK" presents the overview of the mobile money transfer service, its structure and the related work. Section "THE PROPOSED VISUALIZATION-DRIVEN APPROACH" describes the approach suggested, including the analysis workflow and its key elements. Section "USAGE SCENARIOS" outlines the case studies used to demonstrate the proposed approach for fraud detection in mobile payments. Section "EFFICIENCY EVALUATION AND DISCUSSION" considers the results of the efficiency evaluation. Final sections define the directions of the future research, and sum up the authors' contribution.

SUBJECT AREA AND RELATED WORK

MMTS Use Case

Mobile Money Transfer Services, also known as mobile payment, mobile money, and mobile wallet, generally denote to money transfer and micro financing services performed via a mobile device operated by mobile network operator under established financial regulation (Merrit, 2010). In the developing countries, such solution allows compensating the lack of easy access to banking institutions.

This paper is based on the MMTS use case detailed in (Achemlal et al., 2011; Jack et al., 2010). The major points of it are outlined below. The MMTS are managed by a mobile network operator (MNO). The MNO not only provides infrastructure to financial services but emits mobile money, known as *mMoney*, in partnership with a financial institution. The *mMoney* can only be used by subscribers of the MMTS. They enable their users to deposit and withdraw money, transfer money to other users and non-users, pay bills, purchase airtime, and transfer money between service subscribers. Subscribers can have different roles, i.e. retailers, merchants, end users. They hold a prepaid account known as *mWallet*, stored on a platform and accessible via the MNO's network and an application on their mobile device. Some users, such as retailers or service providers, can use computers to access their account. This account contains *mMoney* which can be acquired from the retailers. End-users can either transfer money to other end-users or purchase goods.

Fraud Detection Techniques

The most widely deployed tools used to detect financial frauds are based on rules, linear regression and neural networks. For example, in the Kenyan MMT service M-Pesa (M-Pesa, 2015), fraud detection is carried out by the Minotaur tool which uses neural networks (Neural-technologies, 2017; Cowan, 2012).

Rieke et al. (2013) suggested an approach for financial fraud detection in MMTS based on checking conformance of the current process to a model one using event data. Zhdanova et al. (2014) extended the proposed model-based approach to event-driven detection of the money laundering technique known as micro-structuring of funds or smurfing. In (Coppolino et al., 2015) a fraud detection system based on extension of the Dempster-Shafer theory of evidence is proposed. It allows correlating evidence provided by multiple information sources and computing a belief value to decide whether the transaction is fraudulent. The authors focused on account takeover in MMT services that is one of the most prominent frauds in MMTS.

In other payment systems, several fraud detection techniques have been applied. For example, in the field of credit card fraud detection and market manipulation neural networks, expert systems, case-based reasoning, genetic algorithms, regression, Bayesian networks, decision trees and Hidden Markov Models are used (Al-Khatib, 2012; Golmohammadi et al., 2014; Mohiuddin et al., 2016).

Application of automated analysis techniques for fraud detection assumes that data are clearly structured, complete, correct, does not change over time, and the problem is well defined (Keim et al., 2008). The real life data rarely meet these preconditions. Moreover, in the major cases the automated analysis models are seen as black boxes by the users of the financial fraud detection systems as they do not provide any explanation to the produced results. Leite et al. (2016) highlighted a need to develop visual analytics frameworks for financial fraud detection. The data visualization acts as a mean for the hypothesis generation and verification process that is intuitively clear and does not require explicit application of complex statistical methods (Keim et al., 2008; Kotenko&Novikova, 2013; Novikova&Kotenko, 2013).

Visualization Techniques for Fraud Detection

Due to the complexity of financial data (often with multidimensional attributes), many sophisticated visualization and interaction techniques supporting visually decision making have been proposed, which (Marghescu, 2007-1). Parallel coordinates, scatter-plot matrices, survey plots, special glyphs (Schreck et al., 2007), treemaps and 3D treemaps (Huang et al., 2009), stacked and iconic displays, dense pixel-displays (Ziegler et al., 2010), dendrograms, fish-eye views (Lin et al., 2005) are applied to explore financial data. In the most cases they support such financial tasks as analysis of the financial market as a whole, or single assets in particular, estimation of financial performance of companies, identification of a particular stock producing an unusual trading pattern.

The visualization techniques applied for fraud detection in the commercial fraud detection systems are often limited to the standard 2-dimensional models such as trends, pie charts and histograms and gauge-based glyphs to display characteristics of financial flows, number of registered alerts, their type and criticality, etc. (Neural-technologies, 2017; Fiserv, 2017; Nice Actimize, 2017; SAS, 2017). They are suitable for real-time monitoring as they can communicate the most important information at glance and can be easily included in the reports of any level and purpose. Apart from the standard visual models, geographical maps are often present in fraud detection systems as they allow detecting regions with high financial risk level as well as determining the limits of organization responsibility. Different financial metrics are usually encoded by color or specific icon (Fiserv, 2015; Nice Actimize, 2015). The identification of the hidden relationships between entities based on data from multiple sources, and tracking of the money flows are usually done using tabular methods often supported by the graph-based representation of the financial entities (Nice Actimize, 2017; SAS, 2015). The graph vertexes represent different entities such as accounts, user IDs, phones, credit cards, addresses, organizations, etc usually. The edges between them indicate the usage or participation of the corresponding entity in financial operations, and the line thickness displays the frequency of the transactions between entities. The graph-based presentation of transaction activity helps to discover connections between customers and identify suspicious communication patterns.

Korczak&Łuszczyk (2011) address the problem of graphical representation of sequential financial operations in readable manner. Exploration of transaction chains assists analysts to detect money laundering operations. However, the major concern when designing a visualization algorithm of sequential operations is the complexity of the resulting graph. In order to solve this problem the authors propose the evolutionary algorithm that minimizes the number of edge intersections. In (McGinn, D et al., 2016) the force-directed graph visualization is used to reveal the structure of the algorithmic denial of service attack on the Bitcoin system and disclose some previously hidden insights into the multiple distinct phases of such attack. Di Battista et al (2015) propose a high level visual metaphor of the Bitcoin transactions based on their attributes. The application of the metaphor reduces the potential cluttering generated by standard node-link diagrams, and thus enables better comprehension of flows.

Xie et al. (2014) developed VAET – a visual analytics system for analyzing electronic transactions and discovering temporal trends. A probabilistic decision tree learner is used to estimate a specific saliency value for each transaction which describes relevance of the transaction for a certain analysis task, for example, detection of the suspicious transactions. The variation of this measure over time calculated for a specific group of transactions is displayed using pixel-oriented display. The detailed information on a selected sequence of transactions is outlined using new visual metaphor called KnotLines. In this view, lines reveal the connections among transactions while knots encode the detailed information on the associated transactions.

Chang et al. (2007) present the WireVis tool for the analysis of financial wire transactions for fraud detection. It is based on transaction keyword analysis and built in collaboration with the Bank of America. All the textual elements contained in transaction data records are seen as keywords. WireVis uses a multi-view approach including a keyword heat map and a keyword network graph to visualize the relationships among accounts, time and keywords within wire transactions. The keyword heatmap characterizes the usage frequency of the keyword in the group of users. Authors suggest an interesting modification of the clustering algorithm applied to form groups of similarities. They treat each account as a point in k-dimensional space (where k is the number of keywords), and group the accounts based on their distances to the average point of all accounts. This approach significantly decreases the complexity of the clustering procedure having complexity $O(3n)$. In order to support the visualization of transaction activity over time, authors propose the Strings and Beads view in which the strings refer to the accounts or cluster of accounts over time, and the beads refer to specific transactions on a given day.

Leite et al. (2016) propose a visual analytics approach to the analysis of transactions of the customer that reduces false positive alarms. The authors construct a statistical user profile based on how often the costumer executes operations, how much money he/she usually transfers, and what the geographic locations are involved. Lately such profiles are used to evaluate new transactions. The automated analysis of the transactions is supported by a set of linked views consisting of parallel coordinates, a matrix of scatter plots and a stacked bar. The parallel coordinates and scatter plots are used to represent metrics calculated for each transaction and to visualize the fraudulent level of the transaction.

To the authors' knowledge, there are no publicly available works concerning the study and the adaptation of visual analytics fraud detection methods to mobile payment systems use case. Therefore, the authors cannot easily compare the proposed approach work to existing systems.

THE PROPOSED VISUALIZATION-DRIVEN APPROACH

The MMTS Data

The data from the existing MMTS are not publicly available and in the most cases confidential. The possible solution of the lack of real world MMTS data, necessary for developing and evaluating visual analytics models and techniques, is the usage of artificially generated data. This approach is widely used to train automatic fraud detection techniques based on pattern recognition and machine learning (Gaber et al., 2013). In this work the MMTS log synthetic simulator is used. It models the mobile money transfer platform and simulates legitimate and fraudulent transactions. The MMTS log synthetic simulator was developed within European FP7 project MASSIF and can be used to generate test data containing different fraudulent scenarios. The simulated transactions are based on the properties of the real world transaction events. They also contain a special *ground proof* field which could be used to validate the results obtained during the analysis process.

The generated logs are transaction logs, although different kinds of logs exist in the MMTS system. They contain such information as the phone number of the customer (sender/receiver), their account ID and role (customer, retailer, merchant, operator, etc.), transaction ID, its timestamp, type (money transfer between individuals, cash in or cash out of the mobile wallet, etc.), transaction amount, status as well as sender's and receiver's balance before and after transaction.

To test and evaluate the suggested approach, the authors used the simulator described above to generate different case studies with various fraudulent activities. In generated scenarios each MMTS subscriber has only one account and role associated with him/her.

The Proposed Analysis Workflow and The Tool for Fraudulent Activity Detection

When developing the approach to the analysis of the MMTS data, the authors tried to answer the following questions:

- Are there groups of the MMTS subscribers having similar behavior?
- Do they have similar role in the MMTS?
- How do they use MMTS, i.e. what transactions do they make?
- Are there any outlier or group of outliers having similar or different roles in the MMTS system?
- What are the common contacts, i.e. receivers of the transactions of the selected MMTS subscriber?
- How usually does he/she use the MMTS?
- Are there any particular periods of time with suspicious activity (burst in the number of transactions, changes in the transactions' amounts or number of the transactions receivers)?

Answering these questions, the analyst forms firstly the general understanding of how the MMTS subscribers make use of the services, and then focuses on the particular user or group of users analyzing why they are united in the separate group, what the main characteristics of their transaction activity are, and how their behavior changes in the system over time. Figure 1 demonstrates the main stages of the proposed analysis process. To form the overview, the authors propose to apply the RadViz visualization technique to present MMTS subscribers and a heat map to display their transaction activity during selected period of day. The node-link representation of users transactions is used to analyze the contacts of the selected users. The scatter plot, heat map and table-based graphical presentations of the MMTS logs are applied to detect days with anomalous activity and to get detailed information on the users transactions. Figure 2 shows the interface of the tool for fraudulent activity detection that supports the proposed analysis workflow.

The goal of the RadViz-based *MMTS Subscribers View* (A) is to provide the general overview of the transaction activity in the MMTS during period of time being analyzed. It allows identification of the existing patterns in subscribers' behavior. The scatter plot based *MMTS Subscriber's Days View* (B) helps to identify days with possibly suspicious activity of the selected user. The heat map based *Temporal Profile View* (D) displays how the transaction activity of the selected MMTS user changes over time and is used to explain scatter plot presentation of the periods of time with transactional activity. The graph-based *Contact Graph View* (C) helps to focus on the links of a particular user or a group of users. At last the table *Property View* (E) gives detailed information on the selected MMTS entity (subscriber or transaction) presented graphically. The heat map based visualizations of transaction parameters are also

Figure 1.The analysis workflow proposed to investigate MMTS logs

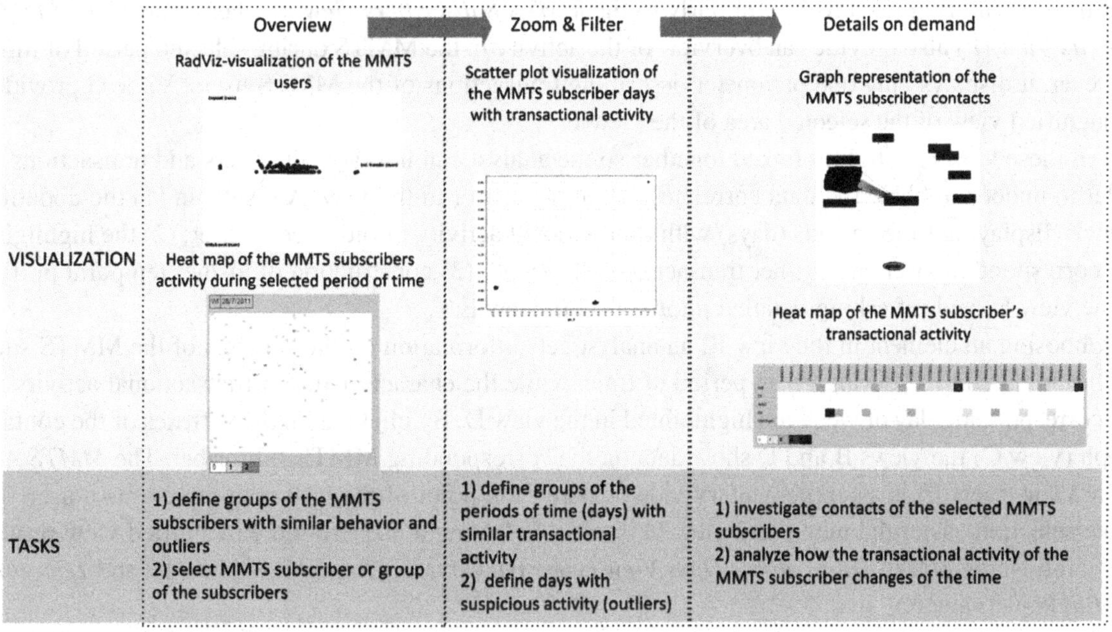

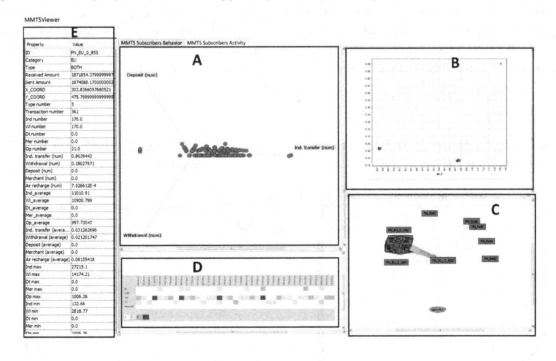

Figure 2. The design of the dashboard supporting the analysis workflow

used to trace changes in transaction activity of the MMTS users when conducting historical analysis or real time monitoring (Figure 3). Similarly to the *MMTS Subscribers view*, the heat map based *MMTS Activity View* (F) also provides an overview of the activity in the MMTS during selected period of time, however, it displays only one parameter describing transactions of the MMTS users. View G provides a magnified view of the selected area of the view F.

All these views are tightly linked together so the analyst can interact with users and transactions in order to understand how the data correlates. Selecting a user in the view A results in (1) the updating view B displaying time periods (days) with transactional activity of the selected user; (2) the highlighting corresponding user and his/her transactions in view C; (3) construction of his/her temporal profile in the view D; and refreshing detailed information in view E.

Choosing an element in the view B, an analyst gets information on the contacts of the MMTS subscriber in the view C during given period of time, while the characteristics of transactional activity of the corresponding day of week are highlighted in the view D. By clicking on the vertexes of the contact graph (view C) the views B and D show data on the corresponding MMTS subscriber. The *MMTS Activity View* (view F) is a supplementary view, where each point of the heat map represents a user. To make selection easier the magnified view G is used. Selecting a user from the magnified view results in the refreshing *MMTS Subscriber's Days View* (view B), *Contact Graph View* (view D) and *Temporal Profile View* (view C).

To support visual exploration of the data, the following interaction techniques are implemented. Flexible filtering mechanism allows specifying different complex logical expressions to filter data. It supports the usage of different attributes of the MMTS entity, i.e. transactions and users, i.e. subscriber's id, role, number of transactions, total amount, minimum and maximum transaction amounts, transaction time, transaction type, its sender and receivers, amount of the transaction, etc. The *tooltip* gives only brief information on selected object. Apart from filtering mechanism and tooltip, each view supports specific interaction techniques which are discussed in detail in the corresponding sections.

Figure 3. The MMTS Activity panel consisting of two views displaying transaction activity of all MMTS subscribers during selected period of time (F) and a set of the selected MMTS users (G)

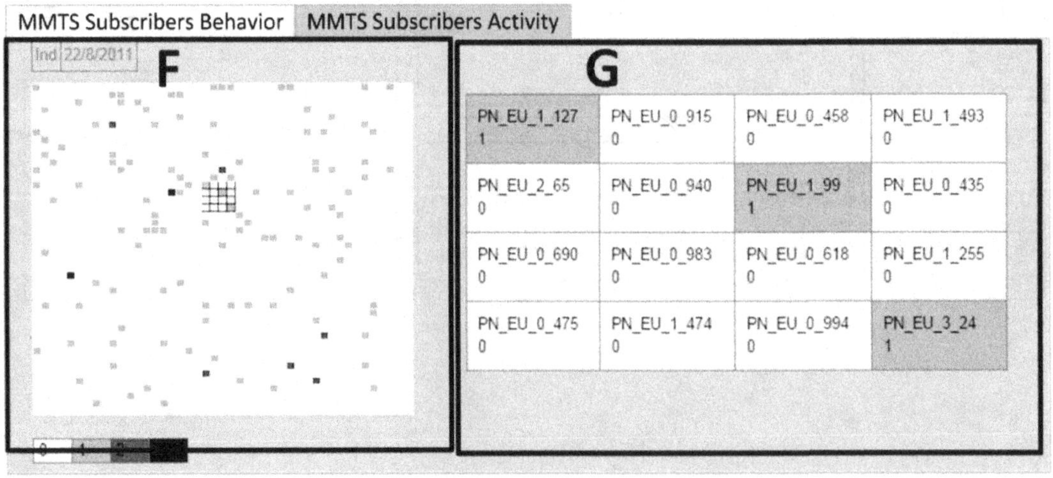

The MMTS Viewer, a tool implementing proposed approach, is written in Java. All visual models are implemented using Prefuse Toolkit (Prefuse, 2017), which allows development of interactive visualizations. It can be easily integrated into Swing applications or Java applets.

The Visualization of the MMTS Subscribers' Behavior

The key element of the proposed approach is the RadViz visualization of the MMTS subscribers' behavior (Ankerst et al., 1998). Its goal is to highlight groups of users and to detect outliers. The RadViz is a non-linear multi-dimensional visualization technique that maps n-dimensional data into 2-dimensional space and can be considered as a visual clustering tool. The visual clustering techniques do not require a prior knowledge about the number of clusters, thus they are suitable to implement the initial analysis of the data without any prior information on their internal structure. The examples of such techniques are algorithms implementing principal component analysis (Abdi, 2010), multidimensional scaling (Hout, 2013), some machine learning algorithms such as t-SNE (Maaten&Hinton, 2008) and self-organizing maps (Kohonen&Honkela, 2007). They perform projection of the original multidimensional space into a space of a smaller dimension, preserving initial distances between objects. The results of scaling are usually presented using 2D- or 3D scatter plots. However, they do not provide any explicit information about the values of the objects' attributes explaining the way they were grouped.

One of the important features of the RadViz visualization is that the layout of the data objects in the 2-dimentional space can be intuitively understood. This ability is explained by the application of the metaphor from physics that is used to determine the position of the object. The analyzed attributes are represented as dimension nodes placed around the perimeter of a circle. Then the objects are displayed as points inside the circle, each point is connected by imaginary n springs to the n respective dimension nodes. The stiffness of each spring is proportional to the value of the corresponding attribute. Thus, the point is located at the position where the spring forces are in equilibrium. Obviously, the objects having higher value for some attribute than for the others are set closer to the corresponding dimension node. If all n coordinates have the same value, the data point lies exactly in the center of the circle.

Another important feature of the RadViz technique is its low computational complexity equal $O(n)$. To compare, the complexity of the currently popular t-SNE clustering technique equals to $O(n^2)$.

For example, analyzing Figure 4, it is possible to conclude that *end user 1* uses MMTS to make individual transfers and recharge his/her mobile account to make calls only, furthermore, the frequencies of the individual transfers are prevailing over air recharge operations, while the *end user 2* makes use of all available types of financial services with equal frequency.

The attributes, describing user activity in the MMTS, that are proposed to use as dimension anchors, are given in the Table 1.

The MMTS subscribers are displayed as colored points inside the unit circle. The color is used to encode user role in the MMTS. The authors assume that the users having similar role in the MMTS should merge in clusters, meaning that they have comparable transactional activity. For example, retailers who are mainly involved in operations of cashing in/out customers mobile wallet should form a cluster, merchants participate only in the merchant payments, therefore they may constitute also a separate cluster. The location of end users is difficult to predict as they can expose rather diverse behavior, nevertheless, they also expected to form groups. Taking into consideration these assumptions it is possible to say that following visual patterns may be signs of the potential fraud: a user does not belong to any cluster or is included in the group of the users having another role; location of a small group of users significantly

Figure 4. Schema of the RadViz-based visualization of user transaction-based behavior

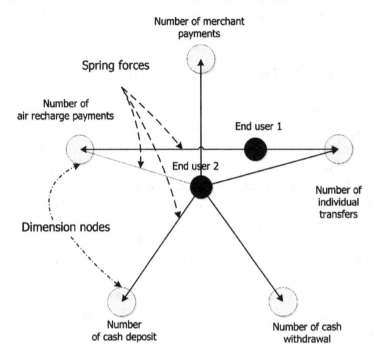

Table 1. Attributes of the MMTS transaction activity proposed to use as RadViz dimension anchors

#	Attribute Name
1	number of individual transfers for a given period of time
2	number of cash deposit operations for a given period of time
3	number of cash withdrawal operations for a given period of time
4	mean amount of individual transfers for a given period of time
5	mean amount of cash deposit operations for a given period of time
6	mean amount of cash withdrawal operations for a given period of time
7	minimum and maximum amount of individual transfers for a given period of time
8	minimum and maximum amount of cash deposit operations for a given period of time
9	minimum and maximum amount of cash withdrawal operations for a given period of time

differs from the rest. Thus, the RadViz visualization of the MMTS users is a good starting point in the analysis of the transactional activity of the users in the MMTS that provides an overall understanding how the subscribers use the MMTS.

However, there are two major problems in usage of the RadViz visualization. The *first one* is an appropriate selection of the layout of dimension nodes which determines the quality of the posterior visualization and ability to detect clusters. It is shown that for *n* variables there are $(n - 1)! / 2$ possible RadViz projections (Di Caro et al., 2010). This means it is possible to produce one non-trivial RadViz projection selecting three transactions attributes only as dimension nodes,. The *second problem* of the

RadViz visualization consists in that many users can be mapped to the same position because they have comparable values for the selected attributes. This RadViz property could hide bursts of anomalous activity in transactions of a certain type.

The experiments showed that the first problem can be easily overcome by selecting only attributes describing deposit, withdrawal operations and individual transfers of the MMTS subscriber, because the most of the fraudulent scenarios assume usage of these operations (Novikova et al, 2014). For this reason the default layout of dimension nodes implemented in the MMTSViewer tool includes the following attributes: number of individual transfers for a given period of time; number of deposit operations for a given period of time; number of withdrawal operations for a given period of time. However, an analyst is provided a possibility to adjust the layout of dimension nodes by selecting them from the predefined list and setting their order.

The second problem is much harder. Let us consider the following example. The *user 1* actively uses MMTS to make individual transfers, that's why he/she regularly cashes in his/her mobile wallet, while withdrawal operations are not typical for this user. Let us assume that the *user 1* usually makes *m* individual transfers, *m* deposit transactions and null withdrawal transactions during a given time unit. Then at the end of the analyzed period of time the system registers *m* withdrawal operations made using mobile account of the *user 1*. This sudden burst of withdrawal activity could remain unspotted by the analyst when exploring transaction activity using RadViz visualization only as the *user 1* will be mapped at the same position with users who make individual transfers, cash in/cash out operations with equal frequencies, i.e. one individual transfer, one cash in and one cash out operations per a selected time unit.

To solve this problem, Novikova et al, (2014) suggested using the heat map based visualization of the MMTS users that allows quantitative analysis of the transactional activity. However, in order to find periods of time with suspicious activity the analyst had to study the whole available period of time spending significant amount of time. To overcome this problem, authors introduced two solutions: (1) scatter plot visualization of the time periods with transactional activity (usually days) used to present results of the multidimensional scaling of the vectors describing activity of the user during the selected period of time, and (2) the mechanism accessing all period being investigated on the presence of time intervals with suspicious activity.

Detection of the Time Periods With Suspicious Transactional Activity

While the RadViz visualization is used to assess the activity of the MMTS subscribers during all time period being analyzed, the scatter plot is applied to detect the time intervals with anomalous financial activity of the user. The authors suggest considering periods of transactional activity as suspicious if they are characterized by (1) sudden change (increase or decrease) in the number of the transactions of any type, (2) sudden change in the mean (maximum or minimum) transaction amount, (3) sudden change in the number of the transaction receivers regardless their type. In order to detect suspicious bursts of the activity, the available period of time is split into time intervals of the equal length. It is suggested to calculate for each interval and each user the attributes describing transactional activity of the user, i.e. attributes listed in the Table 1 and supplemented with the parameters that describe the number of the unique receivers of the transactions of each type. Thus, the activity of the MMTS subscriber is presented as a set of vectors with numeric values.

Analyzing the transactional activity of the MMTS subscribers, the authors came to the conclusion that the users in general case rarely implement more than two financial operations per day, except users having specific roles such as merchants, retailers or operators. Therefore, it is natural to choose time interval equal to a day. Choosing time interval longer than a day may hide possible short-term changes in the MMTS user behavior. The same problem occurs if to apply RadViz to visualize the time intervals with transactional activity. In this case the time intervals presented by the vectors with attributes, proportional to each other, would be mapped to the same position of the unit circle, hiding thus high or low level of the activity. Therefore, it is suggested to apply firstly the multidimensional scaling technique (MDS) in order to map the source data space into two dimensional space, and then visualize the obtained results using scatter plot. Like RadViz, the MDS is used for exploratory data analysis and dimension reduction. The MDS estimates the similarities between vectors and constructs a "map" that conveys, spatially, the relationships among items. Similar items are located close to each other, and dissimilar items are located proportionately further apart (Hout et al., 2013). Thus, it allows detecting time intervals with similar behavior and outliers. Therefore, possible visual sign of fraudulent activity may be formulated as "a day does not belong to any cluster".

Figure 5 shows the results of the multidimensional scaling of the days with transactional activity presented using scatter plot. To construct it, the following attributes were calculated: number of the air recharge operations; number of the unique receivers of the air recharge transactions; number of the deposit operations; number of the unique agents of the deposit transactions; number of the withdrawal operations; number of the unique agents of the withdrawal transactions. From Figure 5 it is clearly seen four visual clusters, two of them are rather numerous, while the third and fourth ones constitute of three days and one day only, correspondingly. A central issue about the MDS is that it conveys information only about relationships, and, therefore, the interpretation of the items' placement in the two-dimensional space is done regardless of the particular meaning of the x-coordinate and y-coordinate. This means that for interpreting the results of the MDS it is necessary to consider only the similarity level, i.e. distances between data objects. Thus, according to the Figure 5 the cluster #4 is more alike to the cluster #2 than to the cluster #1. To explain the scatter plot visualization, authors use a heat map visualization of the MMTS subscriber's temporal profile that is described in the next subsection. According to it, the clusters #1 and #2 constitute of the days during which the MMTS user implemented one individual transfer or one air recharge operation only. The cluster #3 consists of the days during which the user implemented both individual transfer and air recharge operation. The cluster #4 is characterized by two operations of the air recharge. That is why this cluster is located closer to the cluster #2.

The scatter plot helps to detect suspicious changes in the transactional activity of the MMTS user. However, using this approach directly requires checking all subscribers in order to find signs of the fraudulent activity. This may be a very time-consuming analysis task. Therefore, to enhance the process of the anomaly detection in the users' behavior, the authors proposed to assess the level of the fraudulent activity of the subscriber. The estimation of the fraudulent level is done by detecting presence of the outliers in the dataset describing user activity. In the approach proposed the outliers are detected using the local outlier factor (LOF) (Breunig et al., 2000).

This algorithm finds anomalous data points by measuring the local deviation of a given data point with respect to its neighbors, the obtained values are normalized then using z-score. The z-score indicates the proportion of the standard deviation the current value is greater or less than the average one. The values lying in the range [-2.58, 2.58] indicate that the corresponding data points are not the result of a random process (Caldas&Singer, 2006). Thus all objects having scores [-4; -2.58] and [2.58; 4] can be

Figure 5. Scatter plot visualization of the days with transactional activity of the MMTS user

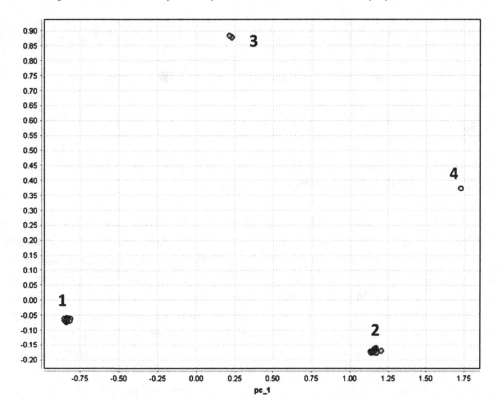

considered anomalous, and if they are present in the data set describing user's behavior, the transactional behavior of the corresponding MMTS subscriber is considered fraudulent. The RadViz visualization is used to display the fraudulent level of the MMTS behavior. The analyst can filter out the users having benign behavior and focus only on those who have suspicious behavior. The calculated z-score for each time interval with transactional activity is displayed using color on the scatter plot.

The Temporal Profile of the MMTS Subscriber

The graphical presentation of the temporal profile of the MMTS subscriber is a key element explaining the results of the RadViz visualization of his/her behavior and the scatter plot of the time intervals with transactional activity. The temporal profile is presented using a heat map which is constructed in the following way. The x-axis represents time, the time unit is equal to the time interval used to construct scatter plot. The y-axis represents the attributes used to assess transactional activity of the user, i.e. number of the transactions of each type, number of the unique receivers of transaction of each type, mean, maximum and minimum amount of the transaction of each time, etc.

Figure 6 shows the number of different transactions implemented by a selected user. According to it the subscriber uses the MMTS to implement deposit (Dt) and withdrawal (Wl) operations as well as individual transfers (Ind), at that he/she periodically makes 3 individual transfers per day. Here air recharge operations (ArRC) are not used.

Figure 6. Temporal profile of the MMTS subscriber represented using heat map

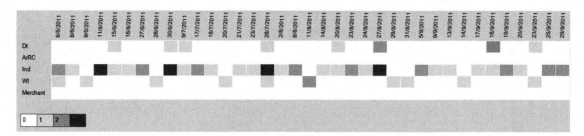

The Graph of the MMTS Subscriber's Contacts

A graph-based approach is applied to visualize contacts of the MMTS subscribers, where the vertexes represent users, and edges – transactions between them. The graph-based visualization technique is a common way for representing transactions in the financial systems. The main advantage of the graph view is that it emphasizes structural properties of the connectivity between users (Novikova et al, 2014).

As mentioned above in the MMTS case study being analyzed, a user has only one mobile account associated with him, therefore, the authors do not display mobile account as a separate vertex connected with the user. However, if the user has several accounts, it is suggested to aggregate them into one meta-node preserving all input and output links in order to improve readability of the generated image.

Color is used to encode the role of the user in the MMTS as well as transaction types. Both color schemes were created using Color-Brewer2 (2015). The transaction types that are frequently used in detection of suspicious activity such as cash withdrawal, deposit and individual money transfers are encoded with color-blind safe colors. The shape of the vertex depends on whether the user is only transaction sender (diamond), receiver (ellipse) or both (rectangle). This feature simplifies the detection of subscribers whose accounts are used only for cash withdrawal or deposit operations. If the user is linked with another user by a set of transactions of the same type then they are displayed as one edge, whose thickness depends on their quantity. The size of the graph vertex could be determined by a sum of received and sent amounts for a given period of time. This option helps to discover subscribers participating in large cash flows.

Apart from interaction techniques implemented for all views, the authors applied *linking* and *brushing* effect to highlight contacts of the MMTS user. When switched to this mode, it is possible to select the user by clicking on the corresponding node. This will make all input and output links visible while the rest will be hidden. The combination of this technique with the *filtering* mechanism allows focusing on transactions with given characteristics of the particular user. The analyst can also set up the vertex layout. Currently two layouts are implemented: radial and custom one, based on scatter plot (Figure 7). In order to construct the latter, the total number of all transactions made by the MMTS subscriber and the number of different transaction types used are calculated for each subscriber. These two attributes define the position of the corresponding node on the plane: x-coordinate is determined by the total number of all transactions, and y-coordinate is determined by number of different transaction types. This layout helps to reveal the most active users. The heightened activity can be a sign of potential fraud.

Figure 7. Graph-based representation of the financial contacts of the MMTS user

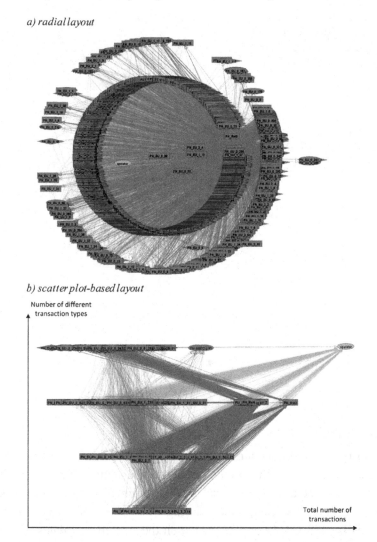

a) radial layout

b) scatter plot-based layout

The Heat Map Representation of the Transaction Activity in the MMTS

The heat map is used to display the activity of the MMTS subscribers during selected time unit and denotes to a complementary view. It can be used to assess existing activity patterns in the MMTS. Each heat map element represents a MMTS user, and the color of the point is determined by the numeric value of the parameter describing MMTS user activity (Figure 8).

The set of parameters available to the analyst matches the set of parameters used to detect time interval with suspicious transactional activity. The analyst has an option to choose what parameter to display. The darker color of the matrix element – the higher the value of the corresponding parameter. Figure 8 shows the number of the individual transfers made in the MMTS during one day. It is clearly seen that only few users made more than four withdrawal transactions (darker elements) during that day, while the rest users made one or two withdrawal operations or did not withdraw electronic money at all.

Figure 8. Heat map representation of the individual transfers during one day

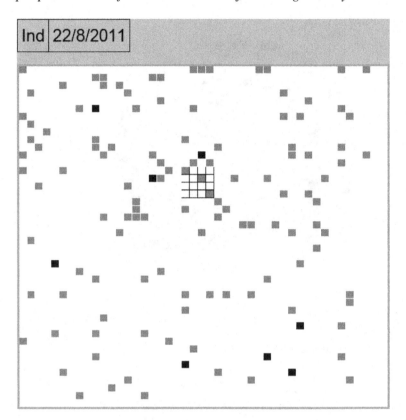

Very often the security analyst is not interested much in what occurred during a particular time unit but rather what changed across a range of time units. Therefore, the analyst may select any set of time units and see not a depiction of the actual values for each subscriber but a variance of the selected parameters. Suppose, for instance, the analyst selects a week, during which the user made 4 individual transfers of electronic money (18539, 1482, 22112 and 21174), then the system calculates the variance of this set of values and displays it. The significant amount of variance may be a sign of mobile bot activity as the infected mobile devices periodically send small sums of electronic money. The color saturation is used to highlight the amount of the variance. White color means no variance, meaning that the corresponding users had the same level of activity in all selected time units. Light grey users have a very small level of variance, grey users have a larger amount of variance, and dark grey users have the most variance.

The heat map visualization of sufficiently large number of MMTS users on one screen poses certain difficulties in the analysis of their behavior due to illegible details of the generated image. To solve this problem, a special zooming mechanism called *magnification area selector* is implemented. The analyst could select an arbitrary set of heat map elements and display them in a separate view. This view is also a heat map visualization of the selected MMTS users which provides account ID and value of the parameter being examined. Each element of this heat map is selectable. Mouse clicking on it results in displaying the temporal profile of the selected user.

USAGE SCENARIOS

The authors tested the proposed approach on a set of benchmark datasets containing different fraudulent activity. These datasets were generated using synthetic MMTS log simulator. The following malicious scenarios were investigated: money laundering scheme, theft of mobile phone and mobile botnet infection.

Money Laundering Misuse Case

Several money laundering schemes exist (FATF, 2010). Most of them assume usage of chains of mules that are used to hide the fraudulent origin of money. Usually fraudsters having a certain amount of money to be laundered divide this amount and send it to several mules. Later on, they withdraw this money from a complicit retailer. In the reality, they would then send the cash obtained to another fraudster, but this money stream is not captured by the MMTS. Chains of mules may be composed of several layers. In the scenario generated by the MMTS log simulator the chain of mules having only one layer of mules is implemented. However, this limitation does not restrict the proposed approach to fraud detection as far as the authors are focusing on determining anomalous activity in MMTS but not on determining particular malicious scheme.

The fraudulent scenario is composed of 500 legitimate users, 10 mules and 4 retailers and described by 5317 transactions implemented during one hour.

When detecting this kind of anomalous activity, the following assumptions were made: the amount of fraudulent transactions is smaller than the average amount of the users; the mules also perform legitimate transactions; a sudden change in transferred *mMoney* amounts corresponds to an anomaly. Thus, the mule could be described by greater number of individual money transfers and withdrawal operations and smaller average amount of these transactions. Basing on these assumptions, the following attributes were selected as anchors: number of individual transfers, and mMoney withdrawals and deposits to detect possible mules participating in the money laundering scenario.

The result of MMTS users' visualization using RadViz technique is shown in Figure 9. It is seen that there is a group of retailers, lying separately from other MMTS subscribers because they are involved only in withdrawal and deposit operations. It is possible to spot two end users located apart from the rest end users. They are marked in Figure 9 by *End users 1* label. Their location is explained by prevailing of the individual money transfers over other mobile money services. Additionally, from the graph-based view it is seen that one of these users only sends money, while another - only receives them.

Further analysis of their contacts shows that these two subscribers are connected with each other via a set of users (Figure 10). According to this, one can conclude that the users *PN_FR1* and *PN_FR2* could be potential fraudsters, and the subscribers connected with them are the mules.

The fraudulent transactions can be also spotted using a scatter plot view of the time periods with transactional activity. As the period being analyzed is very short and equal only to one hour, the time interval is set to one minute. To detect suspicious transaction, the following attributes describing the transactional activity were calculated to: number of individual transfers, unique receivers of the individual transfers and mean value of the individual transfer, number of deposits, unique agents of deposit operations and mean value of the deposits, number of withdrawals, unique agents of the withdrawal operations and mean value of the withdrawals.

Figure 9. RadViz visualization of MMTS users in money laundering scenario

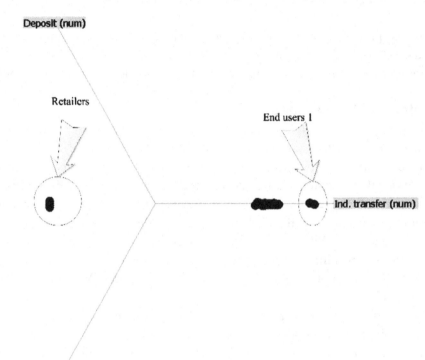

Figure 10. The structure of the examined money laundering scenario

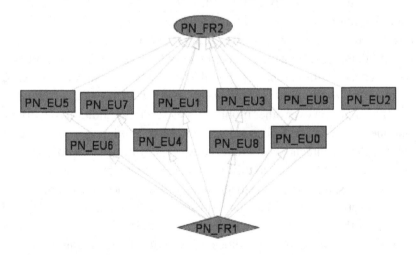

Behavioral Fraud Case Study

The behavioral frauds are the fraudulent scenarios in which the behavior of the fraudster is superimposed on the legitimate user's one. They can occur when the mobile account is taken over after mobile phone is stolen or infected by malware. In this case both actors – legitimate and malicious – use the mobile device to carry out transactions in the same window of time.

The generated dataset contained two types of such fraud. The first one corresponds to a botnet which infected several mobile devices. The malicious program carries out several transfers towards mules who withdraw the money within 72 hours after its reception. This scheme is rather similar to the money laundering scheme except that the amounts involved are not the same, there is no complicit retailer, and the mules are used here to hide the destination of the stolen money and not its origin. Moreover, the fraudulent transactions are initiated by the malicious programs. The second case corresponds to a theft. The mobile device is stolen and the fraudster then tries to withdraw money several times during a short time interval before the phone's theft is reported and the phone is deactivated.

The generated scenario is made of 2 merchants, 6 retailers, 1 operator and 2450 users, 4 of which are mules, and 54850 transactions made during 4 months. There are 4 thieves and 60 cases of the mobile phone theft, and 39 mobile devices infected by the mobile botnet.

In general case when behavior fraud occurs, the following behavior changes are observed: changes in the user's average transaction amount and transaction frequency. That is why when detecting mobile botnet, the following assumptions were made: the amount of fraudulent transactions is slightly inferior than the average amount of the regular users transactions; the time elapsed between two transactions is similar to the average interval between two legitimate transactions; and the legitimate and fraudulent behavior occur during the same window of time. The following transaction attributes were chosen as RadViz dimension nodes: number of individual transfers, *mMoney* withdrawals and deposits, number of merchant and air recharge payments.

Figure 11 shows the results of the RadViz visualization of the MMTS subscribers. It is seen that there are groups of merchants and retailers that lie separately due to the peculiarities of their roles in the MMTS. The majority of the end users are located near the center of the RadViz visualization meaning that they use different types of mobile money services rather uniformly. However, there is a group of four end users (*End users* 2) that have money transfers significantly prevailing over transactions of other types.

The contact analysis of these users shows that the sets of MMTS subscribers sending them individual transfers are intersecting. These two facts allow concluding that these four users are mules whose accounts are used to cash out *mMoney* from the *mWallets*.

In order to detect a set of subscribers with infected mobile devices, the authors filtered out all transactions that are not sent to the mules and this enabled us to detect the botnet. Its structure is presented in Figure 12, the users with infected devices regularly made individual transfers to all mules.

The authors analyzed the activity of the mules and the users with infected devices in the botnet. In general case, the activity of the user with infected device is rather structured, the clusters usually consist of the days with financial operations of different type, and the days with suspicious individual transfers form a separate rather numerous cluster (Figure 13a). However, as this activity has a long-term character the mechanism assessing fraudulent level of the MMTS user activity does not highlight such users as suspicious ones. The mules mostly receive individual transfers and implement withdrawal and air

Figure 11. RadViz visualization of MMTS users in behavior fraud scenario

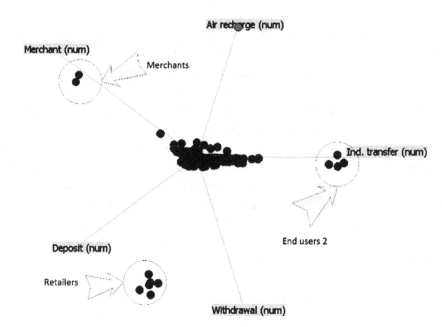

Figure 12. Detection of mobile botnet using the graph-based representation of the MMTS user' contacts

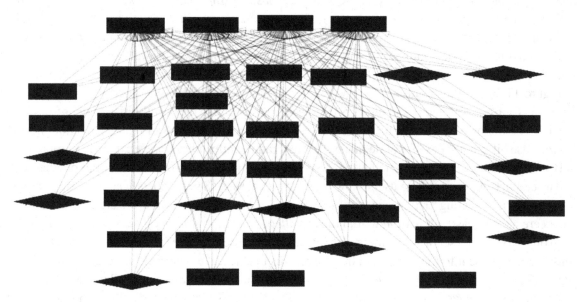

recharge operations to stay online, the intervals with transactional activity differ only in the amount of such operations and unique senders, thus the generated visualization of the time intervals with transaction activity of the mule is rather scattered (Figure 13b).

The temporal profiles of the victim and the mule are shown in Figure 14a and Figure 14b correspondingly, and they easily explain the results of the scatter plot visualizations.

Figure 13. Scatter plot visualization of the time intervals (days)

a) transaction activity of the user with the infected device

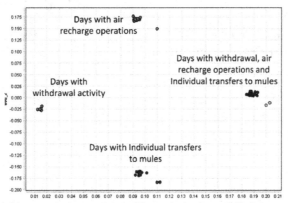

b) transaction activity of the mule in the mobile botnet

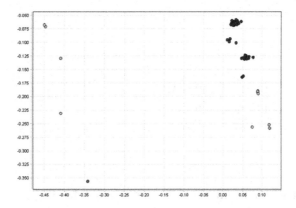

Figure 14. Temporal profiles

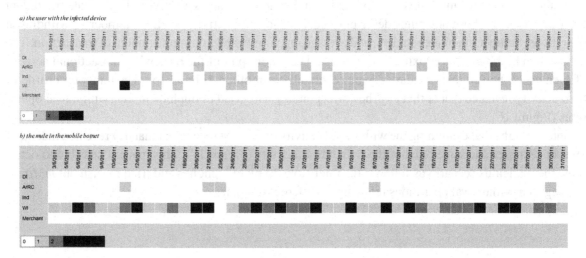

Figure 15. Days of the transactional activity of the user whose phone was stolen

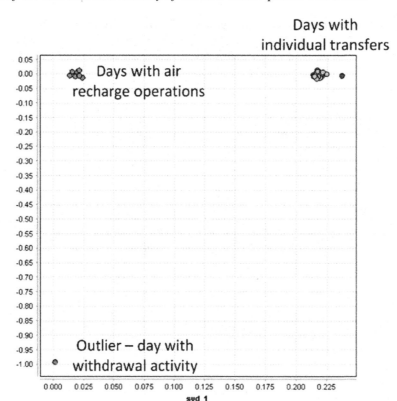

Though the RadViz visualization of the MMTS users was helpful when revealing mules in the money laundering and mobile botnet scenarios, it did not highlighted cases with activity of the mobile phone thieves. As the thieving activity is characterized by a sudden burst of the attempts to withdraw money the authors applied the mechanism assessing fraudulent level of the user activity. To evaluate fraudulent level, the authors chose attributes describing only withdrawal activity, such as number of withdrawal operations, number of unique retailers, and average, maximum and minimum amount of transaction. The mechanism highlighted approximately 60 users. The scatter plot of the days with transactional activity of such users showed that one day is lying separately from others (Figure 15). The temporal profile of their activity revealed that at that day the corresponding user tried to implement four withdrawal operations (Figure 16).

Interestingly, that examining the withdrawal activity of all MMTS users at that day revealed that other 19 users exposed the same burst of the withdrawal activity (Figure 17). This made it possible to determine the malicious scenario as follows: the thief stole 20 mobile phones and tried to cash out *mMoney* making four withdrawal operations referring to different retailers.

Figure 16. Temporal profile of the user whose mobile device was stolen

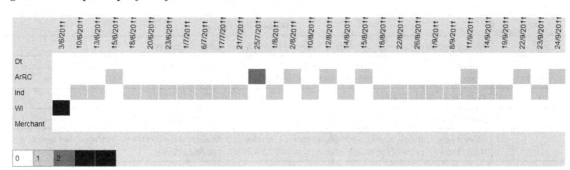

Figure 17. The withdrawal activity represented using the heat map with bursts in the amount of operations

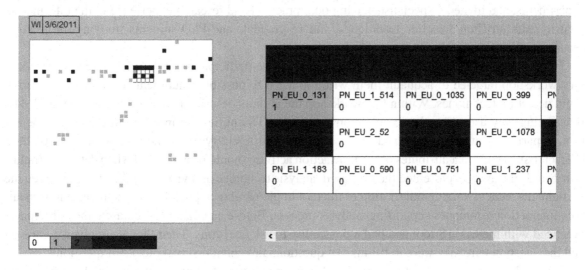

EFFICIENCY EVALUATION AND DISCUSSION

To assess the efficiency of the proposed visualization-driven technique of the analysis of the MMTS transactions, the authors adopted effectiveness metrics proposed by the ISO/IEC 9126-1 (2001) standard purposed to evaluate the software quality. They are task effectiveness, task completion and error frequency which are used to assess the proportion of the correctly achieved goals of the task, the proportion of the completed tasks and the frequency of errors. To estimate these metrics, the authors applied a widely used evaluation method known as inspection designed to assess visualization techniques. It assumes the usage of benchmark (real or artificial) datasets and experts "to examine subjectively the effectiveness of different visualization techniques for different tasks, by visually inspecting the output of the visualization techniques" (Marghescu, 2007-2, p.46). When implementing the evaluation process, the authors were interested in the estimation of the introduced enhancement of the technique in comparison with the technique described in (Novikova et al., 2014) as to the best authors' knowledge there are no visualization driven approaches targeted to support analysis of the fraudulent activity in the mobile money transfer services. In order to do this the authors used 5 datasets previously used in the evaluation process. Their brief description is presented in the Table 2.

Table 2. Description of the test datasets

#	Dataset Description
	Money laundering activity (500 regular users, 10 mules, 6 merchants, 4 retailers)
2	Mobile botnet infection and mobile phone theft (2 merchants, 6 retailers and 4010 users, 4 of which are mules; 4 thieves and 60 cases of the mobile phone theft, and 39 infected mobile devices)
3	Only legitimate activity
4	Money laundering activity (500 regular, users, 5 mules, 8 merchants, 4 retailers)
5	Mobile botnet infection and mobile phone theft (2 merchants, 5 retailers and 4000 users, 5 of which are mules; 2 thieves and 45 cases of the mobile phone theft, and 25 infected mobile devices)

The authors tried to preserve the structure of the focus group engaged in the previous evaluation experiments, and invited 6 specialists having both practical and research experience in the information security and intrusion detection techniques, and 15 graduate and PhD students studying information security and computer science.

The evaluation process was organized as follows. The participants firstly were given an introductory workshop targeted to acquaint them with the MMTS, possible fraudulent scenarios and existing fraud detection techniques. Within this workshop the authors described basic principles of the MMTS functioning: key user roles, basic operations, what MMTS entities are involved in their implementation. A particular attention was paid on description of different types of the fraudulent activity, their characteristics and what attributes of the transaction activity should be examined when detecting frauds. Afterwards the authors presented the proposed analysis technique and visual models used, and then the participants were given a demonstration of the tool implementing approach with the focus on the available interaction techniques supporting analysis process. Before analyzing test datasets the participants practiced with the MMTS tool and had a possibility to ask questions. Then they were given 5 datasets and the questionnaire they were to fill in. The questionnaire contained two blocks of questions.

The goal of the first one is to assess the effectiveness of the technique. It contained the following questions:

1. Determine how people usually use MMTS, what the most popular transaction type is, how many merchants and retailers are, how often subscribers usually do withdrawal and deposit operations, individual transfers, and how many receivers of the individual transfers subscribers usually have.
2. Determine whether dataset contains any suspicious activity.
3. If the fraudulent activity is present, determine what type of fraud is present. Note that data set may contain several fraudulent scenarios.
4. If the mobile botnet is detected, determine its structure: users-mules, users, whose accounts are used to withdraw money. Describe malicious scenario (average transaction sum, transaction frequency, MMTS entities involved).
5. If the case/cases of mobile phone theft determine a list of victims, and date of the theft; describe the malicious scenario (average transaction sum, number of malicious transaction, MMTS entities involved).
6. If the money laundering activity is present, determine the structure of the money laundering scheme: fraudsters who send or withdraw money, mules, whose accounts are used to transfer money.

The goal of the second block of questions is to evaluate the usability of the proposed techniques. Answering these questions, they had to score the system from 1 to 5 (1 = very easy, 5 = very difficult). The questions were as follows:

1. Is it easy or difficult to interpret results of the RadViz visualization in order to describe the behavior of the MMTS user?
2. Is it easy or difficult to interpret results of the scatter plot based visualization of the days with transactional activity of the MMTS user?
3. Is it easy or difficult to interpret results of the graph of the MMTS subscriber's contacts?
4. Is it easy or difficult to explore transactions using the heat map based temporal profile of the MMTS subscriber?
5. Is it easy or difficult to explore data using MMTSViewer?

After the completion of the analysis tasks the participants were also asked to give a feedback about the tool and encouraged to tell their suggestions or critics.

The first question of the first part of the questionnaire has a descriptive character, and almost all participants gave correct answer. Lately, when discussing the obtained results, they marked that combination of the RadViz representation of the users with the heat map presentation of the temporal profile made it easy to understand how the MMTS subscribers use the services, while the scatter plot visualization of the time intervals with transaction activity gave information on how many patterns of transaction activity the user has. The rest questions of the first part of the questionnaire allowed calculating numeric assessment of the accuracy rate, and the obtained results are shown in Figure 18. The accuracy rate is higher in the comparison to the results obtained when evaluating the approach described in (Novikova et al., 2014).

All participants correctly determined the character of the transaction activity in the dataset being examined and answered all questions relating to the mobile botnet infection and money laundering scenarios. The focus group marked that in major cases the analysis of the RadViz visualization in the conjunction with the graph of contacts was enough to detect these types of the fraudulent activity and describe their scenarios.

Figure 18. The accuracy rate evaluation

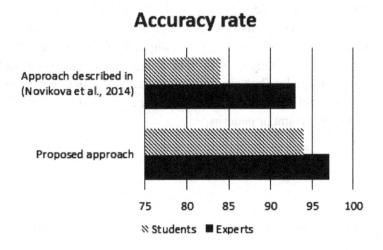

In (Novikova et al., 2014) the detection of mobile phone thefts caused a certain difficulties for both experts and students. The evaluation process showed significant improvement in discovering both cases of the mobile phone thieves and their victims. However, two students failed to determine correctly all victims for one case of mobile phone thief; though they correctly described the patterns of the attack and days with fraudulent activity. Lately, when discussing the results with them, the authors found out that the problem was in the choice of the attributes used to assess the fraudulent level of the users' behavior. Some MMTS users were marked as suspicious ones because they once made more than once air recharge operation per day, and the students did not checked their temporal profile to examine implemented transactions. The rest participants focused on the analysis of withdrawal, deposit operations and individual transfers only, and thus correctly determined victims of the mobile thieves as their number was almost equal to number of the users with suspicious activity.

Interestingly, that experts marked that the visualization of the MMTS subscribers activity using the heat map was helpful in revealing the set of victims of the mobile phone thief, as by determining the correct date it was possible to choose the corresponding date in the *MMTS Subscribers' Activity View* and see the users who also exposed high withdrawal activity at the same period of time. This option decreased the time spent on defining victims of each mobile phone thief. One expert and one student mistakenly classified several users as victims because they once exhibited high withdrawal activity and were marked by the ranking mechanism as suspicious ones. However, the analysis of their temporal profile showed that they expose comparatively high withdrawal activity, and moreover acted as mules in mobile botnet activity, the latter means that benign withdrawal operations were superimposed with fraudulent ones. Thus, it is possible to conclude that some mistakes were made due to lack of attention to details, for example, when interpreting the temporal profiles in conjunction with user contact analysis, and lack of experience when selecting attributes for assessing suspicious level of MMTS behavior.

Most of the experts marked the flexibility in the setting up the visual models, i.e. possibility to choose anchor attributes of the RadViz visualization and attributes used to assess suspicious level of users' behavior and construct the scatter plot. To their opinion, this option allows getting comprehensive understanding of data being analyzed. All participants liked the visual models selected to explore transactions activity in the MMTS, they noted that all views are rather easy to interpret. However, the possibility to get temporal profile of the MMTS user by clicking on the elements of the RadViz or graph visualizations significantly simplifies the analysis process and helps to avoid misunderstanding by providing detailed explanation on generated graphical representation of the data. The experts believed that the combination of the flexibility in the setting up of the visual models and implemented interaction techniques enables revealing novel schemes of the fraudulent activity. To make the usage of the tool more valuable, the participants advised to add a possibility to form formal description of the detected fraudulent scenarios that could be lately used in training automated analysis models or construction of the rules applied to detect frauds. They also recommended implementation of a *search-by-example* mechanism for the temporal profile of the MMTS subscriber that could significantly simplify to detect cases of fraudulent activity with similar patterns.

Summarizing the results of the efficiency evaluation of the proposed visualization technique for MMTS transaction activity, the authors can conclude that the set of visual models and suggested interaction techniques are useful in detection of fraudulent scenarios of different type. The RadViz visualization is helpful when detecting fraudulent scenarios that make use of mules - users whose behavior significantly differs from the behavior of the other MMTS subscribers. It allows also detecting long-term frauds which cause shifts in user behavior with cumulative effect and thus could be revealed when choosing relatively long time unit, i.e. month, to explore MMTS transactions. The graph of the MMTS users' contacts is especially helpful in detecting the set of mules and their contacts. That is why these two techniques were effective when detecting the structure of the money laundering scheme and the mobile botnet. If attacks are characterized by short-term sudden change of the transaction activity of the MMTS user the usage of the fraudulent level assessment mechanism and the scatter plot of the intervals of the transaction activity is preferable. In both cases the heat map visualization of the historical activity of the user supports explanation of the produced visualizations and is an essential part of the analysis process.

FUTURE RESEARCH DIRECTIONS

One of the primary tasks of the future work is the implementation of the interaction techniques supporting similarity search functionality and constructing formal description of the fraudulent scenarios in the format allowing training of the automated analysis models and forming rules for detection of the fraudulent activity. The authors plan to test the proposed approach on the data sets with more complex scenarios of the malicious activity, for example, involving several layers of mules. Another important research direction is to construct the behavior patterns of the MMTS subscribers as in many cases the application of the patterns enhances the detection level of the frauds.

CONCLUSION

The analysis of the state-of-art in fraud detection techniques in the mobile money transfer services showed that the most widely technique to explore electronic transactions is interactive graph-based data representation. It supports link analysis of the user's contacts visually and enables application of graph-theoretic algorithms in order to discover structural peculiarities such as bridges and cliques. The authors proposed to form the metaphoric representation of the MMTS subscriber behavior according to his/her transaction activity. The user's activity is assessed basing on mean amount of transactions, their frequency, and type of operations. The RadViz visualization technique supports graphical representation of the MMTS users according to their behavior and allows determining clusters of users with similar behavior and outliers. It is considered as a starting point of transaction analysis supported by the traditional graph-based representation of subscribers' transactions. To be able to analyze changes in users' behavior the authors developed the mechanism assessing fraudulent level of the MMTS user activity and visualized time periods with transactions using scatter plot. The interpretation of the RadViz visualization and the scatter plot is supported with the heat map representation of the transactions. The authors described a usage scenarios that could be applied to detect fraudulent activity and implemented the efficiency evaluation of the proposed visualization technique. The experts, who participated in the efficiency evaluation process, highlighted the ability of the tool to detect different patterns of the fraudulent activity in MMTS and gave recommendations how to improve interaction mechanisms for searching transaction patterns.

ACKNOWLEDGMENT

This work was partially supported by grants of RFBR (projects No. 16-29-09482, 18-07-01488), by the budget (the project No. AAAA-A16-116033110102-5), and by Government of Russian Federation (Grant 08-08).

REFERENCES

Abdi, H., & Williams, L. J. (2010). Principal component analysis. *WIREs Comp. Stat, 2*, 433–459.

Achemlal, M., et al. (2011). *Scenario requirements*. (Tech. rep.) MASSIF FP7-257475 project.

Al-Khatib, A. (2012). Electronic Payment Fraud Detection Techniques. *World of Computer Science and Information Technology Journal, 2*, 137–141.

Ankerst, M., Berchtold, S., & Keim, D. A. (1998). Similarity Clustering of Dimensions for an Enhanced Visualization of Multidimensional Data. In *IEEE Symposium on Information Visualization (INFOVIS '98)* (pp. 52-60). Washington, DC: IEEE Computer Society. 10.1109/INFVIS.1998.729559

Breunig, M. M., Kriegel, H.-P., Ng, R. T., & Sander, J. (2000). LOF: identifying density-based local outliers. In *Proceedings of the 2000 ACM SIGMOD International Conference on Management of Data (SIGMOD '00)* (pp. 93-104). ACM. 10.1145/342009.335388

Caldas de Castro, M., & Singer, B. (2006). Controlling the False Discovery Rate: A New Application to Account for Multiple and Dependent Test in Local Statistics of Spatial Association. *Geographical Analysis, 38*(2), 180–208. doi:10.1111/j.0016-7363.2006.00682.x

Chang, R., Ghoniem, M., Kosara, R., Ribarsky, W., Yang, J., & Suma, E., … Sudjianto, A. (2007). WireVis: Visualization of Categorical, Time-Varying Data From Financial Transactions. In *IEEE Symposium on Visual Analytics Science and Technology (VAST 2007)* (pp.155-162). Washington, DC: IEEE Computer Society. 10.1109/VAST.2007.4389009

ColorBrewer2. (2015). Retrieved June 13, 2015, from http://colorbrewer2.org

Coppolino, L., D'Antonio, S., Formicola, V., Massei, C., & Romano, L. (2015). Use of the Dempster–Shafer theory to detect account takeovers in mobile money transfer services. *Journal of Ambient Intelligence and Humanized Computing*, 1–10.

Corkin, L. (2016). Kenya's mobile money story and the runaway success of M-Pesa. *Digital Frontiers*. Retrieved September 08.09.2017, from http: http://www.orfonline.org/expert-speaks/kenyas-mobile-money-story-and-the-runaway-success-of-m-pesa

Cowan, J. (2012). Kenya's Safaricom deploys new fraud management system for its mobile payment service. *IoT NOW*. Retrieved September 10, 2017, from https://www.iot-now.com/2012/08/21/7080-kenyas-safaricom-deploys-new-fraud-management-system-for-its-mobile-payment-service/

Delamaire, L., Abdou, H., & Pointon, J. (2009). Credit card fraud and detection techniques: a review. *Banks and Bank Systems*, *4*(2), 57-68.

Di Battista, G., Di Donato, V., Patrignani, M., Pizzonia, M., Roselli, V., & Tamassia, R. (2014). BitConeView: visualization of flows in the bitcoin transaction graph. *2015 IEEE Symposium on Visualization for Cyber Security (VizSec)*, 1-8.

Di Caro, L., Frias-Martinez, V., & Frias-Martinez, E. (2010). *Analyzing the Role of Dimension Arrangement for Data Visualization in Radviz. In Advances in Knowledge Discovery and Data Mining. LNCS* (Vol. 6119, pp. 125–132). Berlin: Springer-Verlag. doi:10.1007/978-3-642-13672-6_13

FATF. (2010). *Money Laundering using New Payment Methods*. Retrieved June 13, 2015, from http://www.fatf-gafi.org/topics/methodsandtrends/documents/moneyla

Fiserv. (2017). *Financial Crime Risk Management solution*. Retrieved September 11, 2017, from https://www.fiserv.com/risk-compliance/financial-crime-risk-management.aspx

Gaber, C., Hemery, B., Achemlal, M., Pasquet, M., & Urien, P. (2013). Synthetic logs generator for fraud detection in mobile transfer services. In *Int. Conference on Collaboration Technologies and Systems (CTS 2013)* (pp.174-179). New York: IEEE. 10.1109/CTS.2013.6567225

Golmohammadi, K., Zaiane, O. R., & Díaz, D. (2014). Detecting stock market manipulation using supervised learning algorithms. *2014 International Conference on Data Science and Advanced Analytics (DSAA)*, 435-441. 10.1109/DSAA.2014.7058109

Hout, M. C., Papesh, M. H., & Goldinger, S. D. (2013). Multidimensional scaling. *Wiley Interdisciplinary Reviews: Cognitive Science, 4*(1), 93–103. doi:10.1002/wcs.1203 PMID:23359318

Huang, M. L., Liang, J., & Nguyen, Q. V. (2009). A Visualization Approach for Frauds Detection in Financial Market. *2009 13th International Conference Information Visualisation*, 197-202.

ISO/IEC 9126-1. (2001). *Product quality. Part 1: Quality model.* Retrieved June 13, 2015, from http://www.iso.org/iso/iso_catalogue/catalogue_tc/catalogue_detail.htm?csnumber=22749

Jack, W., Tavneet, S., & Townsend, R. (2010). Monetary Theory and Electronic Money: Reflections on the Kenyan Experience. Economic Quarterly, 96(1), 83-122.

Kamana, J. (2014). *M-PESA: How Kenya took the lead in mobile money.* Retrieved September 09, 2017, from https://www.mobiletransaction.org/m-pesa-kenya-the-lead-in-mobile-money/

Keim, D., Andrienko, G., Fekete, J.-D., Goerg, C., Kohlhammer, J., & Melancon, G. (2008). Visual Analytics: Definition, Process, and Challenges. In Information Visualisation, LNCS (vol. 4950, pp.154-175). Berlin: Springer-Verlag.

Kohonen, T., & Honkela, T. (2007). Kohonen network. *Scholarpedia, 2*(1), 1568. doi:10.4249cholarpedia.1568

Korczak, J., & Łuszczyk, W. (2011). Visual Exploration of Cash Flow Chains. In *The Federated Conference on Computer Science and Information Systems* (pp.41–46). New York: IEEE.

Kotenko, I., & Novikova, E. (2013). VisSecAnalyzer: a Visual Analytics Tool for Network Security Assessment. In *8th International Conference on Availability, Reliability and Security (ARES 2013). LNCS* (vol. 8128, pp. 345-360). Berlin: Springer-Verlag. 10.1007/978-3-642-40588-4_24

Leite, R. A., Gschwandtner, T., Miksch, S., Gstrein, E., & Kuntner, J. (2016). Visual analytics for fraud detection: focusing on profile analysis. In *Proceedings of the Eurographics / IEEE VGTC Conference on Visualization: Posters* (pp. 45-47). New York: IEEE.

Lin, L., Cao, L., & Zhang, C. (2005). The fish-eye visualization of foreign currency exchange data streams. In *Asia-Pacific Symposium on Information Visualisation (APVis)* (vol. 45, pp. 91-96). Darlinghurst: Australian Computer Society, Inc.

Maaten, L. J. P., & Hinton, G. E. (2008). Visualizing High-Dimensional Data Using t-SNE. *Journal of Machine Learning Research, 9*, 2579–2605.

Marghescu, D. (2007a). *Multidimensional Data Visualization Techniques for Financial Performance Data: A Review* (TUCS Tech. Rep. No 810). Turku, Finland: University of Turku.

Marghescu, D. (2007b). *Evaluating Multidimensional Visualization Techniques in Data Mining Tasks.* TUCS Dissertations, issue 107. Turku: Turku Centre for Computer Science.

McGinn, D., Birch, D., Akroyd, D., Molina-Solana, M., Guo, Y., & Knottenbelt, W. J. (2016). Visualizing Dynamic Bitcoin Transaction Patterns. *Big Data*, 4(2), 109–119. doi:10.1089/big.2015.0056 PMID:27441715

Merrit, C. (2010). *Mobile Money Transfer Services: The Next Phase in the Evolution in Person-to-Person Payments (Tech. rep.)*. Atlanta, GA: Federal Reserve Bank of Atlanta.

Mohiuddin, A., Abdun, N. M., & Rafiqul, Md. I. (2016). A survey of anomaly detection techniques in financial domain. Future Gener. Comput. Syst., 55, 278-288.

Neural-technologies. (2017). *Minotaur Fraud Management Solution*. Retrieved September 10, 2017, from https://www.neuralt.com/73/393/optimus-fraud

Nice Actimize. (2017). *Fraud Prevention & Cybercrime Management Solutions Integrated Fraud Management*. Retrieved September 10, 2017, from http://www.niceactimize.com/fraud-detection-and-prevention

Novikova, E., & Kotenko, I. (2013). Analytical Visualization Techniques for Security Information and Event Management. In *21th Euromicro International Conference on Parallel, Distributed and network-based Processing (PDP 2013)* (pp.519-525). New York: IEEE. 10.1109/PDP.2013.84

Novikova, E., Kotenko, I., & Fedotov, E. (2014). Interactive Multi-view Visualization for Fraud Detection in Mobile Money Transfer Services. *International Journal of Mobile Computing and Multimedia Communications*, 6(4), 73–97. doi:10.4018/IJMCMC.2014100105

Orange Money. (2016). *Orange launches Orange Money in France to allow money transfers to three countries in Africa and within mainland France*. Press release. Retrieved September 09, 2017, from https://www.orange.com/en/Press-Room/press-releases/press-releases-2016/Orange-launches-Orange-Money-in-France-to-allow-money-transfers-to-three-countries-in-Africa-and-within-mainland-France

Prefuse. (2017). *Information Visualization toolkit*. Retrieved September 16, 2017, from http://prefuse.org/

Report, C. G. A. P. (2017). *Brief Fraud in the Mobile Financial Services*. Retrieved September 10, 2017, from http://www.cgap.org/sites/default/files/Brief-Fraud-in-Mobile-Financial-Services-April-2017.pdf

Rieke, R., Zhdanova, M., Repp, J., Giot, R., & Gaber, C. (2013). Fraud Detection in Mobile Payments Utilizing Process Behavior Analysis. In *The 2nd International Workshop on Recent Advances in Security Information and Event Management (RaSIEM 2013)* (pp. 662-669). New York: IEEE.

SAS. (2017). *Fraud Security Intelligence*. Retrieved September 13, 2017, from https://www.sas.com/en_us/solutions/fraud-security-intelligence.html

Schreck, T., Tekusova, T., Kohlhammer, J., & Fellner, D. (2007). Trajectory-based visual analysis of large financial time series data. *ACM SIGKDD Explorations Newsletter*, 9(2), 30–37. doi:10.1145/1345448.1345454

Shneiderman, B. (2003). Dynamic queries for visual information seeking. The Craft of Information Visualization: Readings and Reflections, 14-21.

Wattenberg, M. (1999). *Visualizing the stock market. In CHI Extended Abstracts on Human Factors in Computing Systems* (pp. 188–189). New York: ACM. doi:10.1145/632716.632834

Xie, C., Chen, W., Huang, X., Hu, Y., Barlowe, S., & Yang, J. (2014). VAET: A Visual Analytics Approach for E-Transactions Time-Series. *IEEE Transactions on Visualization and Computer Graphics, 20*(12), 1743–1752. doi:10.1109/TVCG.2014.2346913 PMID:26356888

Zhdanova, M., Repp, J., Rieke, R., Gaber, C., & Hemery, B. (2014). No Smurfs: Revealing Fraud Chains in Mobile Money Transfers. In *2014 Ninth International Conference on Availability, Reliability and Security* (pp. 11-20). New York: IEEE. 10.1109/ARES.2014.10

Ziegler, H., Jenny, M., Gruse, T., & Keim, D. A. (2010). Visual Market Sector Analysis for Financial Time Series Data. In *IEEE Symposium on Visual Analytics Science and Technology (VAST)* (pp.83-90). New York: IEEE. 10.1109/VAST.2010.5652530

Chapter 10
Visualizing Pathway on 3D Maps for an Interactive User Navigation in Mobile Devices

Teddy Mantoro
Sampoerna University, Indonesia

Media Anugerah Ayu
Sampoerna University, Indonesia

Adamu Ibrahim
International Islamic University Malaysia

ABSTRACT

3D maps have become an essential tool for navigation aid. The aim of a navigation aid is to provide an optimal route from the current position to the destination. Unfortunately, most mobile devices' GPS signal accuracy and the display of pathways on 3D maps in the small screen of mobile devices affects the pathway architectural from generating accurate initial positions to destinations. This chapter proposed a technique for visualizing pathway on 3D maps for an interactive user navigation aid in mobile devices. This technique provides visualization of 3D maps in virtual 3D workspace environments which assists a user to navigate to a target location. The Bi-A* path-finding algorithm was used for establishing dynamic target location in Voronoi diagram/Delaunay triangulation. This approach could navigate more than two users in a 3D walk-space and at the same time showing their whereabouts on 3D projections mapped. The map shows the users' location in the scene to navigate from source to the target and the target also moves to the source to meet on the same physical location and image plane.

DOI: 10.4018/978-1-5225-5693-0.ch010

INTRODUCTION

There are misconceptions in scenarios where an individual found himself/herself asking Where am I? Where can I find a certain location? These are the typical questions that might arises whenever a person visited a new environment. Even though individual will be able to know the name of the location of interest, but how to go there from certain location, for some new comers are not that easy, especially in finding the shortest path. 3D maps provide a full range of fundamental requirements of an effective visualization of the scene, built around a geometric representation of an area. Therefore, its representation in handheld GPS tools will enable people to identify the precise latitude and longitude of their present locations, even in the most remote of places. Evidently, the introduction of GPS technology for in-car navigation offers new solutions for finding one's way in urban areas and on the highway. Consequently, the social dynamics involved in traveling on the road have been transformed (Leshed et al., 2008). However, people still get lost or are unable to follow given directions to reach a particular destination. In certain unfortunate situations, a wrong turn can cause a serious accident (Ellard, 2009). Technically, the role of 3D maps is to offer more detailed information than what is available on conventional maps. Although geographic maps represent any space, real or imagined, without regard to context or scale, they have certain drawbacks. The information these conventional maps contain is limited due to fixed representation ratio and lack of interaction with the user. They usually require the translation of added symbols and legends, which may call for a certain level of expertise on the part of the user. The perception and interaction of 2D representations are limited to the interpretation of symbols explained in the legend which not every user knows how to use efficiently. Some works offered to use landmarks to assist in guiding users in their navigation (Ohm et al., 2016). In the case of a 3D map, proper reading is generally much easier and symbols are more straightforward. 3D representations are able to produce more realistic visualizations of navigation fields. A realistic 3D representation has strengths its 2D counterparts do not possess, for example, they are much more precise. The key benefit of a 3D representation is the higher potential for accuracy in presenting spatial data. Besides that, it offers a better platform for multiple cues and small-scale features, which are better suited for pedestrians to locate and identify unknown places. Therefore, a 3D map representation downloadable on a mobile device which represents a certain area in more detail helps the user to identify locations and decide which course to navigate to, at an instant. The accuracy of the generated model by sampling points on the reconstructed realistic 3D model and measuring each point's distance to the closest point in the ground truth models. The accuracy of a reconstruction can be defined as the percentage of points whose distance is in a given threshold (Schöps, 2017).

This paper presents pathway analysis for 3D mobile interactive navigation aid. Bent function, Voronoi diagram and its dual Delaunay triangulation are the algorithms used for establishing user positions and paths to a target destination. The motivation for providing the most favorable locations and path determinations is due to the condition of today's world, where mobility and communication have become essential. People may frequently find themselves asking: 'Where am I right now?', 'How do I get from X to Y?', and 'How can I tell Z from A?" Well-defined path determination and position are deemed as a method that can provide an accurate answer to these questions. This work is part of an integrated application in developing an Intelligent Environment. The goal is to make user interaction with the computer easier in a smart environment where technology spread throughout (pervasive), computers are everywhere at the same time (ubiquitous), and technology is embedded (ambient) in the environment. The technology development needs not to be difficult, tedious or need hard learning to the user. It should potentially be safe, easy, simple, and enable new functionality without a need to learn for a new technology.

3D MAPS IN MOBILE DEVICES

3D representations for navigation aid in mobile devices over a GSM network were experimented in Raposo et al. (1997) as one of the earliest attempts at the development of such a system. Later, the practical implementation by Rakkolainen et al. (1998) was seen to face a measure drawback where its first field experiments were confined to pre-rendered images on a static web page on a laptop computer, not on a mobile device, during their development of 3D City Info Project that was primarily meant to create mobile interactive 3D maps and visualize real-time GPS data. Furthermore, the frame rate with a Cortona CR VRML browser was one frame for every 8 seconds (0.125fps), with severe problems in running Java 2. However, the possibility of rendering large detailed 3D cities in mobile devices at interactive rates over 5 frames per second (without hardware acceleration) and over 30 frames per second with 3D hardware acceleration was presented in the m-LOMA project developed by Nurminen, A. (2006). Hence, the technology stands firm and more of such system started appearing.

There are quite a number of applications for mobile devices that use 3D maps for navigation aid. What are common to all those applications are the 3D graphics renderer and the 3D engine. Their credibility is mostly reflected by accuracies required in communication and recognizing locations, and their alignment to the display on the mobile device's screen (Evangelidis et al., 2018). All of these applications depend exclusively on continuous location determination with positioning accuracies from both GPS and RFID signals. However, the location and positioning's degree of accuracy might be reasonable even if it had an error margin of a few meters. Being in a scene directed by GPS or RFID signals assisted with 3D representation will cause the user to be more aware of the scene because the scene in the assisted device will be compared with the present location. The number of 3D navigation aided applications for mobile devices has steadily increased. A selection of available applications includes lol@ (local location assistance), GiMoDig (Geospatial info-mobility service by real-time data-integration and generalization), m-Loma (*mobile LOcation-aware Messaging Application*), and Mapper (Map personalization). Others are found in the M3I (mobile multi-modal interaction) project, which is a project for pedestrian navigation aids proposed by Wasinger (2003). It provides 2D and 3D graphics, and synthesized speech used to present useful information on routes and places. Embedded speech and gesture recognition allows for situated user interaction. The system is based on a navigation server and a Pocket PC. The 2D and 3D graphics are generated via the embedded Cortona VRML1-browser. A magnetic compass provides the direction the user is (i.e. the PDA's) current facing direction. GPS provides further sensor information, such as velocity and direction, and is also required to locate the user when outside. Infrared beacons are used to locate a user when inside. PDA communication with the server is via a standard HTTP connection, for example through a Bluetooth capable GPRS/UMTS mobile phone, WLAN, or a USB desktop connection. Dead reckoning by the GPS signal or Internet connections means loss of position and destination. Aslan, & Krüger (2004) proposed a pedestrian navigation framework in the Bun Bag Navigator (BNN) Project. The system is based on a light and wearable mobile navigation platform that sends information to the user and allows the user to interact with it. Thus, their platform provides multiple displays to convey information. It is modular constructed and distributed to two PDAs which can share resources and solve arising tasks in a dialogue with each other. The communication between separated modules is realized via WLAN and client-server architecture. The current position and the search destination depends on the wearable devices and is set by their system.

Chakraborty and Hashimoto (2010) have provided a pedestrian navigation framework that is user oriented. The implementation of their framework is planned by using fuzzy set theory, for analysis of the user's subjective preferences, and genetic algorithm, for multiple alternate route generation. This is very closely related to this paper's consideration of current position and desired destinations. While theirs uses fuzzy set theory and genetic algorithms for determining positions, ours uses the Voronoi diagram and its dual Delaunay triangulation. Other frameworks that consider the use of RFID can be found in Okamoto and Uchida (2004). They designed a framework for a specification and configuration policy of an RFID tag system which serves as an infrastructure for pedestrian navigation applications. The RFID tag system transmits signals to the RFID tag of which the communication distance is about 20m in a high-density urban pedestrian road network. The geographic information database which stores the location of the RFID has been prepared. A position can be pinpointed from the received tag ID. When RFIDs have been arranged densely, receiving one tag ID leads to the pinpointing of a position immediately by narrowing the communication range of a receiver. Firstly, the system requires a user's present position and walking direction to start navigation. Therefore, a user has to walk until two adjacent tag IDs are received. Using these tag IDs, the system determines the present position and walking direction. After this procedure a user adds the desired destination. Then, by referring to a path table from the present position to the targeted destination and an incidence table of relationship between adjacent nodes, the system provides the user with the correct walking directions through sound signal and screen information. Every time a new tag ID is received, the same procedure is repeated until the user arrives at the destination. When a user deviates from a designated route considerably, the user is alerted by a warning signal. The positioning and route determination was presented by stored information in a database by which there are no means of generating any information that could not be found in database.

3D MAPS FOR MOBILE NAVIGATION

The purpose of maps is to visualize geographical data in a way that is easy to understand. A 3D map is a two or three-dimensional visualization of a three-dimensional representation of a physical environment, emphasizing the three-dimensional characteristics of this environment intended for navigational purposes (Nurminen, 2006). There are quite a number of applications for navigation guidance with GPS or RFID support or with both GPS and RFID which uses 3D maps in mobile or fixed computing devices for both indoor and outdoor use, such as in-car GPS navigation system, and in-mobile phone GPS navigation applications. All of these applications depend exclusively on continuous location determination with positioning accuracies from both GPS and RFID signals. However, the location and positioning degree of accuracy of a virtual environment presented in 3D form might be reasonable even with an error margin of a few meters. Being in a scene directed by GPS or RFID signals assisted with 3D representation will cause the user to be more aware of the scene because the scene in the assisted device is compared with the present location. Although 3D representation on a small screen mobile device is deemed more suitable for a very big environment like New York, the problem of scaling may arise because recognizing large crowded high population density areas on a small screen tends to be rather difficult.

The number of 3D navigation aided applications for mobile device has steadily increased. A selection of some available applications are presented in Table 1 which include lol@ (local location assistance), GiMoDig (Geospatial info-mobility service by real-time data-integration and generalization), m-Loma (*mobile LOcation-aware Messaging Application*), Mapper (Map personalization), and MONA 3D (Mobile Navigation using 3D City Models).

Table 1. Features of existing 3D mobile navigation systems

Name	Features
lol@ (local location assistance)	Uses remote rendered data from the server and stored data in the device. It is meant for tourist/pedestrian for navigation aid and provides more text information of places and restricts presentation of multiple frames in 3D bird's eye view to give the illusion of 3D scenes.
m-Loma	The application is capable of rendering photorealistic 3D city models with augmented location-based information in a smart phone. It does not use photo images for representing an environment. It can perform textual searches to location-based content and navigate with GPS assisted signal. It is primarily designed for pedestrians, and use depends on data rendered by the server.
Google Maps 5.0	Uses vector graphics for its 2D maps in a mobile device. The vectors enable a two finger swipe to "tilt" the map to get a 3D view of the landscape. It enables offline caching of maps for frequently visited locations and entire trips that have been routed in navigation, including potential reroutes. It is suitable for pedestrian navigation aid.
Mapper	The application uses online data obtained from a remote server and provides route information to one or more destinations for pedestrians in a 3D format.
UpNext	Uses a 3D map/model to explore cities with an interactive touch interface for search and provides offline mode of recent map areas (pre-cached) on its device so that it can still be used offline.
Navigon	It provides a 3D bird's eye view and uses text-to-speech directions, real-time traffic updates, and information about points of interest along the route. It sets anyone from the address book as 'destination'.
GiMoDig	A project aimed at delivering geospatial data to a mobile user by means of real-time data-integration of national primary geo-databases. The application uses online data information acquired from a remote server through wireless connectivity.
Live Search Maps 3-D now Bing Maps	Uses bird's eye view maps and provides photo realistic textures that look like a relatively realistic representation of the environment. Although initially in 3D format, since November 2010, Microsoft decided to drop their 3D feature to focus on other aspects of Bing Maps.
EveryScape	Uses bird's eye view 3D photorealistic representations of cities and location to create (mostly) street-level 3D scenes and also offers: indoor panoramic scenes.
MONA 3D	This project provides mobile navigation aided system with a 3D view to serve for both in-car and pedestrian usage. It has been implemented for the city of Stuttgart and the city of Heidelberg, and addressed drawbacks of limited processing power and bandwidth of mobile devices and enhance navigation efficiency by compression of 3D and smart handling of textures for building facades using and setting up a 3D spatial data infrastructure (3D-SDI) based on OGC open web services (OWS).

POSITION AND PATHWAY SCENARIO

The objective of this paper is to present an optimized pathway technique assessment. The techniques used presently in most mobile 3D maps, as for example indicated in Figure 1, Lehtinen et al., (2007) offers easy shifting from the viewport-coupled mode to a top-down view where movement is POI-based. Path way generation that is suitable for path finding, in this paper, lie within the application of Voronoi diagrams and the Delaunay triangulation. Voronoi diagram is being defined as a collection of geometric objects that partition the plane into cells, each of which consists of the points closer to one particular object than to any other (Eppestein, D. 1990). As a rule, it states that when lines bisect the lines between a center point and its surrounding points, the bisecting lines and connections lines are perpendicular to each other (Zeki et al., 2004).

Since its initial introduction in the early 20th century, the Voronoi diagram has been used in many disciplines of science and engineering due to its natural descriptive and manipulative capability (Kim & Kim, 2006). There are many properties of Voronoi diagrams that apply to many situations, however, in the interest of establishing position and pathway, we consider a property that says if there is a connection

Figure 1. Path oriented 3D view in the mobile device (Lehtinen et al., 2012)

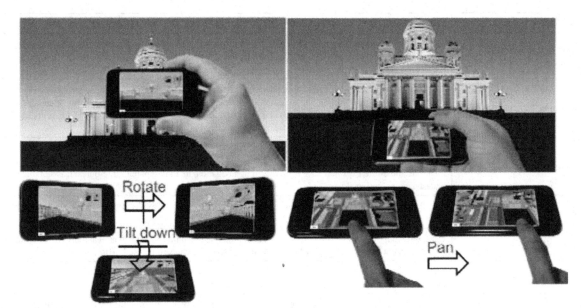

of all the pairs of region whose Voronoi cells are adjacent, then the resulting set of segments will be a triangulation of the node set, called the Delaunay triangulation. Hence, it's called the Voronoi duals. The idea of presenting locations and paths in way finding is shown in Figure 2 by Jimison et al. (2011), who said "there are discrepancies not only in nomenclature, but also in precision of location when user generated content shown in Figure 2; Nearly all physical objects encompass space greater than a singular point of latitude and longitude, and this is reflected in the input of the user."

Thus, the position and pathway determination is for a concept of multiple user interactive navigation with 3D maps in mobile devices where many users can interact together while moving across the perimeter of an unknown location and try to locate each other remains an integral part of pathway analysis. Consequently, the boundary of which the design of the application is considering is fixed. Therefore, the Voronoi diagram will now be employed to divide the entire boundary into appropriate disjointed data points (nodes) and sites (regions). The result of this is a separate layer which establishes the known points (nodes) and their closest neighbors within the different regions as shown in Figure 3.

The design of the implementation of the interactions between multiple users is done by taking into consideration every node within a certain perimeter and their closest neighbors. The users' inputs are the disjointed nodes as shown in Figure 5 where each node represents a stationary point of a mobile device registered on the dedicated link, and the output are the results of the creation of connection to all regions or the tessellation (Voronoi) of the regions within the same perimeter as shown in Figure 3. The node's distances apart are the Delaunay triangulation, established within each neighboring region as shown in Figure 3. The blue lines are the boundary created by the Voronoi diagram with blue points labeled from A to L as the nodes within each region. The nodes are connected to their closest neighbors by red lines forming a Delaunay triangulation. There are lemmas which show the minimum and maximum number of nodes nearest neighbor of each region within any other point in the Voronoi diagram,

Figure 2. Interactive pathways, stages involved in searching for a destination (Jimison et al., 2011)

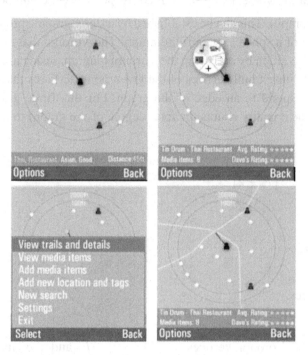

Figure 3. Voronoi diagram and Delaunay triangulation set for path and position determination

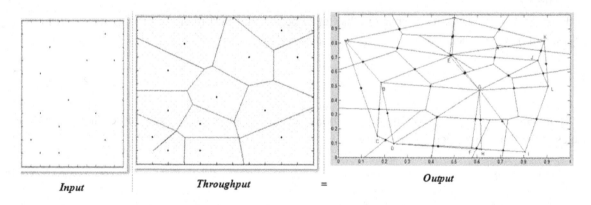

for example: *The number of Voronoi vertices and edges are* $2(n-1) - h$ *and* $3(n-1) - h$ *respectively, where n is the number of sites and h the number of sites on the convex hull of S. The Voronoi diagram V (S) has O(n) many edges and vertices. The average number of edges in the boundary of a Voronoi region is less than 6.* These two dilemmas have been proven by Euler's formula (Mantoro, et al., 2011). The first dilemma proof is by considering a finite graph which satisfies Euler's formula

$$2(n-1) - h \quad and \quad 3(n-1) - h \tag{1}$$

For a convex polyhedron, where V, E *and* F are respectively the number of vertices, edges, and faces of the polyhedron (graph). This can be established by a stereographic projection of the polyhedron, embedded on the surface of a sphere, into a finite graph. The Voronoi diagram of a set of n sites can be transformed into a finite graph by enclosing the Voronoi diagram inside a very large circle and treating each of the h (= # of convex hull vertices of the given point set) arcs that lie inside an unbounded Voronoi polygon to correspond to an edge of the graph. For this finite graph, we use the fact that $2E = 3V$ (each side of the equality counts the total degree of the graph) to deduce, using Euler's formula, that

$$V = 2\left(F - 2\right) \quad and \quad E = 3\left(F - 2\right) \tag{2}$$

$V = h + \#$ Voronoi vertices

$E = h + \#$ Voronoi edges.

Hence the number of Voronoi vertices $= 2\left(F - 2\right) - h$ and the number of Voronoi edges $= 3\left(F - 2\right) - h$. Since $F = n + 1$, these counts are therefore $2\left(n - 1\right) - h$ *and* $3\left(n - 1\right) - h$ respectively.

From Euler's formula for planar graphs, let G be a connected plane graph having n vertices, e edges, and f faces, then

$$n - e + f = 2 \tag{3}$$

From equation (3), the following relation holds for the numbers v, e, f, and c represented as vertices, edges, faces, and connected components.

$$v - e + f = 1 + c: \tag{4}$$

Thus, this relation when applied to the finite Voronoi diagram, that is for a certain boundary. Each vertex has at least three incident edges; by adding up we obtain

$$e \geq \frac{3v}{2} \tag{5}$$

because each edge is counted twice. Substituting this inequality together with $c = 1$ and $f = n + 1$ yields

$$v \leq 2n - 2 \quad and \quad e \leq 3n - 3 \tag{6}$$

Adding up the numbers of edges contained in the boundaries of all $n+1$ faces results in $2e \leq 6n - 6$ because each edge is again counted twice. Thus, the average number of edges in a region's boundary is bounded by

$$\left(6n - 6\right)\Big/\left(n + 1\right) < 6 \tag{7}$$

The Voronoi vertices are of degree at least three, by the proof of the second dilemma above. Vertices of any degree higher than three do not occur if no four point sites are co-circular. The Voronoi diagram $V(S)$ is disconnected if all point sites are collinear; in this case it consists of parallel lines (Aurenhammer & Schwarzkopf, 1991). From Figure 3, the Euclidean distances between one node to another is represented by the red line, when we refer to node A and B, the distance d between them is evaluated as follows:

$$d\left(A, B\right) = \sqrt{\left(A_x - B_x\right)^2 + \left(B_y - A_y\right)^2} \tag{8}$$

$$= \left[\left(A_x - B_x\right)\left(A_x - B_x\right) + \left(B_y - A_y\right)\left(B_y - A_y\right)\right]^{\frac{1}{2}}$$

$$= \left[\left(A_x^2 - A_x B_x - A_x B_x + B_x^2\right) + \left(B_y^2 - B_y A_y - B_y A_y + A_y^2\right)\right]^{\frac{1}{2}} \quad =$$

$$\left(\left(A_x^2 - 2A_x B_x + B_x^2\right) + \left(B_y^2 - 2A_y B_y + A_y^2\right)\right)^{\frac{1}{2}}$$

$$= \left(A_x^2 + A_y^2 - 2A_x B_x - 2A_y B_y + B_x^2 + B_y^2\right)^{\frac{1}{2}}$$

$$d(A, B) = \left(A_x^2 + A_y^2 - 2\left(A_x B_x + A_y B_y\right) + B_x^2 + B_y^2\right)^{\frac{1}{2}}$$

Ignoring other nodes from C to L, and asuming there are no other nodes available. Let $A = \left\{a_1, a_2, a_3, ..., a_n\right\}$ be the set of the n distinct nodes. These nodes are all the regions within the entire perimeter of the Euclidean plane. Thus, the Voronoi diagram of A will be the decomposition of the entire perimeter into n number of regions with a property that node B lies in the region corresponding to region a_i if and only if $d\left(b, a_i\right) < d\left(b, a_j\right)$ for each $a_j \in A$ with $j \neq i$. Series of regions emerge from Figure 4 which comprises of 12 nodes each inside his own region around a given triangulation vertex which encloses the entire node. Within that, they are closer to that node than any other node.

Where 2 nodes are not close to any other node, then the midpoint of the redline joining those nodes falls on the separating line between the Voronoi region for the node.

In another case, where there is a third node that is close to the red line joining 2 given nodes, then the Voronoi region for the third node may extend between the 2 given nodes, crossing the redline joining them and surrounding that lines midpoint. This third node can therefore be considered the closest neighbor to the 2 given nodes than the 2 nodes are to each other. There is a dilemma that shows a case where two nodes may not be considered to be closest neighbors even though their Voronoi region are next to each other and their boundaries touch. As described above, the straight red line dual of Voronoi diagram A for a set $A \subset \mathbb{R}^2$ of $n \geq 3$ points is the Delaunay triangulation of A. However, the triangulation for the set of $A \subset \mathbb{R}^2$ of n nodes can be computed in $O(n)$ time complexity. This indicates that the generators points of the Voronoi diagran are not collinear and when their number is finite and more than 3 closest points join together, they produce a tessilation that forms a triangle. This is shown in Figure 4, except for DFH. In this case its called Delaunay pretriangulation, while those that form the triangle are the called Delaunay triangulation. From the asumption that the area we want to decompose and the established position and pathway are a boundary of a certain perimeter, for example, $\theta(p)$ the solution for pathway and position will be the computation of all the possible nodes that will be created within the perimeter and all the nearest neighbors of each node within. Thus, this will be denoted as follows:

Let the set $\Omega \subseteq \mathbb{R}^{n+1}$ be defined by the set of all possible nodes and regions, and the closest nodes between each other by

$$\Omega = \left\{ (p, \theta) \mid x_i(x, \theta) \leq 0, \forall_i \in \mathbb{N}_q \right\} \tag{9}$$

Figure 4. The GPS tracks data within a given area

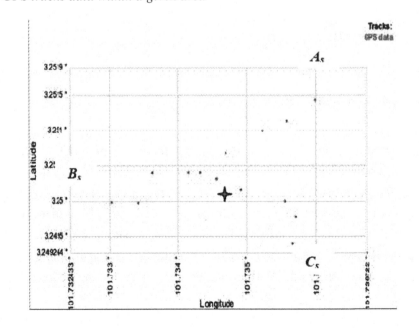

$$\Omega = \left\{ (p, \theta) \mid A_x + B_\theta \leq p \right\} \tag{10}$$

This set provides random entries of all possible nodes and regions within a certain perimeter p. Suppose the perimeter increases, which may lead to the possibility of an increase in each node and region. Thus, this could be resolved by a linear programming approach as follows:

$$A_x = \begin{bmatrix} a_{x1} \\ a_{x2} \\ \cdot \\ \cdot \\ \cdot \\ a_{xn} \end{bmatrix} \quad B_x = \begin{bmatrix} b_{x1} \\ b_{x2} \\ \cdot \\ \cdot \\ \cdot \\ b_{xn} \end{bmatrix} \quad A_y = \begin{bmatrix} a_{y1} \\ a_{y2} \\ \cdot \\ \cdot \\ \cdot \\ a_{yn} \end{bmatrix} \quad B_y = \begin{bmatrix} b_{y1} \\ b_{y2} \\ \cdot \\ \cdot \\ \cdot \\ b_{yn} \end{bmatrix} \tag{11}$$

From equation 2;

$$\text{Let } k = A_x^2 A_y = \begin{bmatrix} a_{x1} \\ a_{x2} \\ \cdot \\ \cdot \\ \cdot \\ a_{xn} \end{bmatrix} \begin{bmatrix} a_{x1} \\ a_{x2} \\ \cdot \\ \cdot \\ \cdot \\ a_{xn} \end{bmatrix} \begin{bmatrix} a_{y1} \\ a_{y2} \\ \cdot \\ \cdot \\ \cdot \\ a_{yn} \end{bmatrix} \tag{12}$$

$$w = A_x B_x + A_y B_y = \begin{bmatrix} a_{x1} \\ a_{x2} \\ \cdot \\ \cdot \\ \cdot \\ a_{xn} \end{bmatrix} \begin{bmatrix} b_{x1} \\ b_{x2} \\ \cdot \\ \cdot \\ \cdot \\ b_{xn} \end{bmatrix} + \begin{bmatrix} a_{y1} \\ a_{y2} \\ \cdot \\ \cdot \\ \cdot \\ a_{yn} \end{bmatrix} \begin{bmatrix} b_{y1} \\ b_{y2} \\ \cdot \\ \cdot \\ \cdot \\ b_{yn} \end{bmatrix} \tag{13}$$

$$m = B_x^2 + B_y^2 = \begin{bmatrix} b_{x1} \\ b_{x2} \\ \cdot \\ \cdot \\ \cdot \\ b_{xn} \end{bmatrix} \begin{bmatrix} b_{x1} \\ b_{x2} \\ \cdot \\ \cdot \\ \cdot \\ b_{xn} \end{bmatrix} + \begin{bmatrix} b_{y1} \\ b_{y2} \\ \cdot \\ \cdot \\ \cdot \\ b_{yn} \end{bmatrix} \begin{bmatrix} b_{y1} \\ b_{y2} \\ \cdot \\ \cdot \\ \cdot \\ b_{yn} \end{bmatrix} \tag{14}$$

Therefore $d(A, B) = \sqrt{k - 2w + m}$

Querying nodes within the perimeter surrounded by a specific region generate other information on the closest neighbors of each bounded nodes, for example in Figure 3, nodes

$$E\hat{B}G, \quad A\hat{B}C, \quad C\hat{B}D, \quad A\hat{B}E, \quad D\hat{B}G$$

are the only possible nodes which are bounded within node *B*. This is the reason why the structure of a Voronoi diagram and Delaunay triangulation are being used for pathway and position determination in the development of navigation guides for mobile devices with a 3D view. The implementation of the Algorithm considering the entire perimeter under consideration in Voronoi of A is as follows:

Input $= A$, which is the set of all the n distinct nodes $A = \left\{a_1, a_2, a_3, ..., a_n\right\}$ within the parameter.

Output $= A_{ij}$ as the decomposition of the entire perimeter into n number of regions connected by the their closest neighboring nodes.

Initialize the procedure with the selected random node a within the entire perimeter.

while*edge index and midpoint along near index* $A a_n$ is established

do Find the vertex closest to the midpoints and the edges

if the midpoint is not closer to another vertex within a region occurring at $\left\{a_n\right\}$

then find the generator of a_n, whose Voronoi polygon contains $a_1, ..., a_n$

else update the triangulation edge indices and edge midpoints

do Find the vertex nearest the midpoints

then update the triangulation edge indices and edge midpoints until the edges are exhausted

Initialize another random node to repeat the algorithm.

PATHWAYS EVALUATION

The individuals' current positions, which is the starting position for navigation, is represented in Figure 4 as A_s, B_s and C_s signifying starting points for A, B, and C respectively, intended for three mobile devices. The symbol Cs indicates the destination for A, B, and C respectively. This position is acquired through the integration of 3D engine application via Voronoi diagram/Delaunay triangulation techniques and satellite navigation system, in particular GPS. The 3D engine provides a symbol (as a node) which only displays based on individual current positions obtained from the satellite navigation system as the starting point, and for each movement, there is always an update on the changes of position. The display symbol which indicates an individual's current position and the change in position update is transformed into a map coordinate system. The position accuracy for example in some navigation devices is achieved via on-board GPS signal receiver strength. However in this case, it is the combination of it and the computation of the nodes initiated by the integration of Voronoi diagram nodes and distance between the neighboring nodes evaluated by Delaunay triangulation. This case happens only for the nodes within a perimeter of a given area. For example, an individual wishes to move from point A to point H. Point A is the reference point which will be initiated by Voronoi diagram/Delaunay triangulation and GPS, while point H is the destination which is calculated by the techniques depending on the given algorithm. The pathway is through neighboring points from A to H. However, where individuals decide to stop after taking some distances away from the starting point, they are able to capture the GPS update of the points where they stopped and any movement along the line, but that does not mean that those points are evaluated by Voronoi diagram and Delaunay triangulation.

There is a dilemma provided above that says that points within a Voronoi diagram are finite for a given area, but for GPS update, any change of movement will be updated as long as there is signal connection between the devices and the satellite navigation system. Within a given area, as represented in Figure 6, 7 and 8, are the same points combined in Figure 5. However, those points were recorded during the time taken at halt while individuals are navigating. The nodes in Figure 6, 7, and 8 were indicating the latitude update of the navigation orientations for the three devices, which are North/South orientation. The whole directional orientation North, South, East and West are presented in Figure 5.

There is an advantage with combining the GPS received signal in position accuracy with computing locations of established points. However, the position accuracy from GPS, where it does not exceed more than 10m, is highly reliable. The change of position via GPS is calibrated to North, as a result, the symbol which will indicates the nodes is an arrow. However, for a system like ours, since it is providing for dedicated point/nodes, it is not necessary to use the arrow sign. That is why a marker differentiated by colors is used (see Figure 5 and 6).

The implementation of this looks similar to the point of interest in some navigation devices, where remote rendering services have to be provided for the client to make a request of the point of interest to the service provider who has a list of all points of interest which the client can select and resend to the service provider for those points, thus the service provider responds to those points. Those points can now be evaluated using bent functions in order to know which paths are used at certain times and which

Figure 5. Latitude update for the given position

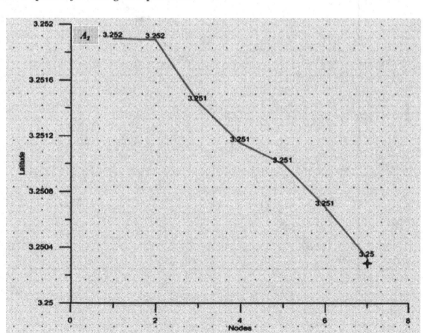

Figure 6. Latitude update for the given positions

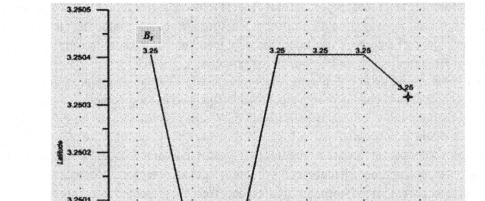

Figure 7. Latitude update for the given positions

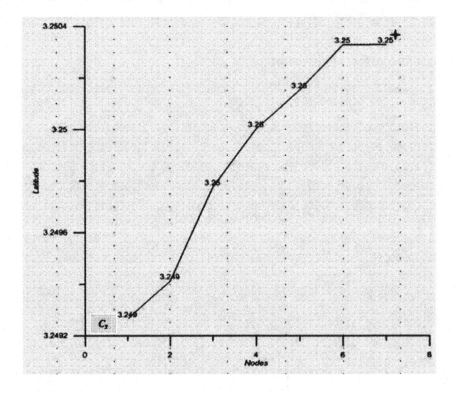

Figure 8. Position of nodes within a given perimeter

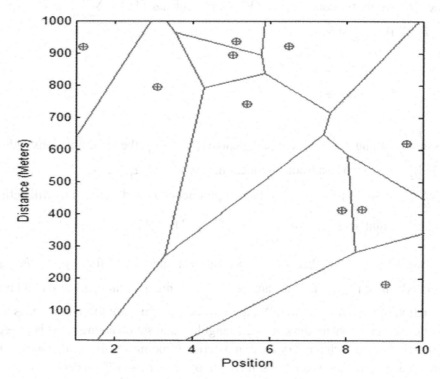

paths are not and how the change of using one path affects the change of using another path. Let A be a set of m elements of R defining a polytope $P = \{x \in \mathbb{R}^n : a^T x \leq 1, a \in A\} \in$ having m facets. For $y \in \mathbb{R}^n$, we define

$$D_y := \left\{ (S,T) : S, T \subseteq A \quad and \quad y \in convex\left((-S) \cup T \right) \right\},$$

where $convex(X)$ denotes the convex hull of X, and where $(-X)$ denotes the set $\{-a : a \in A\}$ For each subset $s \subseteq \{1, 2, \cdots, n\}$, there exists a corresponding vector $s = (s_1, s_2, \cdots, s_n)$ of dimension n by letting $s_i = 0$. Thus, this assumption satisfies a Galois field with two elements denoted by $F_2 = \{0, 1\}$ which also can be supported by F_2^n in vector spaces over F_2. Hence, the algebraic of the entire function $F_2^n \rightarrow F_2$ can be designated by

$$p_n = \frac{F_2[x_1, x_2, \cdots, x_n]}{\left(x_1^2 - x_1, \cdots, x_n^2 - x_n \right)} \tag{15}$$

where for each subset $s \subseteq \{1, 2, \cdots, n\}$, denote $\prod_{i \in s} x_i \in p_n$ by x^s.

Thealgebraicnormalformofa Boolean function $F_2^n \rightarrow F_2$ written as $f(x) = \sum_{s=0}^{2^n-1} a_s x^s,\ \ where\ a_s \in F_2^n$. The degree of $f(x)$ is defined by

$$H(s)\ \ s \in \{0, 1, \cdots, 2^n - 1\}, a_s \neq 0 \tag{16}$$

where $H(s)$ is the Hamming weight of vector s, denote by $AF(n)$, the set of functions with $\deg(f) \leq 1$ Thus, if $f(x) : F_2^n \rightarrow F_2$ is to be a Boolean function, where $x = (x_1, x_2, \cdots, x_n)$, $w = (w_1, w_2, \cdots, w_n)$ and $w.x = w_1 x_1 + x_2 w_2 + \cdots + x_n w_n \in F_2$ is the dot product of w and x, then this will define the Walsh spectrum of $f(x)$ at point w as $s_{(f)}(w) = \sum_{x \in F}(-1)^{f(x)}(-1)^{w.x}$ (17)

Moreover, if we let $f(x)$, such that $x \in F_2^n$ is a Boolean function, if for any $w \in F_2^n, |s(f)| = 2^{n/2}$, then $f(x)$ is called a bent function. Thus, any positions within a certain part of the 3D built model or 3D map is a function of the bent function. What we do here to compare those systems is to established finite points within a region, where three points triangulate and set up connections between them. This will make it possible to give each point or the combination of points attribute information, for example, those belonging to a certain class share the same path, or those ranges from certain neighborhoods are at the mid-valley or those that are far apart require special paths. In Figure 9, there is a representation of the Voronoi nodes for which each node has its closest neighbor generated within a given perimeter. In the design of the system, the class for the positions of each are set (see Table 2). Two or more nodes may be in the same position but distances apart differentiate them. Where two nodes happen to be in the same position, one of them may be below or above the other.

All the distances between those nodes where presented in Table 3 and the interaction with the closest neighbor is presented in Figure 9. The information which is being presented indicates the distances and position interactions from all the given points within that region.

For example, A and B are neighbors (see Figure 9 A), however the system indicates that A connects and forms a triangulation with H and C (see Figure 9 B) and it does not share any boundary with C and H. This does not mean that the closest neighbors do not connect.

The idea only suggests that position and pathway can be established through selected points and connections, however, this indicates a low computing mobile device that computes multiple pathways and renders the displayed results which could be prioritized in order to enhance the computing time. This also suggests that multi-user interactions between two or more connecting devices could be accomplished. Furthermore, there is always a chance for an expansion. This should be done by ignoring the computation of the pathways that are not connected and computing the connected pathways.

Another fact is for implementation of pedestrians. They have the capability to acquire route knowledge, which encompasses the sequence of landmarks in an area coupled with navigation instructions which requires a collection of procedural as well as topological knowledge illustrated in the form of a graph of nodes and edges that are constantly growing as more nodes and edges are added (Nurminen, A. 2008).

Figure 9. Neighborhood relation and poistion connection

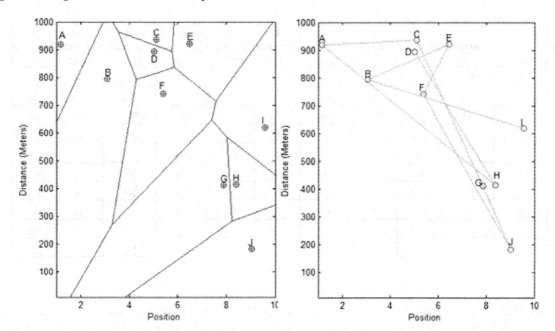

Table 2. Presentation of the position of nodes

Nodes	Positions(m)
A	0.8
B	2.7
C	5.6
D	5.5
E	7.2
F	5.7
G	8.0
H	8.2
I	9.8
J	9.2

DISCUSSION

Voronoi diagram/Delaunay triangulation and A* forward and reverse techniques are the navigation objects used in the design of pedestrian navigation aided system with 3D view in mobile devices. They generate points (nodes) and display the closest neighbors' connections between the sets of points within a given perimeter. The result of any given input is presented in a 3D dataset layer where the viewer's perception of the 3D model, 3D bird's eye view and 2D map, are displayed depending on the user's input choice. The idea of using Voronoi diagram and Delaunay triangulation comes from the fact that they

Table 3. Presentation of the distance apart from each of the nodes in meters

Nodes	A Distance apart	B Distance apart	C Distance apart	D Distance apart	E Distance apart	F Distance apart	G Distance apart	H Distance apart	I Distance apart	J Distance apart
A	No	No	18	No	No	No	No	550	No	No
B	No	No	No	No	76	65	No	No	203	No
C	18	No	No	No	No	No	602	570	No	No
D	No	No	No	No	No	No	No	477	No	No
E	No	76	No	No	No	197	No	No	No	No
F	65	No	No	No	197	No	No	No	130	654
G	No	No	602	No	No	No	No	No	No	300
H	550	No	550	No	No	No	No	No	No	No
I	No	203	No	No	No	130	No	No	No	No
J	No	No	700	No	No	No	300	No	No	No

provide finite points and all the possible connections of the points within a certain boundary. They only establish well defined positions and pathways. Finding the optimal shortest path or even any path to the desired destination requires using pathfinding algorithms. That is why in our case we use A* forward and reverse search technique. The dedicated links provided ensure that only a certain number of devices registered on the links can exhibit "multi-user interactivity". The contribution of this technique for the design of navigation aid is the provision of the solution to people asking: 'Where am I right now?', 'How do I get from X to Y?', and 'How can I tell Z from A?". This is because a well-defined path determination and position are computed by the user's input within a certain boundary and delivered on request. The challenge faced when designing a 3D model for mobile devices for navigation aid is posed by the huge amount of datasets it contains. Thus, data reduction is critical, especially if the mobile device runs on limited memory and storage space. Data reduction is achievable through 3D model simplification or 3D compression. This reduces the model to a more reasonable size, keeps original appearance, and realizes the localization of data storage for display (Nurminen, 2008). 3D digital map system construction considered as a very complex research project. Lots of studies should been done, including in data fusion, modeling, network transmission,data management and analysis of 3D space and geographic coordination (Zhongyuan, 2012).

PEDESTRIAN NAVIGATION IN 3D

An early prototype based on a scenario for pedestrian navigation were developed. The scenario is for a two-way direction of user navigation. Each user has the same capability to see each other from their own angle.

Figure 10. 3D walk-space transitions: user 'blue' will meet user 'green' from different angle

The scenario for two-way direction of user navigation is as the following: Assume we have user blue and user green and they are in different area and they make appointment to meet face to face in certain spot/landmark. Figure 10 shows two 3D images from user blue's and user green's side. Each of them can see their current location in the 3D maps (blue dot) and also in 3D walk-space. They can easily know his location by comparing the ornament/landmark surrounding them and the 3D walk-space natural image in his smartphone. At the same time, both users have to know where the location of their counterpart is, so they can have a feeling/estimation about the distance and time they need to meet. The last one is they need to decide where they are going to meet.

3D transitions or movement process between locations which takes too long runs the risk of becoming tedious, and in the end, annoying to users. One solution to the problem of abrupt transitions is to make the transition gradually by either introducing a brief but smooth animated transition of the viewpoint, or if the visualization is object centric, an animated transformation of the object or objects.

Figure 11 shows the transition/movement of both users. User 'blue' and user 'green' can see that they are moving toward meeting with each other. The facing of the 3D map image related with the facing of 3D walk-space based on the GPS reading. In this case, both diagrams follow the direction of the mobile user, when they are on the move with their smartphones.

Figure 11. The distance of user blue and user green is getting closer and they can estimate based on time and distance between them

Figure 12. User 'blue' in a meeting with user 'green'

Animating transitions between 3D maps and 3D walk-spaces may have two possible benefits. First, users can see the maps directory in 3D, enlarging and moving from its previous location into the new location. If the user were to click on and enter the wrong location, it should be easier to realize the error with animation and compare to the real environment. If, on the other hand, the transition had been abrupt the user might easily continue walking in the location space. Second, the animation emphasizes the relation between the current location and the previous user location and surrounding. This may simplify learning the location structure of the 3D workspace in realism image and the real environment itself.

Figure 12 shows that user 'blue' and user 'green' has finally met as they expected. In Figure 13, the 3D maps and 3D walk-space can be shown in a smartphone by reducing the quality of the image since we are considering the performance of the smartphone to re-draw the image in a small mobile device such as smartphone.

CONCLUSION

As part of a project for designing pedestrian navigation aid with 3D maps in a mobile device, Voronoi diagram and Delaunay triangulation were used for pathway and position determination. This paper presents these techniques and explains why they are necessary. Other researchers used fuzzy set and genetic algorithm for a similar approach. Our approach is promising in the sense that it does not predict position or route but it ensures their presence within a certain perimeter. The success of any navigation aid is to provide an optimal route from the current position; this means that providing precise positions and pathways to the destination is highly critical. The algorithms used are commonly used in robotics and geometrics for route planning. However, its application in mobile devices is also viable. In this paper, we explained the reason for using the techniques and show how it is being implemented in a navigation project, and we explore the mathematics behind it. The recommendation for future work will be on experimental user studies to evaluate its applicability.

Figure 13. 3D walk-space with 3D map inside, tweaked to work in smartphones

REFERENCES

Adamu, A. I., Mantoro, T., Ayu, M. A., & Mahmud, M. (2011). Exploring End-User Preferences of 3D Mobile Interactive Navigation Design. *Proceedings of the 9th International Conference on Advances in Mobile Computing & Multimedia*, 289-292.

Adamu, A. I., Mantoro, T., & Shafi'I, M. A. (2012). Dynamic Interactive 3D Mobile Navigation Aid. *International Journal of Theoretical and Applied Information Technology.*, *37*(2), 159–170.

Aslan, I., & Krüger, A. (2004). The Bum Bag Navigator (BBN): An Advanced Pedestrian Navigation System. *UbiComp Workshop on Artificial Intelligence in Mobile Systems*, 15-19.

Aurenhammer, F., & Schwarzkopf, O. (1991). A simple on-line randomized incremental algorithm for computing higher order Voronoi diagrams. *Proceedings of the seventh annual symposium on Computational geometry*, 142-151. 10.1145/109648.109664

Chakraborty, B., & Hashimoto, T. (2010). A Framework for User Aware Route Selection in Pedestrian Navigation System. *Aware Computing (ISAC), 2010 2nd International Symposium,* 150–153.

Darken, R. P., & Peterson, B. (2001). Spatial Orientation, Wayfinding, and Representation. In Handbook of Virtual Environment Technology. Erlbaum.

Dey, A. K., Abowd, G. D., & Salber, D. (2001). A Conceptual Framework and a Toolkit for Supporting the Rapid Prototyping of Context-Aware Applications. *Journal of Human Computer Interaction*, *16*(2), 97–166. doi:10.1207/S15327051HCI16234_02

Ellard, C. (2009). *Where Am I? Why We Can Find Our Way to the Moon But Get Lost in the Mall*. London: Harper Collins Publishers Ltd.

Eppestein, D. (1990). The Furthest Point Delaunay Triangulation Minimizes Angles. *Computational Geometry*, *1*(3), 143–148. doi:10.1016/0925-7721(92)90013-I

Evangelidis, K., Papadopoulos, T., Papatheodorou, K., Mastorokostas, P., & Hilas, C. (2018). 3D geospatial visualizations: Animation and motion effects on spatial objects. *Computers & Geosciences*, *111*, 200–212. doi:10.1016/j.cageo.2017.11.007

Helbing, P., & Molnar, P. (1995). Social force model for pedestrian dynamics. *Physical Review. E*, *51*(5), 4282–4286. doi:10.1103/PhysRevE.51.4282 PMID:9963139

Jimison, D., Sambasivan, N., & Pahwa, S. (2007). *Wigglestick: An Urban Pedestrian Mobile Social Navigation System*. Workshop Onpervasive Mobile.

Jul, S., & Furnas, G. W. (1997). Navigation in Electronic Worlds: A CHI 97 Workshop. *ACM SIGCHI Bulletin*, *29*(4), 44-49.

Kim, D., & Kim, D. (2006). Region-expansion for the Voronoi diagram of 3D spheres. *Journal of Computer-Aided Design*, *38*(5), 417–430. doi:10.1016/j.cad.2005.11.007

Leshed, G., Velden, T., Rieger, O., Kot, B., & Sengers, P. (2008). In-Car GPS Navigation: Engagement with and Disengagement from the Environment. *Proceedings of the twenty-sixth annual SIGCHI conference on Human factors in computing systems*, 1675-1684. 10.1145/1357054.1357316

Löhner, R. (2010). On the modeling of pedestrian motion. *Journal of Applied Mathematical Modeling*, *34*(2), 366–38. doi:10.1016/j.apm.2009.04.017

Mantoro, T., Abubakar, A., & Ayu, M. A. (2011). Multi-User Navigation: A 3D Mobile Device Interactive Support. *IEEE Symposium on Industrial Electronics and Applications (ISIEA 2011)*, 545-549. 10.1109/ISIEA.2011.6108772

Nurminen, A. (2006). m-LOMA - a Mobile 3D City Map. *Proceedings of the eleventh international conference on 3D web technology*, 7–18.

Nurminen, A. (2008). Mobile 3D City Maps. *Journal of IEEE Computer Graphics and Applications*, *28*(4), 20–31. doi:10.1109/MCG.2008.75 PMID:19004682

Ohm, C., Bienk, S., Kattenbeck, M., Ludwig, B., & Müller, M. (2016). Towards interfaces of mobile pedestrian navigation systems adapted to the user's orientation skills. *Pervasive and Mobile Computing*, *26*, 121–134. doi:10.1016/j.pmcj.2015.10.006

Okamoto, A., & Uchida, T. (2004). Pedestrian Navigation System Easy-to-Use-in Urban Canyon: Algorithms, system development and a field trial in Osaka. *Memoirs of the Faculty of Engineering Osaka City University*, *45*, 65–70.

Rakkolainen, I., Pulkkinen, S., & Heinonen, A. (1998). Visualizing real-time GPS data with VRML worlds. *Proceedings of the ACM-GIS98 workshop*, 52–56.

Raposo, A., Neumann, L., Magalhaes, L., & Ricarte, I. (1997). Efficient Visualization in a mobile WWW environment. *Proceedings of the WebNet 97, World Conference of the WWW, Internet, and Intranet*.

Schöps, T., Sattler, T., Häne, C., & Pollefeys, M. (2017). Large-scale outdoor 3D reconstruction on a mobile device. *Computer Vision and Image Understanding*, *157*, 151–166. doi:10.1016/j.cviu.2016.09.007

Spence, R. (1999). A Framework for Navigation. *International Journal of Human-Computer Studies*, *51*(5), 919–945. doi:10.1006/ijhc.1999.0265

Ville Lehtinen, V., Nurminen, A., & Oulasvirta, A. (2012). Integrating Spatial Sensing to an Interactive Mobile 3D Map. *IEEE Symposium on 3D User Interfaces*.

Wasinger, R., Stahl, C., & Krüger, A. (2003). Mobile Multi-Modal Interaction in a Pedestrian Navigation & Exploration System. *Proceedings of MobileHCI*, 481-485.

Yan, Y., & Kunhui, L. (2010). 3D Visual Design for Mobile Search Result on 3G Mobile Phone. *Proceedings of the International Conference on Intelligent Computation Technology and Automation*, 12-16. 10.1109/ICICTA.2010.489

Zeki, A. M., Ghyasi, A. F., Mujahid, M., Zainul, N., Cheddad, A., Zubayr, M., & Zakaria, S. M. (2004). Design and Implementation of a Voronoi Diagrams Generator using Java. *International Arab Conference on Information Technology*, 194-198.

Zhongyuan, Z. (2012). Research on 3D Digital Map System and Key Technology. *Procedia Environmental Sciences*, *12*, 514–520. doi:10.1016/j.proenv.2012.01.311

Compilation of References

5GAmericas. (2016, November). *Network Slicing for 5G and Beyond*. 5G Americas White Paper. Retrieved from http://www.5gamericas.org/files/3214/7975/0104/5G_Americas_Network_Slicing_11.21_Final.pdf

Abdi, H., & Williams, L. J. (2010). Principal component analysis. *WIREs Comp. Stat*, *2*, 433–459.

Abdulhussein, M., Abbas, T., Servel, A., Hofmann, F., Thein, C., Bedo, J. S., …Trossen, D. (2015, October). *5G Automotive Vision*. ERTICO, European Commission, & 5G-PPP.

Abdulkafi, Kiong, Sileh, Chieng, & Ghaleb. (2016). *A Survey of Energy Efficiency Optimization in Heterogeneous Cellular Networks*. Academic Press.

Abreu-Sernandez, V., & Garcia-Mateo, C. (2000). Adaptive multi-rate speech coder for VoIP transmission. *Electronics Letters*, *36*(23), 1978–1980. doi:10.1049/el:20001344

Acharya, S., Alonso, R., Franklin, M., & Zdonik, S. (1995). Broadcast Disks: Data Management for Asymmetric Communication Environments. *Proc. of ACM Sigmod International Conference on Management of Data*. 10.1145/223784.223816

Achemlal, M., et al. (2011). *Scenario requirements*. (Tech. rep.) MASSIF FP7-257475 project.

Adami, D., Callegari, C., Giordano, S., Pagano, M., & Pepe, T. (2012). Skype-Hunter: A real-time system for the detection and classification of Skype traffic. *International Journal of Communication Systems*, *25*(3), 386–403. doi:10.1002/dac.1247

Adamu, A. I., Mantoro, T., Ayu, M. A., & Mahmud, M. (2011). Exploring End-User Preferences of 3D Mobile Interactive Navigation Design. *Proceedings of the 9th International Conference on Advances in Mobile Computing & Multimedia*, 289-292.

Adamu, A. I., Mantoro, T., & Shafi'I, M. A. (2012). Dynamic Interactive 3D Mobile Navigation Aid. *International Journal of Theoretical and Applied Information Technology.*, *37*(2), 159–170.

Aksoy, D., Altinel, M., Bose, R., Cetintemel, U., Franklin, M., Wang, J., & Zdonik, S. (1999). Research in Data Broadcast and Dissemination. *LNCS*, *1554*, 194–207.

Alam, M. G. R., Tun, Y. K., & Hong, C. S. (2016). Multi-agent and reinforcement learning based code offloading in mobile fog. *International Conference on Information Networking (ICOIN)*, 285–290. 10.1109/ICOIN.2016.7427078

Alcatel-Lucent. (2009). *The LTE Network Architecture - A Comprehensive Tutorial*. Strategic White Paper.

Alexander, S., & Droms, R. (1997, March). *Dynamic Host Configuration Protocol*. IETF RFC.

Alicherry, M., Bhatia, R., & Li, L. E. (2005, August). Joint channel assignment and routing for throughput optimization in multi-radio wireless mesh networks. In *Proceedings of the 11th annual international conference on Mobile computing and networking* (pp. 58-72). ACM. 10.1145/1080829.1080836

Al-Khatib, A. (2012). Electronic Payment Fraud Detection Techniques. *World of Computer Science and Information Technology Journal*, 2, 137–141.

Al-Manthari, B., Hassanein, H., & Nasser, N. (2007). Packet scheduling in 3.5G high –speed downlink packet access networks. *IEEE Network*, 21(1), 52–57. doi:10.1109/MNET.2007.314537

Alshamrani, A., Shen, X., & Xie, L. L. (2011). QoS provisioning for heterogeneous services in cooperative cognitive radio networks. *IEEE Journal on Selected Areas in Communications*, 29(4), 819–830. doi:10.1109/JSAC.2011.110413

Al-Surmi, I., Othman, M., & Ali, B. M. (2010, February). Review on Mobility Management for Future IP-Based Next Generation Wireless Networks. *International conference on Advanced Communication Technology (ICACT)*, 989–994.

Amento, B., Balasubramanian, B., Hall, R. J., Joshi, K., Jung, G., & Purdy, K. H. (2016). FocusStack: Orchestrating Edge Clouds Using Location-Based Focus of Attention. *2016 IEEE/ACM Symposium on Edge Computing (SEC)*, 179-191. 10.1109/SEC.2016.22

Ankerst, M., Berchtold, S., & Keim, D. A. (1998). Similarity Clustering of Dimensions for an Enhanced Visualization of Multidimensional Data. In *IEEE Symposium on Information Visualization (INFOVIS '98)* (pp. 52-60). Washington, DC: IEEE Computer Society. 10.1109/INFVIS.1998.729559

Anyfi. (2014). *Software-Defined Wireless Networking: Concepts, Principles and Motivations*. Whitepaper, Anyfi Networks.

Arkian, H. R., Diyanat, A., & Pourkhalili, A. (2017). MIST: Fog-based data analytics scheme with cost-efficient resource provisioning for IoT crowdsensing applications. *Journal of Network and Computer Applications*, 82, 152–165. doi:10.1016/j.jnca.2017.01.012

Aslan, I., & Krüger, A. (2004). The Bum Bag Navigator (BBN): An Advanced Pedestrian Navigation System. *UbiComp Workshop on Artificial Intelligence in Mobile Systems*, 15-19.

Aurenhammer, F., & Schwarzkopf, O. (1991). A simple on-line randomized incremental algorithm for computing higher order Voronoi diagrams. *Proceedings of the seventh annual symposium on Computational geometry*, 142-151. 10.1145/109648.109664

Avallone, S., & Akyildiz, I. F. (2008). A channel assignment algorithm for multi-radio wireless mesh networks. *Computer Communications*, *31*(7), 1343–1353. doi:10.1016/j.comcom.2008.01.031

Awad, Mohamed, & Chiasserini. (2016). Dynamic Network Selection in Heterogeneous Wireless Networks. *IEEE Consumer Electronics Magazine.*

Badrinath, B. R., & Phatak, S. H. (n.d.). *An Architecture for Mobile Databases.* Technical Report DCS-TR-351. Department of Computer Science, Rutgers University.

Bailey, S., Bansal, D., Dunbar, L., Hood, D., Kis, Z. L., Mack-Crane, … Varma, E. (2013, December). *SDN Architecture Overview.* Open Networking Foundation, Version 1.0 –draft v08.

Banks, A., & Gupta, R. (2014). *MQTT Version 3.1.1. OASIS standard.* Retrieved from http://docs.oasis-open.org/mqtt/mqtt/v3.1.1/csprd02/mqtt-v3.1.1-csprd02.html

Bansal, M., Mehlman, J., Katti, S., & Levis, P. (2012). OpenRadio: A Programmable Wireless Dataplane. *1st Workshop on Hot topics in Software Defined Networks, HotSDN'2012*, 109-114.

Barbara, D. (1999). Mobile Computing and Databases-A Survey. *IEEE TKDE*, *11*(1), 108–117.

Bari, F., & Leung, V. C. M. (2007). Automated network selection in a heterogeneous wireless network environment. *IEEE Network*, *21*(1), 34–40. doi:10.1109/MNET.2007.314536

Basagni, S., Chlamtac, I., Syrotiuk, V. R., & Barry, W. (1998). A Distance Routing Effect Algorithm for Mobility (DREAM). In *Proceedings of the 4th Annual ACM/IEEE International Conference on Mobile Computing and Networking* (pp.76-84). New York: IEEE. 10.1145/288235.288254

Bayer, R., & McCreight, E. M. (1972). Organization and Maintenance of Large Ordered Indices. *Acta Informatica*, *1*(3), 173–189. doi:10.1007/BF00288683

Bellavista, P., & Zanni, A. (2017). Feasibility of Fog Computing Deployment based on Docker Containerization over RaspberryPi. In *Proceedings of the 18th International Conference on Distributed Computing and Networking (ICDCN '17).* ACM. 10.1145/3007748.3007777

Bellman, R. (1958). On a Routing Problem. *Quarterly of Applied Mathematics*, *16*(1), 87–90. doi:10.1090/qam/102435

Betts, M., Davis, N., Dolin, R., Doolan, P., Fratini, S., Hood, D., … Dacheng, Z. (2014, June). *SDN Architecture.* Issue 1, ONF TR-502.

Bhorkar, A., Naghshvar, M., Javidi, T., & Rao, B. (2012). Adaptive Opportunistic Routing for Wireless Ad-Hoc Networks. *IEEE/ACM Transactions on Networking, 20*(1), 243–256. doi:10.1109/TNET.2011.2159844

Biswas, S., & Morris, R. (2005). ExOR: Opportunistic Multi-Hop Routing for Wireless Networks. In *Proceedings of the 2005 Conference on Applications, Technologies, Architectures, and Protocols for Computer Communications (SIGCOMM '05)* (pp. 133-144). New York: ACM. 10.1145/1080091.1080108

Bitam, S., Zeadally, S., & Mellouk, A. (2017). Fog computing job scheduling optimization based on bees swarm. *Enterprise Information Systems*, 1–25.

Bittencourt, L. F., Diaz-Montes, J., Buyya, R., Rana, O. F., & Parashar, M. (2017). Mobility-aware application scheduling in fog computing. *IEEE Cloud Computing, 4*(2), 26–35. doi:10.1109/MCC.2017.27

Bletsas, A., Dimitriou, A., & Sahalos, J. (2010). Interference-limited Opportunistic Relaying with Reactive Sensing. *IEEE Transactions on Wireless Communications, 9*(1), 14–20. doi:10.1109/TWC.2010.01.081128

Bonomi, F., Milito, R., Zhu, J., & Addepalli, S. (2012). Fog Computing and Its Role in the Internet of Things. In *Proceedings of the First Edition of the MCC Workshop on Mobile Cloud Computing* (pp. 13-16). New York: ACM. 10.1145/2342509.2342513

Brady, P. T. (1969). A model for generating on-off speech patterns in two-way conversation. *The Bell System Technical Journal, 48*(7), 2445–2472. doi:10.1002/j.1538-7305.1969.tb01181.x

Breunig, M. M., Kriegel, H.-P., Ng, R. T., & Sander, J. (2000). LOF: identifying density-based local outliers. In *Proceedings of the 2000 ACM SIGMOD International Conference on Management of Data (SIGMOD '00)* (pp. 93-104). ACM. 10.1145/342009.335388

Bruno, R., & Conti, M. (2010). MaxOPP: A Novel Opportunistic Routing for Wireless Mesh Networks. In *Proceedings of IEEE Symposium on Computers and Communications (ISCC)* (pp. 255-260). Riccione, Italy: IEEE. 10.1109/ISCC.2010.5546793

Bruno, R., Conti, M., & Gregori, E. (2003). Optimal Capacity of *p*-persistent CSMA Protocols. *IEEE Communications Letters, 7*(3), 139–141. doi:10.1109/LCOMM.2002.808371

Cadger, F., Curran, K., Santos, J., & Moffett, S. (2013). A Survey of Geographical Routing in Wireless Ad-Hoc Networks. *IEEE Communications Surveys and Tutorials, 15*(2), 621–653. doi:10.1109/SURV.2012.062612.00109

Caldas de Castro, M., & Singer, B. (2006). Controlling the False Discovery Rate: A New Application to Account for Multiple and Dependent Test in Local Statistics of Spatial Association. *Geographical Analysis, 38*(2), 180–208. doi:10.1111/j.0016-7363.2006.00682.x

Cao, X., Chen, J., Zhang, Y., & Sun, Y. (2008). Development of an integrated wireless sensor network micro-environmental monitoring system. *Elsevier ISA Trans., 47*(3), 247–255. doi:10.1016/j.isatra.2008.02.001 PMID:18355827

Carter, S., & Yasinsac, A. (2002). Secure Position Aided Ad hoc Routing. In *Proceedings of the IASTED International Conference on Communications and Computer Networks (CCN02)* (pp.329-334). Academic Press.

Ceselli, A., Premoli, M., & Secci, S. (2017). Mobile Edge Cloud Network Design Optimization. *IEEE/ACM Transactions on Networking*, 1-14.

Ceselli, A., Premoli, M., & Secci, S. (2015). *Cloudlet network design optimization IFIP Networking Conference* (pp. 1–9). Toulouse: IFIP Networking.

Chachulski, S., Jennings, M., Katti, S., & Katabi, D. (2007). Trading Structure for Randomness in Wireless Opportunistic Routing. In *Proceedings of the 2007 Conference on Applications, Technologies, Architectures, and Protocols for Computer Communication* (pp. 169-180). Academic Press. 10.1145/1282380.1282400

Chakraborty, B., & Hashimoto, T. (2010). A Framework for User Aware Route Selection in Pedestrian Navigation System. *Aware Computing (ISAC), 2010 2nd International Symposium*, 150–153.

Chamola, V., Tham, C. K., & Chalapathi, G. S. S. (2017). Latency aware mobile task assignment and load balancing for edge cloudlets. In *IEEE International Conference on Pervasive Computing and Communications Workshops (PerCom Workshops)*. (pp. 587–592). IEEE. 10.1109/PERCOMW.2017.7917628

Chang, C., Srirama, S. N., & Liyanage, M. (2015). A Service-Oriented Mobile Cloud Middleware Framework for Provisioning Mobile Sensing as a Service. *Proceedings of the 21st IEEE International Conference on Parallel and Distributed Systems (ICPADS 2015)*, 124-131. 10.1109/ICPADS.2015.24

Chang, R., Ghoniem, M., Kosara, R., Ribarsky, W., Yang, J., & Suma, E., … Sudjianto, A. (2007). WireVis: Visualization of Categorical, Time-Varying Data From Financial Transactions. In *IEEE Symposium on Visual Analytics Science and Technology (VAST 2007)* (pp.155-162). Washington, DC: IEEE Computer Society. 10.1109/VAST.2007.4389009

Chen, M., Wan, J., & Li, F. (2012). Machine-to-Machine Communications: Architectures, Standards, and Applications. *KSII Transaction on Internet and Information Systems*, *6*(2), 480-497.

Cheng, T. K., Wu, J. L. C., Yang, F. M., & Leu, J. S. (2014). IEEE 802.16e/m energy-efficient sleep-mode operation with delay limitation in multibroadcast services. *International Journal of Communication Systems*, *27*(1), 45–67. doi:10.1002/dac.2342

Chen, J., Cao, X., Cheng, P., Xiao, Y., & Sun, Y. (2010). Distributed collaborative control for industrial automation with wireless sensor and actuator networks. *IEEE Transactions on Industrial Electronics*, *57*(12), 4219–4230. doi:10.1109/TIE.2010.2043038

Chen, Y. S., Deng, D. J., Hsu, Y. M., & Wang, S. D. (2012). Efficient uplink scheduling policy for variable bit rate traffic in IEEE 802.16 BWA systems. *International Journal of Communication Systems*, *25*(6), 734–748. doi:10.1002/dac.1206

Chlamtac, I., Conti, M., & Liu, J. (2003). Mobile Ad Hoc Networking: Imperatives and Challenges. *Ad Hoc Networks*, *1*(1), 13–64. doi:10.1016/S1570-8705(03)00013-1

Choi, H. H., & Cho, D. H. (2007). Hybrid energy-saving algorithm considering silent periods of VoIP traffic for mobile WiMAX. *2007 IEEE International Conference on Communications,* 1-14, 5951-5956. 10.1109/ICC.2007.986

Choi, H. H., Lee, J. R., & Cho, D. H. (2007). Hybrid power saving mechanism for VoIP services with silence suppression in IEEE 802.16e systems. *IEEE Communications Letters*, *11*(5), 455–457. doi:10.1109/LCOMM.2007.070035

Choi, H. H., Lee, J. R., & Cho, D. H. (2009). On the Use of a Power-Saving Mode for Mobile VoIP Devices and Its Performance Evaluation. *IEEE Transactions on Consumer Electronics*, *55*(3), 1537–1545. doi:10.1109/TCE.2009.5278024

Cisco. (2016, April). *PMIP: Multipath Support on MAG and LMA, IP Mobility*. Mobile IP Configuration Guide, Cisco IOS XE.

Clark, B. N., Colbourn, C. J., & Johnson, D. S. (1990). Unit disk graphs. *Discrete Mathematics*, *86*(1-3), 165–177. doi:10.1016/0012-365X(90)90358-O

Clausen, T., & Jacquet, P. (2003). *Optimized link state routing protocol (OLSR)* (No. RFC 3626).

ColorBrewer2. (2015). Retrieved June 13, 2015, from http://colorbrewer2.org

Comer, D. (1979). The Ubiquitous B-Trees. *ACM Computing Surveys*, *11*(2), 121–137. doi:10.1145/356770.356776

Conan, J. L. V., & Friedman, T. (2008). Fixed Point Opportunistic Routing in Delay Tolerant Networks. *IEEE Journal on Selected Areas in Communications*, *26*(5), 773–782. doi:10.1109/JSAC.2008.080604

Contreras, L. M., Cominardi, L., Qian, H., & Bernardos, C. J. (2016, April). Software-Defined Mobility Management: Architecture Proposal and Future Directions. *Springer Mobile Netw Appl, 21*(2), 226–236. doi:10.100711036-015-0663-7

Coppolino, L., D'Antonio, S., Formicola, V., Massei, C., & Romano, L. (2015). Use of the Dempster–Shafer theory to detect account takeovers in mobile money transfer services. *Journal of Ambient Intelligence and Humanized Computing*, 1–10.

Corkin, L. (2016). Kenya's mobile money story and the runaway success of M-Pesa. *Digital Frontiers*. Retrieved September 08.09.2017, from http: http://www.orfonline.org/expert-speaks/kenyas-mobile-money-story-and-the-runaway-success-of-m-pesa

Cowan, J. (2012). Kenya's Safaricom deploys new fraud management system for its mobile payment service. *IoT NOW*. Retrieved September 10, 2017, from https://www.iot-now.com/2012/08/21/7080-kenyas-safaricom-deploys-new-fraud-management-system-for-its-mobile-payment-service/

Cox, R. V., & Kroon, P. (1996). Low bit-rate speech coders for multimedia communication. *IEEE Communications Magazine*, *34*(12), 34–41. doi:10.1109/35.556484

Cuervo, E., Balasubramanian, A., Cho, D.-k., Wolman, A., Saroiu, S., Chandra, R., & Bahl, P. (2010). Maui: Making smartphones last longer with code offload. In *Proceedings of the 8th International Conference on Mobile Systems, Applications, and Services*. ACM.

Darken, R. P., & Peterson, B. (2001). Spatial Orientation, Wayfinding, and Representation. In Handbook of Virtual Environment Technology. Erlbaum.

Delamaire, L., Abdou, H., & Pointon, J. (2009). Credit card fraud and detection techniques: a review. *Banks and Bank Systems*, *4*(2), 57-68.

Demirkol, I., Ersoy, C., & Alagoz, F. (2006). MAC protocols for wireless sensor networks: A survey. *IEEE Communications Magazine*, *44*(4), 115–121. doi:10.1109/MCOM.2006.1632658

Denazis, S., Koufopavlou, O., & Haleplidis, E. (Eds.). (2015, January). Software-Defined Networking (SDN): Layers and Architecture Terminology. RFC 7426.

Devasia. (2017). Iterative Control for Networked Heterogeneous Multi Agent Systems With Uncertainties. *IEEE Transactions on Automatic Control*.

Dey, A. K., Abowd, G. D., & Salber, D. (2001). A Conceptual Framework and a Toolkit for Supporting the Rapid Prototyping of Context-Aware Applications. *Journal of Human Computer Interaction*, *16*(2), 97–166. doi:10.1207/S15327051HCI16234_02

Di Battista, G., Di Donato, V., Patrignani, M., Pizzonia, M., Roselli, V., & Tamassia, R. (2014). BitConeView: visualization of flows in the bitcoin transaction graph. *2015 IEEE Symposium on Visualization for Cyber Security (VizSec)*, 1-8.

Di Caro, L., Frias-Martinez, V., & Frias-Martinez, E. (2010). *Analyzing the Role of Dimension Arrangement for Data Visualization in Radviz. In Advances in Knowledge Discovery and Data Mining. LNCS* (Vol. 6119, pp. 125–132). Berlin: Springer-Verlag. doi:10.1007/978-3-642-13672-6_13

Draves, R., Padhye, J., & Zill, B. (2004, September). Routing in multi-radio, multi-hop wireless mesh networks. In *Proceedings of the 10th annual international conference on Mobile computing and networking* (pp. 114-128). ACM. 10.1145/1023720.1023732

Dubois-Ferriere, H., Grossglauser, M., & Vetterli, M. (2007). Least-Cost Opportunistic Routing. In *Proceedings of 2007 Allerton Conference on Communication, Control, and Computing* (pp. 1-8). Allerton, UK: Academic Press.

Elkhatib, Y., Porter, B., Ribeiro, H. B., Zhani, M. F., Qadir, J., & Riviere, E. (2017, March). On Using Micro-Clouds to Deliver the Fog. *IEEE Internet Computing*, *21*(2), 8–15. doi:10.1109/MIC.2017.35

Ellard, C. (2009). *Where Am I? Why We Can Find Our Way to the Moon But Get Lost in the Mall*. London: Harper Collins Publishers Ltd.

Elmasri, R., & Navathe, S. B. (2003). *Fundamentals of Database Systems* (4th ed.). Addison Wesley.

Elsadek, W. F. & Mikhail, M. N. (2016, October). SRMIP: A Software-Defined RAN Mobile IP Framework for Real Time Applications in Wide Area Motion. *Int. J. of Mobile Computing and Multimedia Communications, 7*(4), 28-49. DOI: 10.4018/IJMCMC.2016100103

Elsadek, W. F. (2016). Toward Hyper Interconnected IoT World using SDN Overlay Network for NGN Seamless Mobility. *8th IEEE International Conference on Cloud Computing Technology and Science (CloudCom)*. 10.1109/CloudCom.2016.0078

Emara, T. Z. (2016). Maximizing Power Saving for VoIP over WiMAX Systems. *International Journal of Mobile Computing and Multimedia Communications, 7*(1), 32–40. doi:10.4018/IJMCMC.2016010103

Emara, T. Z., Saleh, A. I., & Arafat, H. (2014). Power saving mechanism for VoIP services over WiMAX systems. *Wireless Networks, 20*(5), 975–985. doi:10.100711276-013-0650-5

Ephremides, A., & Mowafi, O. A. (1982). Analysis of a hybrid access scheme for buffered users-probabilistic time division. *IEEE Transactions on Software Engineering, SE-8*(1), 5261. doi:10.1109/TSE.1982.234774

Eppestein, D. (1990). The Furthest Point Delaunay Triangulation Minimizes Angles. *Computational Geometry, 1*(3), 143–148. doi:10.1016/0925-7721(92)90013-I

Erramilli, A. C. V., Crovella, M., & Diot, C. (2008). Delegation Forwarding. In *Proceedings of ACM International Symposium on Mobile Ad Hoc Networking and Computing* (pp. 251-260). Hong Kong: ACM.

Erran, L. L., Mao, Z. Z., & Rexford, J. (2012). Toward Software-Defined Cellular Networks. *European Workshop on Software Defined Networking (EWSDN)*, 7-12.

Esmat, B., Mikhail, M. N., & El Kadi, A. (2000, May). *Enhanced Mobile IP Protocol*. IFIP-TC6/European Commission NETWORKING 2000 International Workshop, MWCN 2000, Paris, France. 10.1007/3-540-45494-2_13

Estepa, A., Estepa, R., & Vozmediano, J. (2004). A new approach for VoIP traffic characterization. *IEEE Communications Letters, 8*(10), 644–646. doi:10.1109/LCOMM.2004.835318

ETSI. (2016). *Small Cell LTE Plugfest*. Retrieved from http://www.etsi.org/about/10-news-events/events/1061-small-cell-lte-plugfest-2016

Evangelidis, K., Papadopoulos, T., Papatheodorou, K., Mastorokostas, P., & Hilas, C. (2018). 3D geospatial visualizations: Animation and motion effects on spatial objects. *Computers & Geosciences, 111*, 200–212. doi:10.1016/j.cageo.2017.11.007

Fan, Tian, Zhang, & Zhang. (2017). Virtual MAC concept and its protocol design in virtualised heterogeneous wireless network. *IET Commun., 11*(1), 53–60.

Fang, X., Yang, D., & Xue, G. (2011). Consort: Device-Constrained Opportunistic Routing in Wireless Mesh Networks. In Proceedings of IEEE INFOCOM (pp. 1907-1915). Shanghai, China: IEEE.

Fang, X., Yang, D., & Xue, G. (2013). Map: Multi-Constrained Any Path Routing in Wireless Mesh Networks. *IEEE Transactions on Mobile Computing, 12*(10), 1893–1906. doi:10.1109/TMC.2012.158

FATF. (2010). *Money Laundering using New Payment Methods.* Retrieved June 13, 2015, from http://www.fatf-gafi.org/topics/methodsandtrends/documents/moneyla

Fernando, N., Loke, S. W., & Rahayu, W. (2013). Honeybee: A Programming Framework for Mobile Crowd Computing. In *Proceedings of the 2012 International Conference on Mobile and Ubiquitous Systems: Computing, Networking, and Services* (pp. 224–236). Springer Berlin Heidelberg. 10.1007/978-3-642-40238-8_19

Fernando, N., Loke, S. W., & Rahayu, W. (2016). Computing with nearby mobile devices: a work sharing algorithm for mobile edge-clouds. *IEEE Transactions on Cloud Computing,* 1–1.

Fiserv. (2017). *Financial Crime Risk Management solution.* Retrieved September 11, 2017, from https://www.fiserv.com/risk-compliance/financial-crime-risk-management.aspx

Gaber, C., Hemery, B., Achemlal, M., Pasquet, M., & Urien, P. (2013). Synthetic logs generator for fraud detection in mobile transfer services. In *Int. Conference on Collaboration Technologies and Systems (CTS 2013)* (pp.174-179). New York: IEEE. 10.1109/CTS.2013.6567225

Gau, R. H., & Cheng, C. P. (2013). Optimal tree pruning for location update in machine-to-machine communications. *IEEE Transactions on Wireless Communications, 12*(6), 2620–2632. doi:10.1109/TWC.2013.040413.112086

Ghadimi, E., Landsiedel, O., Soldati, P., Duquennoy, S. & Johansson, M. (2014). Opportunistic Routing in Low Duty-Cycled Wireless Sensor Networks. *ACM Transactions on Sensor Networks, 10*(4), 67:1-67:39.

Ghazvini, F. K., Ali, M. M., & Doughan, M. (2017). Scalable hybrid MAC protocol for M2M communications. *Computer Networks. Elsevier Publications, 127,* 151–160.

Ghosh, A., & Misra, I. S. (2016). An analytical model for a resource constrained QoS guaranteed SINR based CAC scheme for LTE BWA Het-Nets. *International Conference on Advances in Computing, Communications and Informatics (ICACCI).* 10.1109/ICACCI.2016.7732376

Ghosh, A., Wolter, D. R., Andrews, J. G., & Chen, R. H. (2005). Broadband wireless access with WiMax/802.16: Current performance benchmarks and future potential. *IEEE Communications Magazine, 43*(2), 129–136. doi:10.1109/MCOM.2005.1391513

Giordano, S., & Lu, W. (2001). Challenges in Mobile Ad Hoc Networking. *IEEE Communications Magazine*, *39*(6), 29–129. doi:10.1109/MCOM.2001.925680

Golmohammadi, K., Zaiane, O. R., & Díaz, D. (2014). Detecting stock market manipulation using supervised learning algorithms. *2014 International Conference on Data Science and Advanced Analytics (DSAA)*, 435-441. 10.1109/DSAA.2014.7058109

Gubbi, J., Buyya, R., Marusic, S., & Palaniswami, M. (2013). Internet of things (IoT): A vision, architectural elements, and future directions. *Future Generation Computer Systems*, *29*(7), 1645–1660. doi:10.1016/j.future.2013.01.010

Gudipati, A., Perry, D., Erran, L. L., & Katti, S. (2013). SoftRAN: Software Defined Radio Access Network. *Second ACM SIGCOMM Workshop on Hot topics in Software Defined Networking, (HotSDN)*, 25-30. 10.1145/2491185.2491207

Guerin, J., Portmann, M., & Pirzada, A. (2007, December). Routing metrics for multi-radio wireless mesh networks. In *Proceedings of Australasian Telecommunication Networks and Applications Conference* (pp. 343-348). IEEE. 10.1109/ATNAC.2007.4665270

Gundavelli, S. (Ed.). (2008, August). Proxy Mobile IPv6. IETF RFC 5213.

Guo, P., Lin, B., Li, X., He, R., & Li, S. (2016). Optimal deployment and dimensioning of fog computing supported vehicular network. In *Proceedings of the 2016 IEEE Trustcom/BigDataSE/ISPA*, (pp. 2058–2062). IEEE. 10.1109/TrustCom.2016.0315

Guo, S., Gu, Y., Jiang, B., & He, T. (2009). Opportunistic Flooding in Lowduty- Cycle Wireless Sensor Networks with Unreliable Links. In *Proceedings of IEEE/ACM Annual International Conference on Mobile Computing and Networking (MobiCom)* (pp. 133-144). Beijing, China: IEEE. 10.1145/1614320.1614336

Gupta, R., & Rastogi, N. (2012). *LTE Advanced – LIPA and SIPTO*. White Papers.

Gupta, P., & Kumar, P. R. (2000). The capacity of wireless networks. *IEEE Transactions on Information Theory*, *46*(2), 388–404. doi:10.1109/18.825799

Haas, Z.J., & Pearlman, M.R. (1997). *The Zone Routing Protocol (ZRP) for Ad Hoc Networks*. Internet Draft, Available at hdraft-haaszone-routing-protocol-00.txt.

Hajji, W., & Tso, F. P. (2016). Understanding the Performance of Low Power Raspberry Pi Cloud for Big Data. *Electronics (Basel)*, *5*(2), 29. doi:10.3390/electronics5020029

Hampel, G., Rana, A., & Klein, T. (2013). Seamless TCP Mobility Using Lightweight MPTCP Proxy. *11th ACM Symposium on Mobility Management and Wireless Access (MobiWac)*, 139-146. 10.1145/2508222.2508226

Han, M. K., Bhartia, A., Qiu, L., & Rozner, E. (2011). O3: Optimized Overlay Based Opportunistic Routing. In *Proceedings of the ACM International Symposium on Mobile Ad Hoc Networking and Computing (MobiHoc)* (pp. 2:1-2:11). ACM.

Hasan, M., Hossain, E., & Niyato, D. (2013). Random access for machine-to-machine communication in LTE-advanced networks: Issues and approaches. *IEEE Communications Magazine*, *51*(6), 86–93. doi:10.1109/MCOM.2013.6525600

Hashimoto, M., Hasegawa, G., & Murata, M. (2015). An analysis of energy consumption for TCP data transfer with burst transmission over a wireless LAN. *International Journal of Communication Systems*, *28*(14), 1965–1986. doi:10.1002/dac.2832

Haykin, S. (2007). *Neural Networks: A Comprehensive Foundation* (3rd ed.). Prentice-Hall, Inc.

Helbing, P., & Molnar, P. (1995). Social force model for pedestrian dynamics. *Physical Review. E*, *51*(5), 4282–4286. doi:10.1103/PhysRevE.51.4282 PMID:9963139

Hong, K., Lillethun, D., Ramachandran, U., Ottenwalder, B., & Koldehofe, B. (2013). Mobile fog: A programming model for large-scale applications on the Internet of things. In *Proceedings of the Second ACM SIGCOMM Workshop on Mobile Cloud Computing* (pp. 15–20). New York: ACM. 10.1145/2491266.2491270

Hout, M. C., Papesh, M. H., & Goldinger, S. D. (2013). Multidimensional scaling. *Wiley Interdisciplinary Reviews: Cognitive Science*, *4*(1), 93–103. doi:10.1002/wcs.1203 PMID:23359318

Hsu, C., Liu, H., & Seah, W. (2009). Economy: A Duplicate Free Opportunistic Routing. In *Proceedings of the 6th ACM International Conference on Mobile Technology Application and Systems* (pp. 17:1-17:6). New York: ACM.

Hsu, T. H., & Yen, P. Y. (2011). Adaptive time division multiple access-based medium access control protocol for energy conserving and data transmission in wireless sensor networks. *IET Communications*, *5*(18), 2662–2672. doi:10.1049/iet-com.2011.0088

Huang, M. L., Liang, J., & Nguyen, Q. V. (2009). A Visualization Approach for Frauds Detection in Financial Market. *2009 13th International Conference Information Visualisation*, 197-202.

Huang, J. L., & Chen, M.-S. (2002). Dependent Data Broadcasting for Unordered Queries in a Multiple Channel Mobile Environment. *Proc. of the IEEE GLOBECOM*, 972-976.

Huang, J. L., & Chen, M.-S. (2003). Broadcast Program Generation for Unordered Queries with Data Replication. *Proc. of the 8th ACM Symposium on Applied Computing*, 866-870. 10.1145/952532.952704

Huang, Y., Sistla, P., & Wolfson, O. (1994). Data Replication for Mobile Computers. *Proc. of the ACM SIGMOD*, 13-24.

Huerta-Canepa, G., & Lee, D. (2010). A virtual cloud computing provider for mobile devices. In *Proceedings of the 1st ACM Workshop on Mobile Cloud Computing & Services: Social Networks and Beyon* (pp. 6:1–6:5). New York: ACM. 10.1145/1810931.1810937

Hu, Q., Lee, D. L., & Lee, W. C. (1998). Optimal Channel Allocation for Data Dissemination in Mobile Computing Environments. *Proc. of 18th International Conference on Distributed Computing Systems*, 480-487.

Hu, Q., Lee, W. C., & Lee, D. L. (1999). Indexing Techniques for Wireless Data Broadcast under Data Clustering and Scheduling. *Proc. of the 8th ACM International Conference on Information and Knowledge Management*, 351-358. 10.1145/319950.320027

Hussain, T. H., Marimuthu, P. N., & Habib, S. J. (2014). Supporting multimedia applications through network redesign. *International Journal of Communication Systems, 27*(3), 430–448. doi:10.1002/dac.2371

Hwang, H., Hur, I., & Choo, H. (2009). GOAFR plus-ABC: Geographic routing based on Adaptive Boundary Circle in MANETs. In *Proceedings of the 2009 International Conference on Information Networking* (pp. 1-3). Chiang Mai, Thailand: Academic Press.

IEEE Std 802.11-2012. (2012). Information Technology—Telecommunications and information exchange between systems—Local and metropolitan area networks—Specific requirements—Part 11: wireless LAN medium access control (MAC) and physical layer (PHY) specifications.

IEEE. (2006). IEEE Standard for Local and Metropolitan Area Networks Part 16: Air Interface for Fixed and Mobile Broadband Wireless Access Systems Amendment 2: Physical and Medium Access Control Layers for Combined Fixed and Mobile Operation in Licensed Bands and Corrigendum 1. In IEEE Std 802.16e-2005 and IEEE Std 802.16-2004/Cor 1-2005 (Amendment and Corrigendum to IEEE Std 802.16-2004) (pp. 0_1-822).

Igarashi, Y., Ueno, M., & Fujisaki, T. (2012). Proposed Node and Network Models for an M2M Internet. *World Telecommunications Congress (WTC)*, 1-6.

Imielinski, T., & Viswanathan, S. (1994). Adaptive Wireless Information Systems. *Proc. of SIGDBS (Special Interest Group in Database Systems)*, 19-41.

Imielinski, T., Viswanathan, S., & Badrinath, B. R. (1997). Data on Air: Organisation and Access. *IEEE TKDE, 9*(3), 353–371.

Islam, A., Islam, A. A., M. J., Nurain, N., & Raghunathan, V. (2016). Channel assignment techniques for multi-radio wireless mesh networks: A survey. *IEEE Communications Surveys & Tutorials, 18*(2), 988-1017.

ISO/IEC 9126-1. (2001). *Product quality. Part 1: Quality model*. Retrieved June 13, 2015, from http://www.iso.org/iso/iso_catalogue/catalogue_tc/catalogue_detail.htm?csnumber=22749

ITU-T. (1993). P.59: Artificial conversational speech. In ITU-T P.59 (03/93).

ITU-T. (2006). G.723.1: Dual rate speech coder for multimedia communications transmitting at 5.3 and 6.3 kbit/s. In ITU-T G.723.1 (05/06).

ITU-T. (2012). Coding of speech at 8 kbit/s using conjugate-structure algebraic-code-excited linear prediction (CS-ACELP). In ITU-T G.729

Jack, W., Tavneet, S., & Townsend, R. (2010). Monetary Theory and Electronic Money: Reflections on the Kenyan Experience. Economic Quarterly, 96(1), 83-122.

Jacquet, P., Muhlethaler, P., & Qayyum, A. (1998). *Optimized Link State Routing Protocol*. Internet Draft, Available at draft-ietf-manetolsr-00.txt.

Jiang, J., Li, J. D., & Hou, R. (2013). Network selection policy based on effective capacity in heterogeneous wireless communication systems. *Science China. Information Sciences*, 56.

Jimison, D., Sambasivan, N., & Pahwa, S. (2007). *Wigglestick: An Urban Pedestrian Mobile Social Navigation System*. Workshop Onpervasive Mobile.

Jin, S., Chen, X., Qiao, D. J., & Choi, S. (2011). Adaptive sleep mode management in IEEE 802.16m wireless metropolitan area networks. *Computer Networks*, *55*(16), 3774–3783. doi:10.1016/j.comnet.2011.03.002

Jin, X., Erran, L. L., Vanbevery, L., & Rexford, J. (2013). SoftCell: Scalable and Flexible Cellular Core Network Architecture. *Ninth ACM Conference on Emerging Networking Experiments and Technologies (CoNEXT)*, 163-174. 10.1145/2535372.2535377

Johnson, D., Perkins, C., & Arkko, J. (2004, June). *Mobility Support in IPv6*. RFC 3775.

Johnson, D. B., & Maltz, D. A. (1996). Dynamic Source Routing in Ad Hoc Wireless Networks. *Mobile Computing, Kluwer Academic Publishers*, *353*, 153–181. doi:10.1007/978-0-585-29603-6_5

Jul, S., & Furnas, G. W. (1997). Navigation in Electronic Worlds: A CHI 97 Workshop. *ACM SIGCHI Bulletin*, *29*(4), 44-49.

Kalle, R. K., Gupta, M., Bergman, A., Levy, E., Mohanty, S., Venkatachalam, M., & Das, D. (2010). Advanced Mechanisms for Sleep Mode Optimization of VoIP Traffic over IEEE 802.16m. *2010 IEEE Global Telecommunications Conference Globecom 2010*. 10.1109/GLOCOM.2010.5683895

Kamana, J. (2014). *M-PESA: How Kenya took the lead in mobile money*. Retrieved September 09, 2017, from https://www.mobiletransaction.org/m-pesa-kenya-the-lead-in-mobile-money/

Kang, Z., & Wang, X. (n.d.). Load balancing algorithm in heterogeneous wireless networks oriented to smart distribution grid. *12th International Conference on Natural Computation, Fuzzy Systems and Knowledge Discovery (ICNC-FSKD)*. DOI: 10.1109/FSKD.2016.7603496

Kao, S. J., & Chuang, C. C. (2012). Using GI-G-1 queuing model for rtPS performance evaluation in 802.16 networks. *International Journal of Communication Systems, 25*(3), 314–327. doi:10.1002/dac.1242

Kappler, C., Poyhonen, P., Johnsson, M., & Schmid, S. (2007). Dynamic network composition for beyond 3G networks: A 3GPP viewpoint. *IEEE Network, 21*(1), 74–77. doi:10.1109/MNET.2007.314538

Karp, B., & Kung, H. T. (2000). GPSR: Greedy Perimeter Stateless Routing for Wireless Networks. In *Proceedings of the 6th Annual International Conference on Mobile Computing and Networking* (pp. 243-254). New York: Academic Press. 10.1145/345910.345953

Keim, D., Andrienko, G., Fekete, J.-D., Goerg, C., Kohlhammer, J., & Melancon, G. (2008). Visual Analytics: Definition, Process, and Challenges. In Information Visualisation, LNCS (vol. 4950, pp.154-175). Berlin: Springer-Verlag.

Kemp, R., Palmer, N., Kielmann, T., & Bal, H. (2012). Cuckoo: A Computation Offloading Framework for Smartphones. In *Proceedings of the Second International ICST Conference on Mobile Computing, Applications, and Services* (pp. 59–79). Springer Berlin Heidelberg. 10.1007/978-3-642-29336-8_4

Kempen, A., Crivat, T., Trubert, B., Roy, D., & Pierre, G. (2017). MEC-ConPaaS: An Experimental Single-Board Based Mobile Edge Cloud. *2017 5th IEEE International Conference on Mobile Cloud Computing, Services, and Engineering (MobileCloud)*, 17-24.

Ke, S. C., Chen, Y. W., & Fang, H. A. (2014). An energy-saving-centric downlink scheduling scheme for WiMAX networks. *International Journal of Communication Systems, 27*(11), 2518–2535. doi:10.1002/dac.2486

Khandekar, Bhushan, Tingfang, & Vanghi. (2010). *LTE-Advanced: Heterogeneous Networks*. Qualcomm Inc.

Kibria, M. R., & Jamalipour, A. (2007). On designing issues of the next generation mobile network. *IEEE Network, 21*(1), 6–13. doi:10.1109/MNET.2007.314532

Kim, D., & Kim, D. (2006). Region-expansion for the Voronoi diagram of 3D spheres. *Journal of Computer-Aided Design, 38*(5), 417–430. doi:10.1016/j.cad.2005.11.007

Kim, S., Cha, J., Jung, S., Yoon, C., & Lim, K. (2012). Performance evaluation of random access for M2M communication on IEEE 802.16 network. *Proc. ICACT*, 278–283.

Kodialam, M., & Nandagopal, T. (2005, August). Characterizing the capacity region in multi-radio multi-channel wireless mesh networks. In *Proceedings of the 11th annual international conference on Mobile computing and networking* (pp. 73-87). ACM. 10.1145/1080829.1080837

Kohonen, T., & Honkela, T. (2007). Kohonen network. *Scholarpedia, 2*(1), 1568. doi:10.4249cholarpedia.1568

Kolias, C. (Ed.). (2013, September). *ONF Solution Brief OpenFlow™-Enabled Mobile and Wireless Networks*. ONF White Paper.

Koodli, R. (Ed.). (2009, July). Mobile IPv6 Fast Handovers. RFC 5568.

Korczak, J., & Łuszczyk, W. (2011). Visual Exploration of Cash Flow Chains. In *The Federated Conference on Computer Science and Information Systems* (pp.41–46). New York: IEEE.

Korhonen, J. (Ed.). (2010, February). Diameter Proxy Mobile IPv6: Mobile Access Gateway and Local Mobility Anchor Interaction with Diameter Server. Proposed Standard RFC 5779.

Korhonen. (2015, February). *5G Vision –The 5G Infrastructure Public Private Partnership: Next Generation of Communication Networks and Services*. European Commission.

Kotenko, I., & Novikova, E. (2013). VisSecAnalyzer: a Visual Analytics Tool for Network Security Assessment. In *8th International Conference on Availability, Reliability and Security (ARES 2013). LNCS* (vol. 8128, pp. 345-360). Berlin: Springer-Verlag. 10.1007/978-3-642-40588-4_24

Koutsonikolas, D., Hu, Y., & Wang, C. (2008). XCOR: Synergistic Interflow Network Coding and Opportunistic Routing. In *Proceedings of the ACM Annual International Conference on Mobile Computing and Networking* (pp.1-3). San Francisco, CA: ACM.

Krylovskiy, A. (2015). Internet of Things gateways meet Linux containers: Performance evaluation and discussion. In *Proceedings of the 2015 IEEE 2nd World Forum on Internet of Things (WF-IoT) (WF-IOT '15)*. IEEE Computer Society.

Kumar, S., Cifuentes, D., Gollakota, S., & Katabi, D. (2013, October). Bringing Cross-Layer MIMO to Today's Wireless LANs. *ACM SIGCOMM Computer Communication Review, 43*(4), 387–398.

Lahby, M., Mohammedia, M., & Adib, A. (2013). Network selection mechanism by using MAHP/GRA for heterogeneous networks. *Proceedings of the 2013 6th Joint IFIP IEEE Wireless and Mobile Networking, Conference (WMNC)*, 1-3.

Lampin, Q., Barthel, D., Aug-Blum, I., & Valois, F. (2012). QOS Oriented Opportunistic Routing Protocol for Wireless Sensor Networks. In *Proceedings of IEEE/IFIP Wireless Days* (pp. 1-6). Dublin, Ireland: IEEE. 10.1109/WD.2012.6402804

Laufer, R., Velloso, P. B., Vieira, L. F. M., & Kleinrock, L. (2012). Plasma: A New Routing Paradigm for Wireless Multihop Networks. In *Proceedings of IEEE Conference on Computer Communications (INFOCOM)* (pp. 2706-2710). Orlando, FL: IEEE. 10.1109/INFCOM.2012.6195683

Lee, K., & Shin, I. (2013). User mobility-aware decision making for mobile computation offloading. *IEEE 1st International Conference on Cyber-Physical Systems, Networks, and Applications (CPSNA)*, 116–119.

Lee, G., & Haas, Z. (2011). Simple, Practical, and Effective Opportunistic Routing for Short-Haul Multi-Hop Wireless Networks. *IEEE Transactions on Wireless Communications, 10*(11), 3583–3588. doi:10.1109/TWC.2011.092711.101713

Lee, G., & Lo, S.-C. (2003). Broadcast Data Allocation for Efficient Access on Multiple Data Items in Mobile Environments. *Mobile Networks and Applications*, *8*(4), 365–375. doi:10.1023/A:1024579512792

Lee, J. R. (2007). A hybrid energy saving mechanism for VoIP traffic with silence suppression. *Network Control and Optimization. Proceedings*, *4465*, 296–304.

Lee, J. R., & Cho, D. H. (2009). Dual power-saving modes for voice over IP traffic supporting voice activity detection. *IET Communications*, *3*(7), 1239–1249. doi:10.1049/iet-com.2008.0300

Lee, J., & Cho, D. (2008). An optimal power-saving class II for VoIP traffic and its performance evaluations in IEEE 802.16e. *Computer Communications*, *31*(14), 3204–3208. doi:10.1016/j.comcom.2008.04.029

Lee, W. C., Hu, Q., & Lee, D. L. (1997). Channel Allocation Methods for Data Dissemination in Mobile Computing Environments. *Proc. of the 6th IEEE High Performance Distributed Computing*, 274-281. 10.1109/HPDC.1997.626430

Lei, Z., & Gang, C. (2016). QoS-aware user association for load balancing in heterogeneous cellular network with dual connectivity. *Computer and Communications (ICCC), 2016 2nd IEEE International Conference*. DOI: 10.1109/CompComm.2016.7925224

Leite, R. A., Gschwandtner, T., Miksch, S., Gstrein, E., & Kuntner, J. (2016). Visual analytics for fraud detection: focusing on profile analysis. In *Proceedings of the Eurographics / IEEE VGTC Conference on Visualization: Posters* (pp. 45-47). New York: IEEE.

Leong, H. V., & Si, A. (1995). Data Broadcasting Strategies Over Multiple Unreliable Wireless Channels. *Proc. of the 4th International Conference on Information and Knowledge Management*, 96-104. 10.1145/221270.221339

Leong, H. V., & Si, A. (1997). Database Caching Over the Air-Storage. *The Computer Journal*, *40*(7), 401–415. doi:10.1093/comjnl/40.7.401

Leshed, G., Velden, T., Rieger, O., Kot, B., & Sengers, P. (2008). In-Car GPS Navigation: Engagement with and Disengagement from the Environment. *Proceedings of the twenty-sixth annual SIGCHI conference on Human factors in computing systems*, 1675-1684. 10.1145/1357054.1357316

Li, J., & Mohapatra, P. (2003). LAKER: Location Aided Knowledge Extraction Routing for Mobile Ad Hoc Networks. In Proceedings of 2003 IEEE Wireless Communications and Networking, 2003. WCNC 2003 (pp. 1180-1184). New Orleans, LA: IEEE.

Liao, W. H., & Yen, W. M. (2009). Power-saving scheduling with a QoS guarantee in a mobile WiMAX system. *Journal of Network and Computer Applications*, *32*(6), 1144–1152. doi:10.1016/j.jnca.2009.06.002

Lien, S. Y., Chen, K. C., & Lin, Y. (2011). Toward Ubiquitous Massive accesses in 3GPP Machine- to-Machine Communications. *IEEE Communications Magazine*, *49*(4), 66–74. doi:10.1109/MCOM.2011.5741148

Lien, S. Y., Liau, T. H., Kao, C. Y., & Chen, K. C. (2012). Cooperative access class barring for machine-to-machine communications. *IEEE Transactions on Wireless Communications, 11*(1), 27–32. doi:10.1109/TWC.2011.111611.110350

Li, F., & Wang, Y. (2007). Routing in vehicular ad hoc networks: A survey. *IEEE Vehicular Technology Magazine, 2*(2), 12–22. doi:10.1109/MVT.2007.912927

Li, J., Bu, K., Liu, X., & Xiao, B. (2013). ENDA: embracing network inconsistency for dynamic application offloading in mobile cloud computing. In *Proceedings of the second ACM SIGCOMM workshop on Mobile cloud computing (MCC '13)*. ACM. 10.1145/2491266.2491274

Lin, X. H., Liu, L., Wang, H., & Kwok, Y. K. (2011). On Exploiting the On-Off Characteristics of Human Speech to Conserve Energy for the Downlink VoIP in WiMAX Systems. *2011 7th International Wireless Communications and Mobile Computing Conference (Iwcmc)*, 337-342.

Lin, L., Cao, L., & Zhang, C. (2005). The fish-eye visualization of foreign currency exchange data streams. In *Asia-Pacific Symposium on Information Visualisation (APVis)* (vol. 45, pp. 91-96). Darlinghurst: Australian Computer Society, Inc.

Lin, Y., Li, B., & Liang, B. (2008). CodeOR: Opportunistic Routing in Wireless Mesh Networks with Segmented Network Coding. In *Proceedings of the IEEE International Conference on Network Protocols (ICNP)* (pp. 13-22). Orlando, FL: IEEE.

Lin, Y., Li, B., & Liang, B. (2010). SlideOR: Online Opportunistic Network Coding in Wireless Mesh Networks. In *Proceedings of IEEE Conference on Computer Communications (INFOCOM)*, (pp.171-175). San Diego, CA: IEEE. 10.1109/INFCOM.2010.5462249

Lin, Y., & Shen, H. (2017). Cloudfog: Leveraging fog to extend cloud gaming for thin-client mmog with high quality of service. *IEEE Transactions on Parallel and Distributed Systems, 28*(2), 431–445. doi:10.1109/TPDS.2016.2563428

Liu & Lau. (2017). Joint BS-User Association, Power Allocation, and User-Side Interference Cancellation in Cell-free Heterogeneous Networks. *IEEE Transactions on Signal Processing, 65*(2).

Liu, Yi., Yuen, C., Chen, J., & Cao, X. (2013). A Scalable Hybrid MAC Protocol for Massive M2M Networks. *IEEE conference on Wireless Communications and Networking*, 250 – 255.

Liu, Y., Yuen, C., Cao, X., Hassan, N. U., & Chen, J. (2014). Design of a Scalable Hybrid MAC Protocol for Heterogeneous M2M Networks. *IEEE Internet of Things Journal, 1*(1), 99–111. doi:10.1109/JIOT.2014.2310425

Li, W., & Chao, X. (2007). Call admission control for an adaptive heterogeneous multimedia mobile network. *IEEE Transactions on Wireless Communications, 6*(2), 515–525. doi:10.1109/TWC.2006.05192

Li, Y., Mohaisen, A., & Zhang, Z. (2013). Trading Optimality for Scalability in Large-Scale Opportunistic Routing. *IEEE Transactions on Vehicular Technology, 62*(5), 2253–2263. doi:10.1109/TVT.2012.2237045

Löhner, R. (2010). On the modeling of pedestrian motion. *Journal of Applied Mathematical Modeling*, *34*(2), 366–38. doi:10.1016/j.apm.2009.04.017

Loke, S. W., Napier, K., Alali, A., Fernando, N., & Rahayu, W. (2015). Mobile computations with surrounding devices: Proximity sensing and multilayered work stealing. *ACM Transactions on Embedded Computing Systems, 14*(2), 22:1–22:25.

Low, S. H., Paganini, F., & Doyle, J. C. (2002). Internet congestion control. *IEEE Control Systems, 22*(1), 28-43.

Lu, M., Steenkiste, P., & Chen, T. (2009). Design, Implementation and Evaluation of an Efficient Opportunistic Retransmission Protocol. In *Proceedings of IEEE/ACM Annual International Conference on Mobile Computing and Networking* (pp.73-84). IEEE. 10.1145/1614320.1614329

Lu, M., & Wu, J. (2009). Opportunistic Routing Algebra and its Applications. In *Proceedings IEEE Conference on Computer Communications (IEEE INFOCOM)* (pp. 2374-2382). Rio de Janeiro, Brazil: IEEE.

Lu, R., Li, X., Liang, X., Shen, X., & Lin, X. (2011). GRS: The green, reliability, and security of emerging machine to machine communications. *IEEE Communications Magazine*, *49*(4), 28–35. doi:10.1109/MCOM.2011.5741143

Maaten, L. J. P., & Hinton, G. E. (2008). Visualizing High-Dimensional Data Using t-SNE. *Journal of Machine Learning Research*, *9*, 2579–2605.

Malanchini, I., Cesana, M., & Gatti, N. (2012). *Network selection and resource allocation games for wireless access networks. IEEE Transactions on Mobile Computing.*

Mantoro, T., Abubakar, A., & Ayu, M. A. (2011). Multi-User Navigation: A 3D Mobile Device Interactive Support. *IEEE Symposium on Industrial Electronics and Applications (ISIEA 2011)*, 545-549. 10.1109/ISIEA.2011.6108772

Mao, X., Tang, S., Xu, X., Li, X., & Ma, H. (2011). Energy Efficient Opportunistic Routing in Wireless Sensor Networks. *IEEE Transactions on Parallel and Distributed Systems*, *22*(11), 1934–1942. doi:10.1109/TPDS.2011.70

Marghescu, D. (2007a). *Multidimensional Data Visualization Techniques for Financial Performance Data: A Review* (TUCS Tech. Rep. No 810). Turku, Finland: University of Turku.

Marghescu, D. (2007b). *Evaluating Multidimensional Visualization Techniques in Data Mining Tasks.* TUCS Dissertations, issue 107. Turku: Turku Centre for Computer Science.

Marina, M. K., Das, S. R., & Subramanian, A. P. (2010). A topology control approach for utilizing multiple channels in multi-radio wireless mesh networks. *Computer Networks*, *54*(2), 241–256. doi:10.1016/j.comnet.2009.05.015

Marinelli, E. E. (2009). Hyrax: Cloud Computing on Mobile Devices using MapReduce. *Science, 0389*(September), 1–123.

Marques, V., Aguiar, R. L., Garcia, C., Moreno, J. I., Beaujean, C., Melin, E., & Liebsch, M. (2003). An IP-based QoS architecture for 4G operator scenarios. *IEEE Transactions on Wireless Communications, 10*(3), 54–62. doi:10.1109/MWC.2003.1209596

Marques, V., Xavier, P. C., Aguiar, R. L., Liebsch, M., & Duarte, M. O. (2005). Evaluation of a mobile IPv6-based architecture supporting user mobility QoS and AAAC in heterogeneous networks. *IEEE Journal on Selected Areas in Communications, 23*(11), 2138–2151. doi:10.1109/JSAC.2005.856825

Mase, K. (2011). How to Deliver Your Message from/to a Disaster Area. *IEEE Communications Magazine, 49*(1), 52–57. doi:10.1109/MCOM.2011.5681015

Mauve, M., Widmer, J., & Hartenstein, H. (2001). A Survey on Position-Based Routing in Mobile Ad Hoc Networks. *IEEE Network, 15*(6), 30–39. doi:10.1109/65.967595

McGinn, D., Birch, D., Akroyd, D., Molina-Solana, M., Guo, Y., & Knottenbelt, W. J. (2016). Visualizing Dynamic Bitcoin Transaction Patterns. *Big Data, 4*(2), 109–119. doi:10.1089/big.2015.0056 PMID:27441715

Menon V.G. & Prathap P. M. (2017). Vehicular Fog Computing: Challenges Applications and Future Directions. *International Journal of Vehicular Telematics and Infotainment Systems, 1*(2), 15-23.

Menon, Prathap & Priya. (2016). Ensuring Reliable Communication in Disaster Recovery Operations with Reliable Routing Technique. Mobile Information Systems.

Menon, V. G., & Prathap, P. M. (2016). Routing in Highly Dynamic Ad Hoc Networks: Issues and Challenges. *International Journal on Computer Science and Engineering, 8*(4), 112–116.

Menon, V. G., & Prathap, P. M. J. (2016). Analysing the Behaviour and Performance of Opportunistic Routing Protocols in Highly Mobile Wireless Ad Hoc Networks. *IACSIT International Journal of Engineering and Technology, 8*(5), 1916–1924. doi:10.21817/ijet/2016/v8i5/160805409

Menon, V. G., & Prathap, P. M. J. (2016). Comparative Analysis of Opportunistic Routing Protocols for Underwater Acoustic Sensor Networks. *Proceedings of the IEEE International Conference on Emerging Technological Trends.* 10.1109/ICETT.2016.7873733

Menon, V. G., Prathap, P. M. J., & Vijay, A. (2016). Eliminating Redundant Relaying of Data Packets for Efficient Opportunistic Routing in Dynamic Wireless Ad Hoc Networks. *Asian Journal of Information Technology, 12*(17), 3991–3994.

Menon, V. G., Priya, P. M. J., & Prathap, P. M. J. (2013). Analyzing the behavior and performance of greedy perimeter stateless routing protocol in highly dynamic mobile ad hoc networks. *Life Science Journal, 10*(2), 1601–1605.

Merrit, C. (2010). *Mobile Money Transfer Services: The Next Phase in the Evolution in Person-to-Person Payments (Tech. rep.)*. Atlanta, GA: Federal Reserve Bank of Atlanta.

Mikhail, N. M., Esmat, B., & El Kadi, A. (2001, July). *A New Architecture for Mobile Computing, Mobile and Wireless Computing*. Academic Press.

Mohammad, G., & Khoshkholgh, V. C. M. (2017). Analyzing Coverage Probability of Multi-tier Heterogeneous Networks Under Quantized Multi-User ZF Beam forming. *IEEE Transactions on Vehicular Technology*. DOI: 10.1109/TVT.2017.2780519

Mohiuddin, A., Abdun, N. M., & Rafiqul, Md. I. (2016). A survey of anomaly detection techniques in financial domain. Future Gener. Comput. Syst., 55, 278-288.

Morabito, R., & Beijar, N. (2016) Enabling Data Processing at the Network Edge through Lightweight Virtualization Technologies. *2016 IEEE International Conference on Sensing, Communication and Networking (SECON Workshops)*, 1-6. 10.1109/SECONW.2016.7746807

Na, J., & Kim, C.-K. (2006). GLR: A Novel Geographic Routing Scheme for Large Wireless Ad Hoc Networks. *Computer Networks*, *50*(17), 3434–3448. doi:10.1016/j.comnet.2006.01.004

Namboodiri, V., & Gao, L. X. (2010). Energy-Efficient VoIP over Wireless LANs. *IEEE Transactions on Mobile Computing*, *9*(4), 566–581. doi:10.1109/TMC.2009.150

Nassr, M., Jun, J., Eidenbenz, S., Hansson, A., & Mielke, A. (2007). Scalable and Reliable Sensor Network Routing: Performance Study from Field Deployment. In *Proceedings of IEEE INFOCOM 2007 - 26th IEEE International Conference on Computer Communications* (pp. 670–678). Anchorage, AK: IEEE. 10.1109/INFCOM.2007.84

Naveed, A., Kanhere, S. S., & Jha, S. K. (2007, October). Topology control and channel assignment in multi-radio multi-channel wireless mesh networks. In *Mobile Proceedings of IEEE International Conference on Adhoc and Sensor Systems, 2007*(pp. 1-9). IEEE.

Navidi, W., & Camp, T. (2004). Stationary Distributions for the Random Waypoint Mobility Model. *IEEE Transactions on Mobile Computing*, *3*(1), 99–108. doi:10.1109/TMC.2004.1261820

Nelson, S., Bakht, M., Kravets, R., & Harris, A. F. III. (2009). Encounter-Based Routing in DTNs. *Mobile Computing and Communications Review*, *13*(1), 56–59. doi:10.1145/1558590.1558602

Neural-technologies. (2017). *Minotaur Fraud Management Solution*. Retrieved September 10, 2017, from https://www.neuralt.com/73/393/optimus-fraud

Nga, D. T. T., Kim, M. G., & Kang, M. (2007). Delay-guaranteed energy saving algorithm for the delay-sensitive applications in IEEE 802.16e systems 1339. *IEEE Transactions on Consumer Electronics*, *53*(4), 1339–1347. doi:10.1109/TCE.2007.4429222

Nice Actimize. (2017). *Fraud Prevention & Cybercrime Management Solutions Integrated Fraud Management*. Retrieved September 10, 2017, from http://www.niceactimize.com/fraud-detection-and-prevention

NMC. (2015a, February). *LTE IP Address Allocation Schemes II: A Case for Two Cities*. Netmanias Technical Document.

NMC. (2015b, February). *LTE IP Address Allocation Schemes I: Basic*. Netmanias Technical Document.

Novikova, E., & Kotenko, I. (2013). Analytical Visualization Techniques for Security Information and Event Management. In *21th Euromicro International Conference on Parallel, Distributed and network-based Processing (PDP 2013)* (pp.519-525). New York: IEEE. 10.1109/PDP.2013.84

Novikova, E., Kotenko, I., & Fedotov, E. (2014). Interactive Multi-view Visualization for Fraud Detection in Mobile Money Transfer Services. *International Journal of Mobile Computing and Multimedia Communications*, *6*(4), 73–97. doi:10.4018/IJMCMC.2014100105

Nurminen, A. (2006). m-LOMA - a Mobile 3D City Map. *Proceedings of the eleventh international conference on 3D web technology*, 7–18.

Nurminen, A. (2008). Mobile 3D City Maps. *Journal of IEEE Computer Graphics and Applications*, *28*(4), 20–31. doi:10.1109/MCG.2008.75 PMID:19004682

Nygren, A., Pfa, B., Lantz, B., Heller, B., Barker, C., Beckmann, C., … Kis, Z. L. (2013, October). *The OpenFlow Switch Specification*, Version 1.4.0. ONF, Wire Protocol 0x05, ONF TS-012.

Odini, M., Sahai, A., Veitch, A., Gamela, A., Khan, A., Perlman, B., … Lei, Z. (2015, September). *Network Functions Virtualization (NFV); Ecosystem*. Report on SDN Usage in NFV Architectural Framework, ETSI, Draft ETSI GS NFV-EVE 005 V0.2.0.

Ohm, C., Bienk, S., Kattenbeck, M., Ludwig, B., & Müller, M. (2016). Towards interfaces of mobile pedestrian navigation systems adapted to the user's orientation skills. *Pervasive and Mobile Computing*, *26*, 121–134. doi:10.1016/j.pmcj.2015.10.006

Okamoto, A., & Uchida, T. (2004). Pedestrian Navigation System Easy-to-Use-in Urban Canyon: Algorithms, system development and a field trial in Osaka. *Memoirs of the Faculty of Engineering Osaka City University*, *45*, 65–70.

ONF. (2012, April). *Software-Defined Networking: The New Norm for Networks*. ONF White Paper.

OpenFog Consortium. (2017). OpenFog Reference Architecture for Fog Computing. *OPFRA001, 20817*, 162.

Orange Money. (2016). *Orange launches Orange Money in France to allow money transfers to three countries in Africa and within mainland France.* Press release. Retrieved September 09, 2017, from https://www.orange.com/en/Press-Room/press-releases/press-releases-2016/Orange-launches-Orange-Money-in-France-to-allow-money-transfers-to-three-countries-in-Africa-and-within-mainland-France

Ottenwälder, B., Koldehofe, B., Rothermel, K., & Ramachandran, U. (2013). MigCEP: operator migration for mobility driven distributed complex event processing. In *Proceedings of the 7th ACM international conference on Distributed event-based systems (DEBS '13).* ACM. 10.1145/2488222.2488265

Pahl, C., Helmer, S., Miori, L., Sanin, J., & Lee, B. (2016). A Container-Based Edge Cloud PaaS Architecture Based on Raspberry Pi Clusters. *2016 IEEE 4th International Conference on Future Internet of Things and Cloud Workshops (FiCloudW)*, 117-124.

Park, C. W., Hwang, D., & Lee, T. J. (2014). Enhancement of IEEE 802.11ah MAC for M2M communications. *IEEE Communications Letters*, *18*(7), 1151–1154. doi:10.1109/LCOMM.2014.2323311

Park, V. D., & Corson, M. S. (1997). A Highly Adaptive Distributed Routing Algorithm for Mobile Wireless Networks. In *Proceedings of the Sixteenth Annual Joint Conference of the IEEE Computer and Communications Societies* (pp. 1405-1413). Kobe, Japan: IEEE. 10.1109/INFCOM.1997.631180

Paulson, L. D. (2003). Will Fuel Cells Replace Batteries in Mobile Devices? *IEEE Computer Magazine*, *36*(11), 10–12. doi:10.1109/MC.2003.1244525

Perkins, C. (Ed.). (2002, August). IP Mobility Support for IPv4. RFC 3344.

Perkins, C. E., & Bhagwat, P. (1994). Highly Dynamic Destination-Sequenced Distance-Vector Routing (DSDV) for Mobile Computers. In *Proceedings of the Conference on Communications Architectures, Protocols and Applications (SIGCOMM '94)* (pp. 234-244). New York: ACM. 10.1145/190314.190336

Perkins, C. E., & Royer, E. M. (1999). Ad-Hoc On-Demand Distance Vector Routing. In *Proceedings of Second IEEE Workshop on Mobile Computing Systems and Applications* (pp. 90-100). New Orleans, LA: IEEE. 10.1109/MCSA.1999.749281

Petrolo, R., Morabito, R., Loscrì, V., & Mitton, N. (2017). Article. *Annales des Télécommunications*, 1–10.

Phemius, K., Bouet, M., & Leguay, J. (2014, May). Disco: Distributed Multi-Domain SDN Controllers. *IEEE Network Operations and Management Symposium (NOMS)*, Krakow, Poland.

Pitoura, E., & Samaras, G. (1998). *Data Management for Mobile Computing.* London: Kluwer Academic Publishers. doi:10.1007/978-1-4615-5527-8

Prefuse. (2017). *Information Visualization toolkit.* Retrieved September 16, 2017, from http://prefuse.org/

Rad, A. H. M., & Wong, V. W. (2006, June). Joint optimal channel assignment and congestion control for multi-channel wireless mesh networks. *Proceedings of IEEE International Conference on Communications*, 5, 1984-1989.

Rahaman, M. S., Mei, Y., Hamilton, M., & Salim, F. D. (2017). Capra: A contour-based accessible path routing algorithm. *Information Sciences*, *385*, 157–173. doi:10.1016/j.ins.2016.12.041

Rajandekar, A., & Sikdar, B. (2015). A survey of MAC layer issues and protocols for machine-to-machine communications. *IEEE Internet of Things Journal*, *2*(2), 175–186. doi:10.1109/JIOT.2015.2394438

Rakkolainen, I., Pulkkinen, S., & Heinonen, A. (1998). Visualizing real-time GPS data with VRML worlds. *Proceedings of the ACM-GIS98 workshop*, 52–56.

Ramachandran, K. N., Belding, E. M., Almeroth, K. C., & Buddhikot, M. M. (2006, April). Interference-aware channel assignment in multi-radio wireless mesh networks. *Proceedings of 25th IEEE International Conference on Computer Communications*, 1-12. 10.1109/INFOCOM.2006.177

Ramanathan, S. (1997). A unified framework and algorithms for (T/F/C)DMA channel assignment in wireless networks. *Proceedings - IEEE INFOCOM*, 900–907.

Raniwala, A., & Chiueh, T. C. (2005, March). Architecture and algorithms for an IEEE 802.11-based multi-channel wireless mesh network. *Proceedings of 24th Annual Joint Conference of the IEEE Computer and Communications Societies*, *3*, 2223-2234.

Raniwala, A., Gopalan, K., & Chiueh, T. C. (2004). Centralized channel assignment and routing algorithms for multi-channel wireless mesh networks. *Mobile Computing and Communications Review*, *8*(2), 50–65. doi:10.1145/997122.997130

Raposo, A., Neumann, L., Magalhaes, L., & Ricarte, I. (1997). Efficient Visualization in a mobile WWW environment. *Proceedings of the WebNet 97, World Conference of the WWW, Internet, and Intranet*.

Ravi, A., & Peddoju, S. K. (2014). Mobility managed energy efficient Android mobile devices using cloudlet. In *2014 IEEE Students' Technology Symposium (TechSym)* (pp. 402-407). Kharagpur, India: IEEE.

Ren, T., Koutsopoulos, I., & Tassiulus, L. (2002). QoS provisioning for real time traffic in wireless packet networks. *Proceedings of the IEEE GLOBECOM*. 10.1109/GLOCOM.2002.1188482

Report, C. G. A. P. (2017). *Brief Fraud in the Mobile Financial Services*. Retrieved September 10, 2017, from http://www.cgap.org/sites/default/files/Brief-Fraud-in-Mobile-Financial-Services-April-2017.pdf

Rhee, I., Warrier, A., Aia, M., Min, J., & Sichitiu, M. L. (2008). Z-MAC: A hybrid MAC for wireless sensor networks. *IEEE Trans. Netw.*, *16*(3), 511–524. doi:10.1109/TNET.2007.900704

Rieke, R., Zhdanova, M., Repp, J., Giot, R., & Gaber, C. (2013). Fraud Detection in Mobile Payments Utilizing Process Behavior Analysis. In *The 2nd International Workshop on Recent Advances in Security Information and Event Management (RaSIEM 2013)* (pp. 662-669). New York: IEEE.

Rosario, D., Zhao, Z., Braun, T., Cerqueira, E., Santos, A., & Alyafawi, I. (2014). Opportunistic Routing for Multi-Flow Video Dissemination Over Flying Ad hoc Networks. In *Proceeding of IEEE International Symposium on World of Wireless, Mobile and Multimedia Network* (pp. 1-6). Sydney, Australia: IEEE. 10.1109/WoWMoM.2014.6918947

Rozner, E., Seshadri, J., Mehta, Y., & Qiu, L. (2009). SOAR: Simple Opportunistic Adaptive Routing Protocol for Wireless Mesh Networks. *IEEE Transactions on Mobile Computing*, 8(1), 1622–1635. doi:10.1109/TMC.2009.82

Ruckus Wireless, Inc. (2013). *The Choice of Mobility Solutions Enabling IP-Session Continuity Between Heterogeneous Radio Access Networks*. Interworking Wi-Fi and Mobile Networks White Paper.

Saha, B., Misra, S., & Obaidat, M. (2013). A Web-Based Integrated Environment for Simulation and Analysis with NS-2. *IEEE Wireless Communications*, 20(4), 109–115. doi:10.1109/MWC.2013.6590057

Saini, J. S., & Sohi, B. S. (2016). A Survey on Channel Assignment Techniques of Multi-Radio Multi-channel Wireless Mesh Network. *Indian Journal of Science and Technology*, 9(42).

SAS. (2017). *Fraud Security Intelligence*. Retrieved September 13, 2017, from https://www.sas.com/en_us/solutions/fraud-security-intelligence.html

Satyanarayanan, M., Bahl, P., Caceres, R., & Davies, N. (2009). The case for VM-based cloudlets in mobile computing. *IEEE Pervasive Computing*, 8(4), 14–23. doi:10.1109/MPRV.2009.82

Savic, Z. (2011). *LTE Design and Deployment Strategies*. Cisco Systems Inc.

Schmidt, T. (Ed.). (2014, November). Multicast Listener Extensions for Mobile IPv6 and Proxy Mobile IPv6 Fast Handovers. IETF RFC 7411.

Schöps, T., Sattler, T., Häne, C., & Pollefeys, M. (2017). Large-scale outdoor 3D reconstruction on a mobile device. *Computer Vision and Image Understanding*, 157, 151–166. doi:10.1016/j.cviu.2016.09.007

Schreck, T., Tekusova, T., Kohlhammer, J., & Fellner, D. (2007). Trajectory-based visual analysis of large financial time series data. *ACM SIGKDD Explorations Newsletter*, 9(2), 30–37. doi:10.1145/1345448.1345454

Seeley, D. (1997). *Planimate™-Animated Planning Platforms*. InterDynamics Pty Ltd.

Sehgal, A., & Agrawal, R. (2010). QoS based network selection for 4G systems. *IEEE Transactions on Consumer Electronics*, 56(2), 560–565. doi:10.1109/TCE.2010.5505970

Seite, P., Yegin, A., & Gundavelli, S. (2016, March). *MAG Multipath Binding Option*. Internet-Draft.

Shen, W., & Zeng, Q. A. (2008). Cost-function-based network selection strategy in integrated wireless and mobile networks. *IEEE Transactions on Vehicular Technology*, 57(6), 3778–3788. doi:10.1109/TVT.2008.917257

Shi, C., Li, Y., Zhang, J., & Sun, Y. (2017). A Survey of Heterogeneous Information Network Analysis. IEEE Transactions on Knowledge and Data Engineering, 29(1).

Shi, C., Lakafosis, V., Ammar, M. H., & Zegura, E. W. (2012). Serendipity: Enabling remote computing among intermittently connected mobile devices. In *Proceedings of the Thirteenth ACM International Symposium on Mobile Ad Hoc Networking and Computing* (pp. 145–154). New York: ACM. 10.1145/2248371.2248394

Shi, H., Chen, N., & Deters, R. (2015). Combining mobile and fog computing: Using CoAP to link mobile device clouds with fog computing. In *Proceedings of the 2015 IEEE International Conference on Data Science and Data Intensive Systems* (pp. 564–571). IEEE. 10.1109/DSDIS.2015.115

Shin, S. C. W. Y., & Lee, Y. (2013). Parallel Opportunistic Routing in Wireless Networks. *IEEE Transactions on Information Theory*, 59(10), 6290–6300. doi:10.1109/TIT.2013.2272884

Shitiri, E., Park, I., & Cho, H. (2017). OrMAC: A Hybrid MAC Protocol Using Orthogonal Codes for Channel Access in M2M Networks. *Sensors. MDPI Publications*, 17(9), 1–10.

Shneiderman, B. (2003). Dynamic queries for visual information seeking. The Craft of Information Visualization: Readings and Reflections, 14-21.

Si, A., & Leong, H. V. (1999). Query Optimization for Broadcast Database. *Data & Knowledge Engineering*, 29(3), 351–380. doi:10.1016/S0169-023X(98)00040-8

Siek, J. G., Lee, L. Q., & Lumsdaine, A. (2001). *The Boost Graph Library: User Guide and Reference Manual, Portable Documents*. Pearson Education.

Simoes, M. G., & Bose, B. K. (1995). Neural-Network-Based Estimation of Feedback Signals for a Vector Controlled Induction-Motor Drive. *IEEE Transactions on Industry Applications*, 31(3), 620–629. doi:10.1109/28.382124

Si, W., Selvakennedy, S., & Zomaya, A. Y. (2010). An overview of channel assignment methods for multi-radio multi-channel wireless mesh networks. *Journal of Parallel and Distributed Computing*, 70(5), 505–524. doi:10.1016/j.jpdc.2009.09.011

Skalli, H., Ghosh, S., Das, S. K., & Lenzini, L. (2007). Channel assignment strategies for multiradio wireless mesh networks: Issues and solutions. *IEEE Communications Magazine*, 45(11), 86–95. doi:10.1109/MCOM.2007.4378326

Soliman, H. C., Malki, E. K., & Bellier, L. (2005, August). *Hierarchical Mobile IPv6 Mobility Management (HMIPv6)*. RFC 4140.

Song, W., Zhuang, W., & Cheng, Y. (2007). Load balancing for cellular/WLAN integrated networks. *IEEE Network*, 21(1), 27–33. doi:10.1109/MNET.2007.314535

Soo, S., Chang, C., Loke, S., & Srirama, S. N. (2017). Proactive Mobile Fog Computing using Work Stealing: Data Processing at the Edge. *International Journal of Mobile Computing and Multimedia Communications*, 8(4), 1–19. doi:10.4018/IJMCMC.2017100101

Sozer, E. M., Stojanovic, M., & Proakis, J. G. (2000). Underwater acoustic networks. *IEEE Journal of Oceanic Engineering*, 25(1), 72–83. doi:10.1109/48.820738

Spence, R. (1999). A Framework for Navigation. *International Journal of Human-Computer Studies*, 51(5), 919–945. doi:10.1006/ijhc.1999.0265

Su, J., Lin, F., Zhou, X., & Lu, X. (2015). Steiner tree based optimal resource caching scheme in fog computing. *China Communications*, 12(8), 161–168. doi:10.1109/CC.2015.7224698

Suresh, L., Schulz-Zander, J., Merz, R., Feldmann, A., & Vazao, T. (2012). Towards Programmable Enterprise WLANS with Odin. *First Workshop on Hot Topics in Software Defined Networks*, 115-120. 10.1145/2342441.2342465

Tantayakul, K., Dhaou, R., & Paillassa, B. (2016, March). Impact of SDN on Mobility Management. *30th IEEE Advanced Information Networking and Applications (AINA)*, 260-265.

Teixeira, M. A., & Guardieiro, P. R. (2013). Adaptive packet scheduling for the uplink traffic in IEEE 802.16e networks. *International Journal of Communication Systems*, 26(8), 1038–1053. doi:10.1002/dac.1390

The Network Simulator home page. (n.d.). Retrieved from http://www.isi.edu/nsnam/ns/

Truong-Huu, T., Tham, C. K., & Niyato, D. (2014). To Offload or to Wait: An Opportunistic Offloading Algorithm for Parallel Tasks in a Mobile Cloud. In *2014 IEEE 6th International Conference on Cloud Computing Technology and Science* (pp. 182-189). Singapore: IEEE.

Verba, N., Chao, K.-M., James, A., Goldsmith, D., Fei, X., & Stan, S.-D. (2017). Platform as a service gateway for the Fog of Things. *Advanced Engineering Informatics*, 33, 243–257. doi:10.1016/j.aei.2016.11.003

Verikoukis, C., Alonso, L., & Giamalis, T. (2005). *Cross-layer optimization for wireless systems: A European research key challenge. Global Communications Newsletter.*

Verma. (2017). Pheromone and Path Length Factor-Based Trustworthiness Estimations in Heterogeneous Wireless Sensor Networks. *IEEE Sensors Journal, 17*(1).

Verma, P. K., Tripathi, R., & Naik, K. (2014). A Robust Hybrid-MAC Protocol for M2M Communications. *5th IEEE International Conference on Computer and Communication Technology (ICCCT)*, 267–271. 10.1109/ICCCT.2014.7001503

Verma, P. K., Verma, R., Prakash, A., Agrawal, A., Naik, K., Tripathi, R., ... Abogharaf, A. (2016). Machine-to-Machine (M2M) Communications: A Survey. *Journal of Network and Computer Applications. Elsevier Publications, 66,* 83–105.

Ville Lehtinen, V., Nurminen, A., & Oulasvirta, A. (2012). Integrating Spatial Sensing to an Interactive Mobile 3D Map. *IEEE Symposium on 3D User Interfaces.*

Wakikawa, R., Gundavelli, S. (2010, May). *Support for Proxy Mobile IPv4.* IETF RFC 5844.

Waluyo, A. B., Srinivasan, B., & Taniar, D. (2003). Global Index for Multi Channels Data Dissemination in Mobile Databases. *Proc. of the 18th International Symposium on Computer and Information Sciences (ISCIS'03)* (vol. 2869, pp. 210-217). Springer.

Waluyo, A. B., Srinivasan, B., & Taniar, D. (2004). Optimising Query Access Time over Broadcast Channel in Mobile Databases. In Lecture Notes in Computer Science: Vol. 3207. *Proc. of the Embedded and Ubiquitous Computing* (pp. 439–449). Springer-Verlag. doi:10.1007/978-3-540-30121-9_42

Wang, G., Zhong, X., Mei, S., & Wang, J. (2010). An adaptive medium access control mechanism for cellular based machine to machine (M2M) communication. *Proc. IEEE Int. Conf. Wireless Inf. Technol. Syst. (ICWITS),* 1-4. 10.1109/ICWITS.2010.5611820

Wang, H., Chen, S., Xu, H., Ai, M., & Shi, Y. (2015, April). SoftNet: A Software Defined Decentralized Mobile Network Architecture Toward 5G. *IEEE Network, 29*(2), 16–22. doi:10.1109/MNET.2015.7064898

Wang, J., Wang, Z., Xia, Y., & Wang, H. (2007, September). A practical approach for channel assignment in multi-channel multi-radio wireless mesh networks. *Proceedings of Fourth International Conference on Broadband Communications, Networks and Systems,* 317-319.

Wang, S., Urgaonkar, R., Zafer, M., Chan, T., He, K., & Leung, K. K. (2015). Dynamic service migration in mobile edge-clouds. *IFIP Networking Conference (IFIP Networking),* 1-9.

Wang, Z., Chen, Y., & Li, C. (2012). CORMAN: A Novel Cooperative Opportunistic Routing Scheme in Mobile Ad Hoc Networks. *IEEE Journal on Selected Areas in Communications, 30*(2), 289–296. doi:10.1109/JSAC.2012.120207

Wasinger, R., Stahl, C., & Krüger, A. (2003). Mobile Multi-Modal Interaction in a Pedestrian Navigation & Exploration System. *Proceedings of MobileHCI,* 481-485.

Wattenberg, M. (1999). *Visualizing the stock market. In CHI Extended Abstracts on Human Factors in Computing Systems* (pp. 188–189). New York: ACM. doi:10.1145/632716.632834

Westphal, C. (2006). Opportunistic Routing in Dynamic Ad Hoc Networks: The OPRAH Protocol. In *Proceedings of the 2006 IEEE International Conference on Mobile Ad Hoc and Sensor Systems* (pp. 570-573). Vancouver, Canada: IEEE. 10.1109/MOBHOC.2006.278612

Weyland, A. (2002, December). *Evaluation of Mobile IP Implementations under Linux*. Academic Press.

Witt, M., & Turau, V. (2005). BGR: Blind Geographic Routing for Sensor Networks. In *Proceedings of the Third International Workshop on Intelligent Solutions in Embedded Systems* (pp. 51-61). Academic Press. 10.1109/WISES.2005.1438712

Wu, Yuen, & Cheng. (2016). Energy-Minimized Multipath Video Transport to Mobile Devices in Heterogeneous Wireless Networks. *IEEE Journal on Selected Areas in Communications, 34*(5).

Xiao, M., Wu, J., Liu, K., & Huang, L. (2013). Tour: Time-Sensitive Opportunistic Utility Based Routing in Delay Tolerant Networks. In Proceedings of IEEE INFOCOM (pp. 2085 2091). Turin, Italy: IEEE.

Xiao, Y., Chen, C. L. P., & Wang, B. (2000). *Quality of service provisioning framework for multimedia traffic in wireless/mobile networks*. IEEE IC Computer Communication Networks.

Xie, C., Chen, W., Huang, X., Hu, Y., Barlowe, S., & Yang, J. (2014). VAET: A Visual Analytics Approach for E-Transactions Time-Series. *IEEE Transactions on Visualization and Computer Graphics, 20*(12), 1743–1752. doi:10.1109/TVCG.2014.2346913 PMID:26356888

Yang, M., Li, Y., Jin, D., Su, L., Ma, S., & Zeng, L. (2013). OpenRAN: A Software-Defined RAN Architecture via Virtualization. *Computer Communication Review, 43*(4), 549–550. doi:10.1145/2534169.2491732

Yan, Y., & Kunhui, L. (2010). 3D Visual Design for Mobile Search Result on 3G Mobile Phone. *Proceedings of the International Conference on Intelligent Computation Technology and Automation*, 12-16. 10.1109/ICICTA.2010.489

Yap, K. K., Sherwood, R., Kobayashi, M., Huang, T. Y., Chan, M., Handigol, N., ... Parulkar, G. (2010). Blueprint for introducing innovation into Wireless Mobile Networks. *2nd ACM SIGCOMM Workshop on Virtualized Infrastructure Systems and Architectures*, 25-32. 10.1145/1851399.1851404

Yen, J. Y. (1970). An Algorithm for Finding Shortest Routes from All Source Devices to A Given Destination in General Networks. *Quarterly of Applied Mathematics, 27*(1), 526–530. doi:10.1090/qam/253822

Ye, W., Heidemann, J., & Estrin, D. (2004). Medium access control with coordinated adaptive sleeping for wireless sensor networks. *IEEE/ACM Transactions on Networking, 12*(3), 493–506. doi:10.1109/TNET.2004.828953

Yiping, C., & Yuhang, Y. (2007). A new 4G architecture providing multimode terminals always best connected services. *IEEE Wireless Communications, 14*(2), 33–39. doi:10.1109/MWC.2007.358962

Yousafzai, A., Chang, V., Gani, A., & Noor, R. M. (2016). Directory-based incentive management services for ad-hoc mobile clouds. *International Journal of Information Management, 36*(6, Part A), 900–906. doi:10.1016/j.ijinfomgt.2016.05.019

Yuan, Y., Yang, H., Wong, S., Lu, S., & Arbaugh, W. (2005). ROMER: Resilient Opportunistic Mesh Routing for Wireless Mesh Networks. In *Proceedings of the IEEE Workshop on Wireless Mesh Networks* (pp. 1-9). IEEE.

Yu, R., Zhang, Y., Gjessing, S., Xia, W., & Yang, K. (2013). Toward cloud-based vehicular networks with efficient resource management. *IEEE Network*, *27*(5), 48–55. doi:10.1109/MNET.2013.6616115

Zaslavsky, A., & Tari, S. (1998). Mobile Computing: Overview and Current Status. *Australian Computer Journal*, *30*(2), 42–52.

Zeki, A. M., Ghyasi, A. F., Mujahid, M., Zainul, N., Cheddad, A., Zubayr, M., & Zakaria, S. M. (2004). Design and Implementation of a Voronoi Diagrams Generator using Java. *International Arab Conference on Information Technology*, 194-198.

Zeng, K., Lou, W., Yang, J., & Brown, D. R. III. (2007). On Throughput Efficiency of Geographic Opportunistic Routing in Multihop Wireless Networks. *Mobile Networks and Applications*, *12*(5), 347–357. doi:10.100711036-008-0051-7

Zeng, K., Yang, Z., & Lou, W. (2009). Location-Aided Opportunistic Forwarding in Multirate and Multihop Wireless Networks. *IEEE Transactions on Vehicular Technology*, *58*(6), 3032–3040. doi:10.1109/TVT.2008.2011637

Zhang, C., Addepalli, S., Murthy, N., Fourie, L., Zarny, M., & Dunbar, L. (2015, June). *L4-L7 Service Function Chaining Solution Architecture*. ONF White Paper.

Zhang, C., Sun, Y., Mo, Y., Zhang, Y., & Bu, S. (2016). Social-aware content downloading for fog radio access networks supported device-to-device communications. In *Proceedings of the 2016 IEEE International Conference on Ubiquitous Wireless Broadband* (pp. 1–4). IEEE. 10.1109/ICUWB.2016.7790392

Zhang, D., Chen, Z., Ren, J., & Zhang, N. (2015). *Energy Harvesting-Aided Spectrum Sensing and Data Transmission in Heterogeneous. Cognitive Radio Sensor Network*. IEEE.

Zhang, R., Ruby, R., Pan, J., Cai, L., & Shen, X. (2010). A hybrid reservation/contention-based MAC for video streaming over wireless networks. *IEEE Journal on Selected Areas in Communications*, *28*(3), 389–398. doi:10.1109/JSAC.2010.100410

Zhang, X., & Li, B. (2009). Optimized Multipath Network Coding in Lossy Wireless Networks. *IEEE Journal on Selected Areas in Communications*, *27*(5), 622–634. doi:10.1109/JSAC.2009.090605

Zhang, Y., Niyato, D., & Wang, P. (2015). Offloading in Mobile Cloudlet Systems with Intermittent Connectivity. *IEEE Transactions on Mobile Computing*, *14*(12), 2516–2529. doi:10.1109/TMC.2015.2405539

Zhang, Y., Niyato, D., Wang, P., & Tham, C. K. (2014). Dynamic offloading algorithm in intermittently connected mobile cloudlet systems. *IEEE International Conference on Communications (ICC)*, 4190-4195. 10.1109/ICC.2014.6883978

Zhang, Y., Yu, R., Nekovee, M., Liu, Y., Xie, S., & Gjessing, S. (2012). Cognitive Machine to Machine Communications: Visions and Potentials for the Smart Grid. *IEEE Network*, 26(3), 6–13. doi:10.1109/MNET.2012.6201210

Zhang, Y., Yu, R., Xie, S., Yao, W., Xiao, Y., & Guizani, M. (2011). Home M2M networks: Architecture, standards and QoS improvements. *IEEE Communications Magazine*, 49(4), 44–52. doi:10.1109/MCOM.2011.5741145

Zhao, Z., Rosario, D., Braun, T., & Cerqueira, E. (2014). Context-Aware Opportunistic Routing in Mobile Ad-Hoc Networks Incorporating Device Mobility. In *Proceedings of the IEEE Wireless Communications and Networking Conference* (pp. 2138–2143). Istanbul, Turkey: IEEE.

Zhao, Z., Rosario, D., Braun, T., Cerqueira, E., Xu, H., & Huang, L. (2013). Topology and Link Quality-Aware Geographical Opportunistic Routing in Wireless Ad-Hoc Networks. In *Proceedings of the IEEE International Wireless Communications and Mobile Computing Conference (IWCMC)* (pp. 1522-1527). Sardinia, Italy: IEEE. 10.1109/IWCMC.2013.6583782

Zhdanova, M., Repp, J., Rieke, R., Gaber, C., & Hemery, B. (2014). No Smurfs: Revealing Fraud Chains in Mobile Money Transfers. In *2014 Ninth International Conference on Availability, Reliability and Security* (pp. 11-20). New York: IEEE. 10.1109/ARES.2014.10

Zhongyuan, Z. (2012). Research on 3D Digital Map System and Key Technology. *Procedia Environmental Sciences*, 12, 514–520. doi:10.1016/j.proenv.2012.01.311

Zhong, Z., Wang, J., Lu, G., & Nelakuditi, S. (2006). On Selection of Candidates for Opportunistic Any Path Forwarding. *Mobile Computing and Communications Review*, 10(4), 1–2. doi:10.1145/1215976.1215978

Zhou, G., He, T., Krishnamurthy, S., & Stankovic, J. A. (2004). Impact of radio irregularity on wireless sensor networks. *Proceedings of ACM MobiSys*, 125-138. 10.1145/990064.990081

Ziegler, H., Jenny, M., Gruse, T., & Keim, D. A. (2010). Visual Market Sector Analysis for Financial Time Series Data. In *IEEE Symposium on Visual Analytics Science and Technology (VAST)* (pp.83-90). New York: IEEE. 10.1109/VAST.2010.5652530

Zubair, S., Fisal, N., Abazeed, M. B., Salihu, B. A., & Khan, A. S. (2015). Lightweight distributed geographical: A lightweight distributed protocol for virtual clustering in geographical forwarding cognitive radio sensor networks. *International Journal of Communication Systems*, 28(1), 1–18. doi:10.1002/dac.2635

About the Contributors

Adamu Abubakar received his B.Sc Geography, PGD and M.Sc Computer Science all from Bayero University Kano Nigeria, in 2004, 2006, and 2009 respectively. He received his PhD in IT from International Islamic University Malaysia (IIUM) in 2012. Currently he is in Kulliyyah (Faculty) of Information and Communication Technology and senior member of Intelligence research group (INTEG), in International Islamic University Malaysia. His current research interest is in the field Ubiquitous computing, Location Authority and Navigation.

Media Anugerah Ayu has graduated with a PhD in Information Science and Engineering from School of Engineering, College of Engineering and Computer Science (CECS), the Australian National University (ANU), Canberra, Australia. She currently holds a position as an Associate Professor in Faculty of Engineering and Technology, Sampoerna University, Jakarta, Indonesia. She has published more than 90 research papers in international journals, conferences, book chapters and books in IT related areas. She also has three patent pendings to her credits in IT related research and innovation. Her research interest is set around the area of web application development, ubiquitous computing, mobile applications for intelligent environment, activity recognition, ICT for teaching and learning, and human computer interaction.

Abira Banik completed graduation (B.E.) in Computer Science & Engineering from Tripura Institute of Technology, Agartala, Tripura, India. She is currently pursuing M.TECH in Computer Science & Engineering From Tripura University, Tripura, India. Her area of interests are Wireless Mesh Networks, QoS routing.

Chii Chang is a research fellow at Mobile and Cloud Laboratory (Mobile & Cloud Lab), Institute of Computer Science, University of Tartu, Estonia. He leads the Internet of Things (IoT) research activities in Mobile & Cloud Lab. His research interests involve service-oriented computing, workflow management systems, system integration, mobile peer-to-peer computing, ubiquitous computing, mobile social network in proximity, context-awareness, IoT and Fog computing. He received PhD, Master of Information Technology and Bachelor of Computing at Faculty of Information Technology, Monash University, Australia.

Tamer Z. Emara received the B.Sc. (2005) from Computer Engineering and Automatic Control department, Tanta University, Egypt. He got M.Sc. degree (2015) in the area of Wireless Network at Computers Engineering and Control Systems department, Mansoura University, Egypt. Currently, he is a PhD candidate at Big Data Institute, Shenzhen University, China.

Walaa Farouk received her PhD. in Computer Science and Engineering from the American University in Cairo (AUC), 2017. She holds two master degrees; the first from AUC in CSE, 2011, while the second from the Arab Academy for Science and Technology & Marine Transport in MBA, 2014. She also obtained a diploma in International Arbitration from the joint program European & Cairo University, 2013. Her BS degree in CSE was from Ain Shams University Faculty of Engineering, 2000. She worked as Information Security Solution Manager for designing and consulting several Middle East enterprise telecommunication projects. Her strong technical and practical experiences have actually enriched her researches with industrial practice.

Igor Kotenko graduated with honors from St. Petersburg Academy of Space Engineering and St. Petersburg Signal Academy. He obtained the Ph.D. degree in 1990 and the National degree of Doctor of Engineering Science in 1999. He is Professor of computer science and Head of the Laboratory of Computer Security Problems of St. Petersburg Institute for Informatics and Automation, and Head of the International Laboratory of Information Security of Cyber-Physical Systems of ITMO University. He is the author of more than 350 refereed publications. Igor Kotenko has a high experience in the research on computer network security and participated in several projects on developing new security technologies. For example, he was a project leader in the research projects from the US Air Force research department, via its EOARD (European Office of Aerospace Research and Development) branch, EU FP7 and FP6 Projects, HP, Intel, F-Secure, etc. The research results of Igor Kotenko were tested and implemented in more than fifty Russian research and development projects. The research performed under these contracts was concerned with innovative methods for network intrusion detection, simulation of network attacks, vulnerability assessment, security protocols design, verification and validation of security policy, etc. He has chaired several International conferences and workshops, and serves as editor on multiple editorial boards.

Seng W. Loke is a Professor in Computer Science at the School of Information Technology, Deakin University, Australia. He is an Adjunct Professor at La Trobe University and an Adjunct Professor at the Southern Cross University. He leads the IoT Research Group at Deakin University. He was Reader and Associate Professor in the Department of Computer Science and Information Technology at La Trobe University. He led the Pervasive Computing Lab at La Trobe, and have (co-)authored more than 220 research publications including numerous works on context-aware computing, and mobile and pervasive computing. He has been on the program committee of numerous conferences/-workshops in the area, including Pervasive 2008. He completed his PhD at the University of Melbourne.

Abhishek Majumder received his B.E. degree in Computer Science & Engineering from National Institute of Technology, Agartala and M.Tech. degree in Information Technology from Tezpur University, Assam in 2006 and 2008, respectively. Currently he is pursuing Ph.D from the Department of Computer Science & Engineering, Assam University (A Central University), Silchar, India. His areas of interest are Wireless Mesh Network and Ad-Hoc Networks. He is working as an Assistant Professor in the Department of Computer Science & Engineering, Tripura University (A Central University), Suryamaninagar, India. He is a member of IEEE.

Teddy Mantoro obtained a PhD, an MSc, and a BSc, all in Computer Science. He was awarded a PhD from School of Computer Science, the Australian National University (ANU), Canberra, Australia. Currently, he is an Professor in Faculty of Engineering and Technology, Sampoerna University, Jakarta, Indonesia. He has filed four Malaysian patents in credit to his name and published 5 books and more than 140 papers in book chapters, journals and conference proceedings. His research interest is in pervasive/ubiquitous computing, Information Security, context aware computing, mobile computing and intelligent environment/IoT.

Bhuvaneswari Mariappan is working as Assistant Professor in Electronics and Communication Engineering, Anna University – BIT Campus, Trichy-24. She has nearly two decades of teaching experience. She has more than 15 National and International research publications, presented more than 15 conference papers and published. Her area of interest includes 5G,IoT,Cognitive radios and Wireless Communication.

Varun G. Menon is a Review Board Member of IEEE Transactions on Vehicular Technology and Computer and Communications Journal, Elsevier. He is an Editorial Board member for many reputed international journals. His research interests are in opportunistic routing, mobile ad hoc networks, underwater sensor networks, fog computing.

Mikhail Mikhail is the founder of computer science and engineering department at American University in Cairo (AUC). In 1985, the launched computer science program was the first in Egypt and outside USA borders to be accredited by the American Professional Computer Science Accreditation Board (CSAB) and later by ABET (Accreditation Board of Engineering and Technology). At AUC, he served as Mathematics Head (1983-1985), Computer Science Program Director (1985-1988), Computer Science Department Chair (1988-1994, 2002-2006), and Computer Science and Engineering Department Chair (2006-2008). Prof. Mikhail holds dual B.Sc. degrees both from Cairo University; one in Aeronautical Engineering and other in Mathematics. His Master Degree in Aero-Thermodynamics and his PhD in Aerospace Engineering are from Carleton University, Ottawa, Canada. Moreover, he consulted a number of high tech projects while being in Canada. He worked on mechanical systems designs for NASA's Space Shuttle program. He has a long list of publications and supervision experience in numerous Ph. D. and M.Sc. thesis at Cairo University, Zagazig University, and AUC.

Satish Narayana Srirama is a Research Professor and head of the Mobile & Cloud Lab (http://mc.cs. ut.ee/) at Institute of Computer Science, University of Tartu. He received his PhD in computer science from RWTH Aachen University, Germany. Srirama has co-authored over 115 publications in international conferences and journals. He is an Editor of Wiley Software: Practice and Experience Journal, and he was an AE of IEEE Transactions on Cloud Computing and a PC member of several international conferences. His current research focuses on cloud computing, mobile web services, mobile cloud, Internet of Things and migrating scientific computing and enterprise applications to the cloud.

Evgenia Novikova is an associate professor at the Computer Department of the Saint-Petersburg State Electrotechnical University and senior research fellow at the Laboratory of Computer Security Problems of the Saint-Petersburg Institute for Informatics and Automation of the Russian Academy of Sciences. She received her PhD degree in information security in 2009. Her most recent interests are in developing and application of visual analytics techniques for information security tasks including security information management and intrusion detection, she has also published research papers on information authentication techniques. Evgenia Novikova has co-authored 15 research papers in international conferences and workshops.

Arun Prakash received BE degree from the Department of Electronics and Communication Engineering, MJP Rohilkhand University, India in 2001, MTech degree in Digital Communication from Uttar Pradesh Technical University, India in 2006 and PhD from the Department of Electronics and Communication Engineering, Motilal Nehru National Institute of Technology Allahabad, India in 2011. He was a Visiting Research Scholar at the University of Waterloo, Canada during September 2008 to February 2009. Presently, he is working as an Assistant Professor, Department of Electronics and Communication Engineering at Motilal Nehru National Institute of Technology Allahabad, India. His research interests are in the area of wireless and mobile network, with emphasis on IP level mobility management.

Joe Prathap is working as Associate Professor in Department of Information Technology, RMD Engineering College. He has published many papers in reputed international journals. His research interests are in Network Security, Ad Hoc Networks and Sensor Networks.

Sander Soo recently concluded his Master's studies in the Computer Science curriculum at the Institute of Computer Science, University of Tartu, Estonia. He received his Bachelor's degree in Computer Engineering in 2014 for research conducted at the Institute of Technology, University of Tartu, Estonia in regards to the application of augmented reality within the context of individuals with visual impairment. His current research interests include various aspects of software engineering, distributed processing, Internet of Things, mobile cloud computing, Fog and Edge computing.

Rajeev Tripathi received his BTech, MTech, and PhD in Electronics and Communication Engineering from Allahabad University, India. At present, he is a Professor in the Department of Electronics and Communication Engineering, Motilal Nehru National Institute of Technology Allahabad, India. He worked as a faculty member at the University of The West Indies, St. Augustine, Trinidad, WI, during September 2002 to June 2004. He was visiting faculty at School of Engineering, Liverpool John Moorse University, UK. His research interests are in high speed communication networks, performance of next generation networks: switching aspects, MAC protocols, mobile ad-hoc networking and mobile networks.

Pawan Kumar Verma obtained his B.E. degree from Agra University, India, and the M.Tech. degree from C-DAC, Mohali, India in 2005 and 2009 respectively. He has worked as a Consultant at Cadence Design Systems, Noida, India. He has obtained his PhD degree from Motilal Nehru National Institute of Technology, Allahabad, India in 2016. He was a visiting research scholar at University of Waterloo, Canada during 2012. After completing his PhD, he has also worked as an Assistant Professor in the ECE department at Raj Kumar Goel Institute of Technology, Ghaziabad. Currently, he is working as an Assistant Professor in the department of Electronics and Communication at Dr. B.R. Ambedkar National Institute of Technology, Jalandhar, India. His main research interests are M2M communications, MANETs, VANETs, wireless networks and mobile computing.

Rajesh Verma has worked in the department of Electronics & Communication Engineering in various Institutes/Universities like MNNIT, Allahabad, Galgotias university Greater Noida, AKG Engineering College, Ghaziabad. Currently, he is working with Raj Kumar Goel Institute of Technology and Management, Ghaziabad as Professor & Director of the Institute. His areas of interest include MAC protocol Design for Wireless Networks, Ad-hoc networks, Sensor networks, M2M networks & many more.

Index

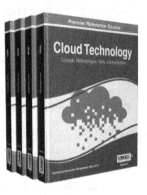

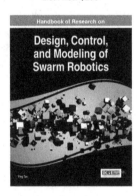

Stay Current on the Latest Emerging Research Developments

Become an IGI Global Reviewer for Authored Book Projects

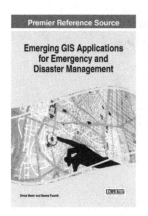

Premier Reference Source

Emerging GIS Applications for Emergency and Disaster Management

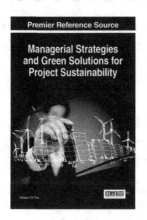

Premier Reference Source

Managerial Strategies and Green Solutions for Project Sustainability

Premier Reference Source

Comparative Approaches to Using R and Python for Statistical Data Analysis

Premier Reference Source

Solutions for High-Touch Communications in a High-Tech World

The overall success of an authored book project is dependent on quality and timely reviews.

In this competitive age of scholarly publishing, constructive and timely feedback significantly decreases the turnaround time of manuscripts from submission to acceptance, allowing the publication and discovery of progressive research at a much more expeditious rate. Several IGI Global authored book projects are currently seeking highly qualified experts in the field to fill vacancies on their respective editorial review boards:

Applications may be sent to:
development@igi-global.com

Applicants must have a doctorate (or an equivalent degree) as well as publishing and reviewing experience. Reviewers are asked to write reviews in a timely, collegial, and constructive manner. All reviewers will begin their role on an ad-hoc basis for a period of one year, and upon successful completion of this term can be considered for full editorial review board status, with the potential for a subsequent promotion to Associate Editor.

If you have a colleague that may be interested in this opportunity, we encourage you to share this information with them.

Printed in the United States
By Bookmasters